T0350921

Noise and Vibration Control in Automotive Bodies

Automotive Series

Series Editor: Thomas Kurfess

Noise and Vibration Control in Automotive Bodies

Jian Pang

This edition first published 2019 by John Wiley & Sons Ltd. under exclusive licence granted by China Machine Press for all media and languages (excluding simplified and traditional Chinese) throughout the world (excluding Mainland China), and with non-exclusive license for electronic versions in Mainland China.

© 2019 China Machine Press

All rights reserved. No part of this publication may be reproduced, stored in a retrieval system, or transmitted, in any form or by any means, electronic, mechanical, photocopying, recording or otherwise, except as permitted by law. Advice on how to obtain permission to reuse material from this title is available at http://www.wiley.com/go/permissions.

The right of Jian Pang to be identified as the author of this work has been asserted in accordance with law.

Registered Offices
John Wiley & Sons, Inc., 111 River Street, Hoboken, NJ 07030, USA
John Wiley & Sons Ltd, The Atrium, Southern Gate, Chichester, West Sussex, PO19 8SQ, UK

Editorial Office
The Atrium, Southern Gate, Chichester, West Sussex, PO19 8SQ, UK

For details of our global editorial offices, customer services, and more information about Wiley products visit us at www.wiley.com.

Wiley also publishes its books in a variety of electronic formats and by print-on-demand. Some content that appears in standard print versions of this book may not be available in other formats.

Limit of Liability/Disclaimer of Warranty
While the publisher and authors have used their best efforts in preparing this work, they make no representations or warranties with respect to the accuracy or completeness of the contents of this work and specifically disclaim all warranties, including without limitation any implied warranties of merchantability or fitness for a particular purpose. No warranty may be created or extended by sales representatives, written sales materials or promotional statements for this work. The fact that an organization, website, or product is referred to in this work as a citation and/or potential source of further information does not mean that the publisher and authors endorse the information or services the organization, website, or product may provide or recommendations it may make. This work is sold with the understanding that the publisher is not engaged in rendering professional services. The advice and strategies contained herein may not be suitable for your situation. You should consult with a specialist where appropriate. Further, readers should be aware that websites listed in this work may have changed or disappeared between when this work was written and when it is read. Neither the publisher nor authors shall be liable for any loss of profit or any other commercial damages, including but not limited to special, incidental, consequential, or other damages.

Library of Congress Cataloging-in-Publication Data

Names: Pang, Jian, 1963– author.
Title: Noise and vibration control in automotive bodies / Jian Pang.
Description: Hoboken, NJ : John Wiley & Sons, 2019. | Includes index. |
Identifiers: LCCN 2018023710 (print) | LCCN 2018033626 (ebook) | ISBN 9781119515517 (Adobe PDF) |
    ISBN 9781119515524 (ePub) | ISBN 9781119515494 (hardcover)
Subjects: LCSH: Automobiles–Noise. | Automobiles–Vibration.
Classification: LCC TL246 (ebook) | LCC TL246 .P35 2018 (print) | DDC 629.2/31–dc23
LC record available at https://lccn.loc.gov/2018023710

Cover Design: Wiley
Cover Images: © 1971yes/iStockphoto; © 3alexd/iStockphoto; © olegback/iStockphoto;
© solarseven/Shutterstock

Set in 10/12pt Warnock by SPi Global, Pondicherry, India

Printed and bound by CPI Group (UK) Ltd, Croydon, CR0 4YY

10 9 8 7 6 5 4 3 2 1

Contents

Preface

I have been working in the field of noise and vibration for more than 30 years. I have been looking for a good book on noise, vibration, and harshness (NVH), to no avail.

With the development of the automotive market, the customers are paying more and more attention to the driving quality. There is almost a consensus in the automotive industry: NVH is the most important indicator to determine the perception of the driving quality. All vehicle systems, such as engine, body, and so on, generate NVH problems. When the serious NVH problems are unraveled, the original minor problems are highlighted. After the minor problems are solved, sound quality becomes the focus of attention. When the sound quality reaches a satisfying level, the customers care about the sound DNA. Almost all the auto companies have invested a lot of resources to develop NVH capabilities and upgrade technologies, and the NVH engineers are hungry for the related knowledges. In the vast NVH world, the knowledges are scattered around. I desire to pick up these scattered pearls, with diligence and wisdom, to weave a string of necklaces and then to dedicate them to a number of peers, which has become a source of motivation for my writing.

The automobile structure is very complex, consisting of body, power train, suspension, and so on. The main systems are "hung" on the body. For example, the power train is connected to the body by mounts, the suspension is linked to the body through bushings or directly to the body, and the exhaust system is attached to the body by hangers. The body carries the drivers and passengers, so its structural characteristics directly affect the passengers' perceptions. Therefore, the body is the core of the vehicle, and its structure determines the vehicle performance. Of course, the body is extremely important to the vehicle NVH performance as well.

The body comprises frames, beams, panels, trimmed parts, etc. The frames determine the overall body stiffness and modes, the panels are related to the local vibration and the sound radiation, and the door and the body together govern the closing door sound quality. The trimmed parts and materials affect the performance of sound package, and the structures of driving points determine the transmission of structure-borne sound.

This book gives a comprehensive picture of the automotive body noise and vibration analysis and control. It has nine chapters, discussing the NVH of overall body structure, NVH of local structure, sound package, sensitivity, wind noise, sound quality, squeak and rattle, and target system.

Working on the front line of product development for many years, every day I encounter a variety of NVH problems, some routine problems, some very fresh problems, and some very difficult problems. After successfully solving the problems,

I am introduced to numerous engineering cases, and I am always curious to explore the theories behind them. On the other hand, I summarize the main points of the engineering problems and then abstract them to the scientific problems by mathematical methods or statistical methods. This kind of approach is prevalent throughout this book, that is, the engineering practice and theory are closely combined.

Due to the combination of theory and practice, I sincerely hope that the book can provide a valuable reference for engineers, designers, researchers, and graduate students in the fields of automotive body design and NVH. Readers may have better ideas, and they are welcome to discuss together.

In the process of writing this book, I have received encouragement and help from many experts and colleagues. They expect me to contribute more to the NVH field. I would like to express my gratitude to them as this expectation is also the driving force for me to write this book.

1

Introduction

1.1 Automotive Body Structure and Noise and Vibration Problems

1.1.1 Automotive Body Structure

An automotive structure, including the body, power train, suspension, and so on, is very complex. The main systems of the vehicle are "hung" on the body – for example, the power plant is connected with the body by mountings, the suspension is connected with the body by bushings or directly connected with the body, and the exhaust system is connected with the body by hangers – so the body is a core of the vehicle and determines the vehicle's performance. However, the body is also a place for carrying passengers, so its structural characteristics directly influence the perception of the vehicle's users.

1.1.1.1 Unitized Body and Body-on-Frame

There are two major forms of automotive structure: the unitized body and the body-on-frame. When the body and the chassis frame are integrated as a whole structure, as shown in Figure 1.1, this is known as a unitized body, also called an integrated body or integral body. The unitized body itself takes the load of vehicle, rather than the load being taken by an independent frame. The advantages of a unitized body include its simple structure, small size, light weight, and low cost, but its disadvantage is that the body's loading capacity is limited. Most passenger vehicles have a unitized body.

Body-on-frame, also called a separate frame structure, non-integrated body, monocoque, or body chassis frame construction, is a body structure in which the chassis frame is separated from the body. The chassis frame, which has high structural strength, is arranged below the body. This structure has the advantages of high stiffness, high strength, strong loading capacity, and strong capacity to resist bending deformation and torsion deformation, but the disadvantages are that the structure is complex, heavy, and expensive. Trucks, buses, off-road vehicles, large sport-utility vehicles (SUVs), and a small number of passenger sedans use a body-on-frame structure.

The noise and vibration problems and control methods described in this book are based on the unitized body structure, so throughout, "body" refers to the unitized body. In automotive engineering, vehicle noise and vibration is usually denoted by "NVH", for noise, vibration, and harshness. Harshness represents the subjective sensation on the human body of vehicle noise and vibration.

Noise and Vibration Control in Automotive Bodies, First Edition. Jian Pang.
© 2019 China Machine Press. All rights reserved. Published 2019 by John Wiley & Sons Ltd.

Figure 1.1 Structure of a united body.

Figure 1.2 Structure of a trimmed body.

1.1.1.2 Body-in-White, Trimmed Body, and Whole Vehicle Body

The stages of construction of a vehicle body are divided into the body-in-white (BIW), the trimmed body, and the whole vehicle body. The BIW refers to a body consisting of frames and panels, including front and rear side frames, rocker frames, cross members, dash panel, floors, roof, and front and rear windshields. Sometimes the BIW is further divided into a BIW without windshields and a BIW with windshields. The BIW without windshields refers to the welded body structure, as shown in Figure 1.1.

The doors, trimmed parts, and seats are installed on the BIW to form the trimmed body, as shown in Figure 1.2. The trimmed body includes the BIW, doors, engine hood, trunk lid, seats, steering system, sound absorptive materials, and insulators.

After the trimmed body and other systems, such as the power plant, exhaust, and suspension, are integrated into the vehicle, the body is called the whole vehicle body. The structures of the whole vehicle body and the trimmed body are the same, but their boundary conditions are different. The body in a whole vehicle is connected with other systems, so it is subjected to constraints from these systems.

1.1.1.3 Classification of Body Structure

Body design involves many performance attributes, such as NVH, crash safety, fatigue and reliability, fuel economy, and handling. The body can be classified by each attribute according its characteristics. In this book, the body is classified from the perspective of NVH. The body is divided into four categories of structure according to its NVH functions, namely the frame structure, panel structure, trimmed structure, and accessory structure, as shown in Figure 1.3. The frame structure refers to a body frame comprising side frames, cross members, and pillars that are connected by the joints. The panel structure refers to the metal plates that cover the body frame, such as the dash panel, roof, floor, side panels, and door panels. The trimmed structure refers to the parts that reduce noise and vibration, such as the dash insulator and damping structure. The accessory structure refers to the accessory parts installed on the body, such as the steering shaft, mirrors, and seats. The make-up and functions of these four structures are briefly described below.

The frame structure, as shown in Figure 1.4, is the foundation of the body. The frame is composed of front and rear side frames, cross members, pillars, and so on. Several side frames, cross members, and pillars intersect, forming a joint. The cross section,

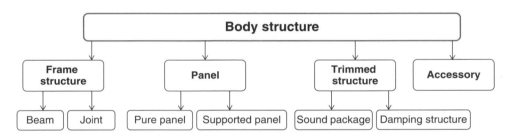

Figure 1.3 Classification of body structure.

Figure 1.4 Body frame structure.

size, and span of a beam determine its stiffness. The joint has significant influence on the body frame stiffness. The frames can only be tightly intersected if the joints have sufficient stiffness. If the frames are stiff enough, but the joint is weak, the stiffness of the body frame is still weak. Therefore, the stiffness of the body frame is determined by both the frame stiffness and the joint stiffness, while the body frame stiffness determines the modal shapes and frequencies of the vehicle body.

The panels are mounted on the frames to form an enclosed body space. The panels are divided into pure panels (or local pure panels) and supported panels. A pure panel is one without support, such as the fender shown in Figure 1.5. Most body panels are supported by metal beams or reinforcement adhesives, or have beaded surfaces, so this kind of panel is called supported panel. Examples of supported panels include the outer door panel (Figure 1.6), where the internal side is supported by the side-impact beam or

Figure 1.5 Fender.

(a)　　　　　　　　　　　　　　　　　　(b)

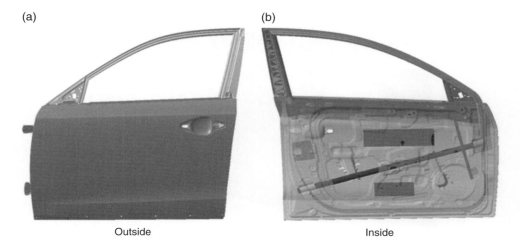

Outside　　　　　　　　　　　　　　　　Inside

Figure 1.6 A door panel. (a) Outside. (b) Internal side.

reinforcement adhesives, and the beaded floor (Figure 1.7), which is supported by the cross members. Sometimes, it is difficult to distinguish between a pure panel and a supported panel, as in the case of the roof shown in Figure 1.8. The roof is a big panel supported by several cross rails, but some area of the roof between two rails is so large that it can be regarded as a pure panel.

The trimmed structure bonded to the panels and frames includes decoration parts that also function to absorb and insulate sound and non-metallic parts that suppress the transmission of noise and vibration. From the NVH perspective, the trimmed structure can be divided into four categories: sound insulation structure, sound absorption structure, damping structure, and barrier structure. The sound insulation structure includes the dash insulator and carpets. The sound absorption structure includes the headliner and the sound absorption layer of the dash insulator. In most cases, the sound insulation structure and the sound absorption structure are integrated to form a sound-absorption-insulation structure, such as the dash inner insulator, as shown in Figure 1.9. The damping structure refers to the damping layer on the panels, including damping material on the floor, as shown in Figure 1.10, and the constrained damping layer installed on panels such as the sandwiched dash panel system shown in Figure 1.11. The barrier structure is a special foaming structure placed inside the frame cavities in order to

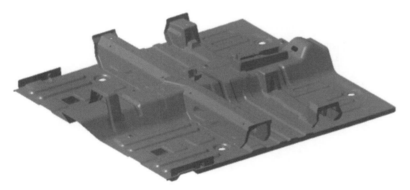

Figure 1.7 A floor.

Pure local panel

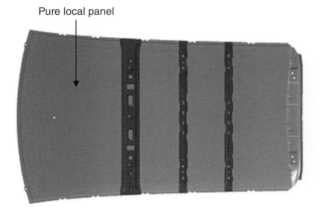

Figure 1.8 A roof.

Figure 1.9 A dash inner insulator.

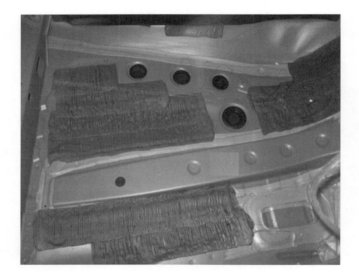

Figure 1.10 Damping material on floor.

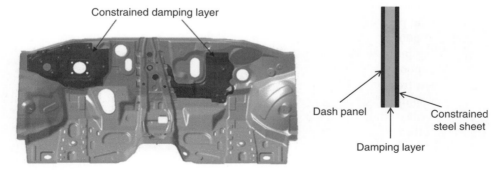

Figure 1.11 A sandwiched dash panel system.

(a)   (b)

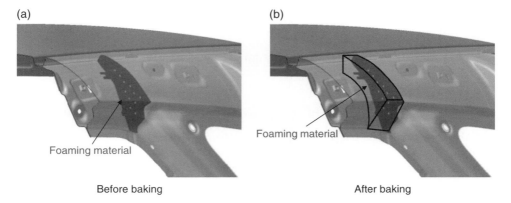

Before baking   After baking

Figure 1.12 Foaming material inside a frame cavity.

prevent sound transmission. The volume of the foaming material is small, but after being baked in high temperature environment, it expands and its volume increases dozens of times, filling a section of the frame cavity, as shown in Figure 1.12.

The accessory structure refers to the other structures installed on the body, such as the steering shaft system, instrument panel, seats, shift system, and mirrors. Occupants may directly perceive vibrations induced by these structures.

1.1.2 Noise and Vibration Problems Caused by Body Frame Structure

The frame structure is the basis of the body, in the same way that a frame is the basis of a building. If the housing framework is not well constructed, its capacity to carry load and resist earthquakes will be deteriorated, and the house could even collapse. A poorly designed and constructed body frame will generate many problems, such as resonance, body deformation, squeak and rattle (S&R), and even safety issues.

The body frame is the supporting structure for other parts, such as the doors, panels, engine hood, trunk lid, and accessory brackets. If the stiffness of the frame structure is insufficient, the door and other components will not be well supported. Under low excitations, friction between these components and the frame could be generated. Under impulse excitations, the components could impact each other. Friction and impact between components are two major causes of S&R.

Over a long period of vibration and shock excitation, the body frame may deform, resulting in poor fitting between the doors and the body frame, and deterioration of body sealing performance. During high-speed driving, the dynamic sealing will be particularly poor, resulting in a huge wind noise.

Excitations from the engine and road directly act on the body frame. These excitations are dominated by low-frequency components. If the frame structure does not have sufficient stiffness, i.e. its modal frequency is low, the body can be easily excited, resulting in the body resonance.

The body frame affects not only the NVH performance, but also crash safety, handling performance, reliability, and so on. For example, the effect of a head-on collision is directly related to the structure of the front frames and cross members.

The frame structure usually induces the low-frequency vibration and noise problems that affect overall vehicle NVH performance. Therefore, the stiffness and modal frequency of the frame structure are very important. Body stiffness can be controlled from two aspects, i.e. the frame and the joint; for both, the stiffness should be as high as possible.

1.1.3   Noise and Vibration Problems Caused by Body Panel Structure

A body panel is similar to a piece of paper or a drum. When the paper is waved in the air, the paper hums because it vibrates and radiates sound. When the drum is hit, the drumstick applies a force to the drum surface, which vibrates and generates sound.

The body is connected to many systems that generate excitations, such as the engine, exhaust, and suspension. In addition, when a vehicle is driven at high speed, the wind excites the body. When subjected the external excitations, the body panels vibrate and generate sound, just like a piece of paper or a drum.

The vibration and noise problems generated by the body panels are divided into two categories: one is the direct radiation of sound, and the other is the interior booming caused by the coupling between a panel and the body cavity.

Under an external excitation, the panel vibrates and radiates sound because of its thin and large surface. To reduce this, the body panel should be designed to be as stiff as possible, and large flat surfaces should be avoided. Usually, there are three ways to increase the frequency of the panel. The first one is to bead the panel to form an intertwined structure. The second one is the use of convex design, i.e. the panel is designed as several planes or arcs. The third one is to add support to the panel. In some cases, if the frequency of the panel cannot be increased by the above methods, damping treatment may be implemented. There are two types of damping treatment: free damping and constrained damping. A layer of damping material is pasted onto the panel to form free damping, such as the damping on the floor, whereas a layer of sandwiched damping structure is added onto the panel to form constraint damping.

The reason for interior booming is resonance between a body panel mode and the acoustic cavity mode. The air inside the body is an enclosed space that forms an acoustic cavity with specific modal shapes and frequencies. For example, the first cavity mode is along the vehicle's longitudinal direction and its frequency is low. When the modal frequency of a panel perpendicular to the direction (such as the trunk lid) is the same as the frequency of the first cavity mode, and the panel is excited by an external excitation with the same frequency, the coupling between the structure and the cavity will generate an annoying low-frequency booming that makes occupants uncomfortable.

Most noise and vibration problems generated by body panels are low- and middle-frequency ones, but they also induce a few high-frequency noise problems. These problems are mainly caused by local body structures, so it is very important to control the stiffness and damping of the local structures in order to suppress NVH problems.

1.1.4   Interior Trimmed Structure and Sound Treatment

The trimmed structure itself does not create NVH problems: in fact, it prevents or attenuates sound transmission. The special trimmed structure includes the sound insulation structure, sound absorption structure, and barrier structure. However, the general trimmed structure represents a combination of the special trimmed structure and the damping structure.

The sound absorption structure is composed of absorptive materials, and its function is to eliminate middle- and high-frequency noise. The sound insulation structure is composed of sound insulation material, and its function is to eliminate low- and middle-frequency noise. Usually, the sound insulation structure and sound absorption structure are combined to form a sound-absorption-insulation structure. When outside sound hits the body, some is reflected, and the rest passes through the body and enters the interior. The function of the sound-absorption-insulation structure is to attenuate the penetrated sound. If the structure is not well designed, this function cannot be realized.

The barrier structure prevents outside sound passing through the frame cavities and into the interior through the use of baffling materials. Most body frames, such as A-pillars, B-pillars, C-pillars, side frames, and rockers, are tube-like structures. There are holes on the frames and pillars that are designed for manufacturing processes or for the installation of other components. The outside sound travels inside the hollow frames or pillars, and then enters the interior through the holes. Therefore, the internal channels of the tubes must be blocked with baffling materials in order to prevent the sound transmission. The baffling material is a foaming material: i.e. the original, small-sized material is inserted into a section of a frame or pillar, then after baking in a high temperature environment, it expands and firmly fills the inside of the tube.

The damping structure is a layer of damping material that is placed on the surface of a metal sheet or sandwiched between two sheets. The function of the damping structure is mainly to reduce vibration and noise in the range of 200–500 Hz.

1.1.5 Noise and Vibration Problems Caused by Body Accessory Structures

Body accessory structures can be divided into three categories: bracket, steering system, and seat.

Many components, such as side mirrors, the internal mirror, the battery, the CPU control unit, and the glove box, are mounted on the body by brackets. If the stiffness of these brackets is insufficient, the components will vibrate. For example, insufficient stiffness of the side mirror bracket gives it a low modal frequency. During cruising, the mirror is easily excited by engine or road vibration, so the mirror could shake, affecting the driver's vision. In another example, a battery bracket with low stiffness causes the bracket-battery system to have a low frequency. The system can be easily excited by road input, and the bracket could generate a low-frequency roaring. Therefore, the frequencies of these brackets must be high enough to separate the systems' frequencies from the excitation frequencies of the engine and the road.

The steering system, consisting of the steering shaft system and the cross car beam (CCB) system, is a large accessory. It is also connected to the body through brackets. If the stiffness of the steering shaft system and/or CCB are insufficient, the modal frequency of the steering system will be too low to be coupled to the engine excitation frequency, causing the steering wheel to vibrate. Even if the steering shaft and CCB are stiff enough, the system modal frequency could still be low if their connections to the body are weak, thus it could fall into the external excitation frequency range and cause vibration on the steering wheel. Therefore, the whole system – the steering shaft,

CCB, and brackets – must be sufficiently stiff to prevent vibrations. In addition, some components, such as the CD box, are connected with the CCB through brackets, so these brackets should have sufficient stiffness.

The seat is an accessory that the occupants directly touch, and the vibration perceived by occupants involves two aspects. The first perception is the seat's overall longitudinal and/or lateral vibration: i.e. the seat's low-frequency longitudinal and/or lateral modes are excited by the engine or the road. The second perception is that the occupants feel uncomfortable because of an unsuitable design of seat cushion and back cushion that results in poor vibration isolation.

1.2 Transfer of Structural-Borne Noise and Airborne Noise to Interior

The process of noise and vibration transfer from the outside to the interior can be described in three stages: source, transfer path, and receiver, as shown in Figure 1.13. The following materials describe the transfer process from the source.

1.2.1 Description of Vehicle Noise and Vibration Sources

The noise and vibration sources are outside the body. The vehicle is subjected to three sources of noise and vibration: power train excitation, road excitation, and wind excitation. The three sources are briefly described below.

The power train system includes the engine, transmission, intake system, exhaust system, and drive shaft. They are directly connected with the body, so the sources of noise and vibration are directly applied to the body. A distinctive feature of the sources is that the noise and vibration is closely related to the engine speed and firing order. They are the most important sources for interior noise and vibration during idling and at low vehicle speed.

The interaction between the tires and the road generates noise that is directly transferred into the interior. Simultaneously, the vibration generated by the action between the road and the tires is transmitted to the body through the suspension system. This type of noise and vibration is related to vehicle speed, and also to the parameters of the tire and suspension system. The road/tire noise is the major interior noise source when a vehicle moves at middling speed, especially on rough roads.

When the vehicle runs at high speed, wind strongly acts on the body. The noise generated by friction between the wind and the body, called wind noise, is transmitted to the interior through the body. At the same time, the wind excites the body panels, and the excited panels radiate noise to the interior. Wind noise is closely related to vehicle speed. Generally, when the vehicle travels at high speed (e.g. above $120\,\mathrm{km\,h^{-1}}$), the wind noise will overwhelm the power train noise and the road noise, making it the largest noise source.

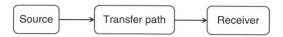

Figure 1.13 Source–transfer path–receiver model.

These three noise and vibration sources are transmitted into the interior in two ways: via the airborne sound transmission path and via the structural-borne sound transmission path.

1.2.2 Structural-Borne Noise and Airborne Noise

1.2.2.1 Airborne Noise and Transmission

As the name suggests, airborne noise refers to sound transmitted through the air and then heard by the occupants. Figure 1.14 shows the transfer process of a drumming sound to people outside and inside a house. When the drum is hit, the sound generated by the drum membrane directly transfers to the person standing outside the house. The sound is called airborne sound because it is generated and transmitted in the air, and then heard by the person.

The person in the house can still hear the drumming sound, but the sound level that the person hears is lower than that heard by the person standing outside the house. When the drumming sound transfers to the house, the sound is partially reflected by the walls, doors, windows, etc., with only a portion of the sound entering the house. The drumming sound wave outside the house is called the incident wave, the sound wave reflected by the walls is called the reflected wave, and the sound wave that passes through the wall and enters the house is called the transmitted wave. The sound wave heard by the person inside the house is the transmitted wave.

Many sound sources heard by a vehicle's occupants are airborne noise, i.e. the sounds transmit directly to the occupants through the air. The vehicle body is similar to the wall of a house. When the outside noise hits through the body, some sound waves are reflected back, while others pass through the body and enter the interior (although some sound waves are also absorbed by the body itself). One difference between a vehicle and a house is that a lot of sound absorptive materials and sound insulators are installed on the body, which absorb and insulate some of the sound waves. Figure 1.15 shows the transfer process of the airborne noise to the occupants' ears.

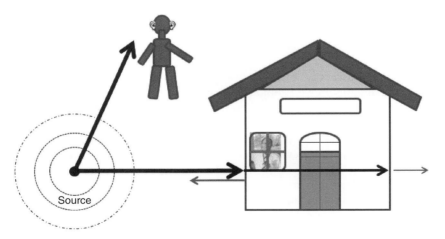

Figure 1.14 Transfer process of drumming to occupants: airborne noise.

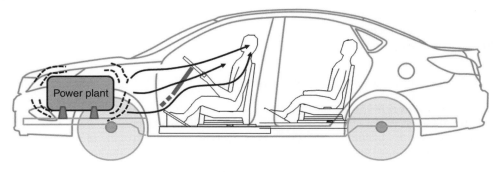

Figure 1.15 Transfer process of airborne noise to occupants' ears.

Figure 1.16 Transmission of structural-borne sound in rails.

The airborne noise sources directly transmitted into the passenger compartment include:

- power plant radiation sound
- intake orifice noise
- exhaust orifice noise
- driveline radiation noise
- noise generated by the cooling fan
- road noise
- wind noise.

1.2.2.2 Structural-Borne Noise and Transmission

Airborne sound is directly transmitted to human ears, whereas structural-borne sound is indirectly transmitted. As the name implies, structural-borne sound refers to sound transmitted in a structure, then radiated to the air, and finally heard by human ears.

If you were to put your ear close to a train track, as shown in Figure 1.16, you might hear a train coming long before you could see it. When a train moves, its vibration is transmitted to the rails through the wheels, and the waves generated by this vibration are transmitted through the structure of the rails. Then the waves in the structure radiate into the air as sound that you can hear. Because the propagation speed of sound waves in the solid rails is much faster than in the air, you can hear the train long before you see it.

There are many transfer paths for structural-borne sound inside the vehicle. For example, engine vibration is transmitted to the subframe, then the vibration waves are transmitted inside the body frames and reach the body panels; finally, the excited panels radiate sound to the interior. This radiated sound is the structural-borne sound. Below are examples of structural-borne sound transmitted in the vehicle.

- Power plant vibration is transmitted to the body through mounting systems, and then the excited panels radiate sound to the interior.
- Exhaust vibration is transmitted to the body through hangers, and then the excited panels radiate sound to the interior.
- Driveshaft vibration is transmitted to the body through bearing supports, and then the excited panels radiate sound to the interior.
- Road excitation is transmitted to the body through the suspension, and then the excited panels radiate sound to the interior.

1.2.3 Transfer of Noise and Vibration Sources to Interior

Noise and vibration sources are outside the body, and the occupants sit in the vehicle and perceive the noise and vibration, so the body is a barrier between the sources and the occupants. In the analysis of vehicle noise and vibration, the source–transfer path–receiver model shown in Figure 1.13 is used. So the body is the transfer path of noise and vibration transmission, whereas the receivers are the occupants who perceive the noise and vibration.

According to the definitions of airborne noise and structural-borne noise, the model expressed in Figure 1.13 can be extended, as shown in Figure 1.17. Figure 1.17 shows the sources of airborne noise and structural-borne noise and the corresponding transfer paths.

The interior noise and vibration are determined by the outside sources and transfer paths, and can be expressed by the following equation:

$$NV = \sum_{i=1}^{N} S_i H_i, \tag{1.1}$$

where NV represents the interior noise or vibration, S_i is i^{th} noise source or vibration source outside the vehicle, and H_i is the ith noise or vibration transfer path.

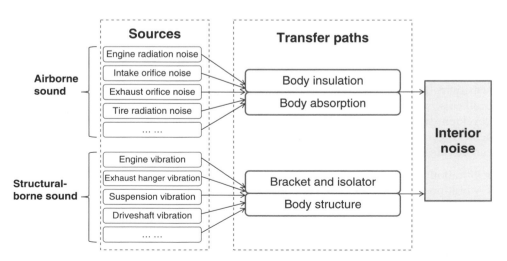

Figure 1.17 Sources and body transfer paths of airborne noise and structural-borne noise.

From Eq. (1.1), it can be seen that the interior noise and vibration can be controlled from two aspects: source and transfer path. Chapter 5 describes the characteristics of major sources in detail. After understanding the characteristics, engineers can control the noise and vibration sources. The characteristics of the airborne and structural-borne transfer paths are briefly described as follows.

For airborne noise, the transfer path is the body sound insulation and sound absorption layers. The sound source is transmitted to the body, some sound waves are reflected, some are absorbed, and the rest pass through the body and enter the interior. The performance of the body sound insulation and absorption is represented by sound–sound transfer function, which means attenuation of the outside sound transmitted into the interior. The higher the sound–sound transfer function, the greater the sound attenuation.

For structural-borne noise, the transfer path refers to the transmission of vibration at a body point to sound in the interior. The performance of the structural-borne sound can be represented by sound–vibration transfer function, which means the interior sound generated by unit force applied on a point. Compared with the airborne noise transfer path, the structural-borne transfer path is more complex. For example, to examine the transfer paths of the engine mountings, three factors must be analyzed: first, the stiffness and modes of mounting brackets; second, the dynamic stiffness of the connected points between the body and the mountings; and third, the interior noise generated by applying force at the connected points.

One of the most important areas of vehicle NVH research is analysis of the characteristics of the transfer paths, and the discovery of methods to control the transfer paths. In this chapter and the following chapters, I focus on how to design the transfer paths to attenuate the transmission of outside noise and vibration sources to the interior.

1.3 Key Techniques for Body Noise and Vibration Control

The frame structure determines the overall modes of the body, and the panel structure determines the noise radiation and its coupling with the cavity mode. The body is the noise and vibration transfer path, and it can be divided into airborne and structural-borne transfer paths, which are related to the body's acoustic sensitivity and vibration sensitivity, respectively. In some special circumstances, the body could have special problems, such as high-speed wind noise, door closing sound quality, or S&R. The development of the body is based on a target system, so clear targets are the key to achieving a successful body development. The key techniques for body NVH control can be categorized as follows:

- vibration and control of overall body structures
- vibration and sound radiation of body local structures
- sound package
- body noise and vibration sensitivity
- wind noise and control
- door closing sound quality and control
- S&R of vehicle body.

In this book, the above seven aspects are covered from Chapter 2 to Chapter 8, respectively. The following sections briefly describe the seven aspects.

1.3.1 Vibration and Control of Overall Body Structure

Controlling the vibration of the overall body structure means controlling the body stiffness and mode from the perspective of the body frame structure in order to achieve good noise and vibration performance. The main research areas include:

- Control of overall body stiffness
- Identification of overall body mode
- Control of overall body mode.

1.3.1.1 Control of Overall Body Stiffness

The stiffness is the basis of the body. Insufficient stiffness brings not only NVH problems such as vehicle shake, interior booming, and S&R, but also safety and reliability problems. The research scope of body stiffness includes measurement, analysis, and control of body stiffness.

Stiffness is divided into bending stiffness and torsional stiffness. Bending stiffness refers to the body's capacity to resist bending deformation under the action of an external force. The connection points between the front shock absorbers and the body and the rear shock absorbers and the body are constrained, and a concentrated force is applied at the installation location of the rear seats. The applied force is divided by the maximum deformation to obtain the body's bending stiffness.

When the loads applied on both sides of the vehicle are different, the body is twisted, resulting in torsional deformation. Torsional stiffness refers to the body's capability to resist torsional deformation. The connection points between the rear shock absorbers and the body are constrained, and a torque is applied at the connection points between the body and the front shock absorbers. The body torsional stiffness is obtained by dividing the torsional angle by the torque.

Using the above boundary conditions and loads, the body's bending stiffness and torsional stiffness can be obtained by testing or by computer aided engineering (CAE) analysis.

Factors affecting the overall body stiffness include the layout of the overall frame, the frame cross-section, and joint stiffness. The overall layout refers to the arrangement of the frames, cross members, and pillars, which must form closed loops. The frame's stiffness depends on its cross-section. The bending stiffness depends on the moment of inertia of the section, and the torsional stiffness depends on the polar moment of inertia of the section. The moment of inertia and the polar moment of inertia of a closed-loop section are much larger than those of an open-loop section, so the frame sections should be designed as closed loops wherever possible. The joint stiffness is defined as the local stiffness at the intersections of the frames, pillars, etc.

High body stiffness can only be achieved through rational layout of the body frame structure, a cross-section with large moment of inertia, and high joint stiffness. When analyzing body stiffness, all three factors must be simultaneously considered.

1.3.1.2   Identification of Overall Body Mode

Vehicle body mode identification involves obtaining the modal frequencies and mode shapes of the vehicle body by testing or analyzing, and then determining the factors that affect the modal characteristics. The most important body modes are the first bending mode and the first torsional mode.

Accelerometers are placed on particular points of the body, and exciters are used to vibrate high-stiffness locations (such as the connection point between a shock absorber and the body). After the excitation signals and responses have been processed, the transfer functions between the outputs and the inputs can be obtained and the modal parameters are extracted. The body modal analysis is usually implemented by finite element (FE) analysis. After the body is discretized into a number of finite grids, and the excitation points and the response points are chosen, the accelerations and forces are calculated, and the transfer functions and modal parameters can be obtained.

In body testing and analysis, the BIW is the most important because it is the most basic structure. After the windshields, seats, doors, and other trimmed structures are added onto the BIW, the body weight increases, and its stiffness changes. Under normal circumstances, the modal frequency of the bending mode of the trimmed body is much lower than that of the BIW. The trimmed parts greatly increase the torsional stiffness of the trimmed body, but although the body weight increases, the torsional modal frequency changes little compared with that of the BIW. The modal frequencies of the BIW, the trimmed body, and the whole vehicle body are related, so after the modal frequency of the BIW has been obtained, the modal frequencies of the trimmed body and whole vehicle body can be roughly calculated.

Modal identification also involves determining the nodes of the dominant modes. The lump masses of the external systems should be placed on the nodes or as close to them as possible so that the systems' influence on the body's modes or the influence of external inputs on the body can be minimized.

1.3.1.3   Control of Overall Body Mode

Control of the overall body mode refers to methods to control the body modes by decoupling systems and excitation, decoupling modes of adjacent systems, and establishing modal tables. Overall body mode control involves three aspects: first, the modal frequency and excitation of each system is analyzed so that adjacent systems can be decoupled from each other and from external excitation; second, a complete mode distribution table is developed; and third, modal decoupling and noise and vibration suppression is achieved by adjusting the stiffness, mass, and structure distribution of the vehicle's systems.

The first task of mode control is to determine the associated systems and their coupling status: i.e. to determine the principles of modal separation and decoupling. Decoupling involves three aspects: decoupling of the overall body modal frequency and the external excitation frequency, decoupling of the overall body modes and the modes of the systems connected with the body, and decoupling of the overall body modes and the local modes.

The second task of mode control is to develop a modal distribution table. After the excitation frequencies, the modal frequencies of the overall body, and the frequencies of the connected systems have been determined, a modal table can be established. This means that the modal frequencies of each system can be planned, which can guide the development of each system.

There are three body modal planning tables. The first is the whole-vehicle modal planning table, in which the modal frequencies of the body and each system are listed in a table. The purpose of the table is to indicate whether each system can be decoupled and to show whether they are separated from the idle excitation frequencies. After the whole-vehicle modal planning table has been determined, the modal frequency targets of each system can be determined, so the development of each system can be relatively independent. The second table is the body modal frequency table, in which the modal frequencies of the overall body and each system/component connected to the body are put together. The purpose of this table is to decouple the overall body modes and the local body modes, and to separate the modal frequencies from excitation frequencies. The third table is the excitation frequency and body mode table, in which the excitation frequencies and the body modal frequencies are placed in a chart or table to illustrate the relationships among body modal frequency, excitation frequency, engine speed, order, and so on. The third table and second table can be used simultaneously, so the relationship between the body mode frequency and excitation can be quickly diagnosed.

The third task of modal control is to separate and control the body modes by modifying the body structure. The overall body mode is mainly determined by the body's stiffness and mass distribution, so the body mode can be altered from two aspects. In addition, some systems connected to the body can be regarded as dynamic vibration dampers that can adjust the body's response at certain modal frequencies. The first body bending mode and first torsional mode should be designed to be as high as possible and far away from the frequencies of the main excitation sources. If a body modal frequency and an excitation frequency are unavoidably overlapped, the excitation should be placed as close as possible to the modal nodes.

1.3.2   Vibration and Sound Radiation of Body Local Structures

The frame structure is the foundation of the vehicle body, and the panels and accessories are connected to the frame structure by welding, riveting, etc. The body local structure has two categories: panels and accessories. Panels are metal sheets covering the frame, such as the dash panel, roof, floor, side panels, doors, engine hood, and trunk lid, which are connected to the frame by welding or other means to form an enclosed body. Accessories are the components installed on the body, such as the steering system, instrument panel, interior rearview mirror, and exterior side mirrors.

A panel is a thin-walled plate. After being excited, it easily radiates noise. The enclosed air inside the body forms a cavity with an acoustic cavity mode. The modal frequency of the panel is relatively low, so its structural mode can easily couple with the acoustic cavity mode, generating interior booming. Many accessories are directly related to the occupants' perceptions. The noise and vibration produced by panels and accessories, such as steering wheel vibration and mirror shake, can be directly perceived by the occupants.

Controlling the vibration and noise of local structures involves the following aspects:

- Control of panel vibration and sound radiation
- Acoustic cavity modes
- Control of accessory vibration.

1.3.2.1 Panel Vibration and Sound Radiation

A panel is like a drum membrane. When the membrane is beaten, its surface vibrates like a ripple across water, as shown in Figure 1.18. Likewise, when a body panel is excited, the vibration wave forms a layered pattern that expands from its center to the edge.

When the panel modal frequency and the external excitation frequency are the same or close to each other, the panel is easily excited and radiates noise to the interior. For example, when vibration from the suspension system is transmitted to the floor, the floor is excited and radiates noise into interior. Another example is the air conditioning tube passing through the dash panel: the engine vibration will be transmitted to the dash panel through the tube, and then the excited panel radiates sound.

The primary task of panel structure research is to establish the relationship between the panel mode and the excitation. The main body panel modes include the dash panel mode, roof mode, floor mode, hood mode, and trunk lid mode. Almost all the excitation sources are likely to excite the panels, so it is necessary to determine the source frequencies and establish a modal planning table that includes the panel modal frequencies and the excitation frequencies. From the table, the modal coupling relationships between the panel modes and the excitations can be clearly discerned, and attempts can be made to decouple them. In addition, adjacent panels should have different modal frequencies, and the panel modes should also be different from the cavity modes.

The second task of panel structure research is to study the vibration characteristics of panels. Body panels are too complex for the use of analytical methods in solving vibration problems. In engineering, the structural modes and vibration responses of body panels are acquired by testing and/or numerical calculation methods. However, some body panels can be simplified as a basic supported plate to allow the use of an analytical method.

The third task of panel structure research is to study the acoustic radiation characteristics of panels. Panel vibration brings two types of noise problems: first, panel vibration can directly radiate noise to the interior, and second, an interior booming noise can be generating by coupling between the panel mode and the cavity acoustic mode. The sound level of the direct radiation is given as the radiated sound power (W_{rad}), and expressed as

$$W_{rad} = \sigma \rho_0 c S \langle \overline{u}^2 \rangle, \tag{1.2}$$

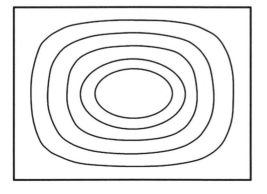

Figure 1.18 Structural wave pattern of a vibrated panel: first mode.

where σ is the sound radiation coefficient, ρ_0 is the air density, c is the sound speed, S is the panel area, and $\langle \bar{u}^2 \rangle$ is the square of the average sound speed.

The study of panel sound radiation includes the radiation mechanism of the panel structure, sound radiation efficiency, and analysis of the contribution of each panel to sound radiation. Studying the radiation mechanism involves analyzing the characteristics of bending wave propagation inside the panel, including the wave pattern, speed, and frequency, and finding the sound radiation process of the bending waves to the interior. The radiation capacity of the bending waves depends on their frequencies. The bending wave frequencies can only radiate sound when they are higher than a critical frequency. According to Eq. (1.2), the radiated sound power is proportional to the square of the panel vibration speed. The panel radiation efficiency represents its radiation capacity: i.e. sound energy radiated to the air per unit of time. The greater the radiated sound energy, the higher the efficiency. Panel radiation contribution analysis involves determining the ratio of the interior sound pressure contributed by each panel to that contributed by the panels overall in order to find the main panels that contribute to the interior sound.

The fourth task of panel structure research is to control panel vibration and sound radiation, which can be done by adjusting the stiffness, damping, and weight of local panels, and by using vibration dampers.

The panel stiffness perpendicular to the panel determines its modal frequencies. For a panel with low frequency, the best method to increase its frequency is to add support onto its surface. For example, the frequencies of most dash panels are between 50 and 150 Hz, and a dash panel can be excited by components that pass through it, such as the steering shaft, air conditioning tubes and clutch cable. The excitation frequencies may be coupled with the panel frequency. To separate this coupling, the panel structure must be modified. There are several ways to reinforce the panel: the first way is to punch a flat panel to form a beaded panel, the second is to force the flat panel in several different planes, and the third is to weld supporting beams onto the panel or to coat it with reinforcement adhesives. In some locations where it is hard to place supporting beams, reinforcement adhesives can be used, such as the adhesive used on outer door panels. Figure 1.19 shows a flat panel and a bead panel of the same size; the frequency of the bead panel is much higher than that of the flat panel.

When supporting beams cannot be used to reinforce a panel's stiffness, damping treatment is the most commonly used method to suppress panel vibration and sound radiation. Damping material is pasted onto the panel surface, a sandwiched damping

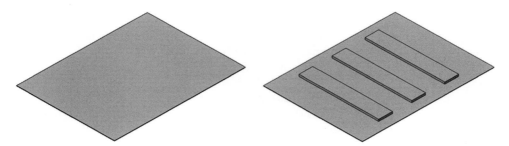

Figure 1.19 Flat plate and bead plate.

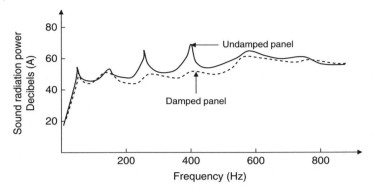

Figure 1.20 Comparison of magnitude of sound radiation of an undamped panel and a damped panel.

structure is installed on the panel, or a multi-layer damping plate is used directly. Figure 1.20 shows a comparison of the magnitude of the sound radiation of an undamped panel and a damped panel.

There are three methods of damping treatment: free damping, constrained damping, and laminated steel. In free damping, the damping material is directly pasted onto the surface of a panel, such as the floor, wheelhouse, or luggage compartment, to attenuate the road noise. In constrained damping, a damping layer is sandwiched between a body panel and a constrained metal sheet to form a "sandwich" structure: for example, a sandwiched panel used on the rear wheelhouse can attenuate noise from the road and splashing water. Laminated steel is an independent "sandwiched" damping panel that is also a form of constrained damping. In some vehicles, laminated steel is for the dash panel.

Damping material is used where the strain energy of the panel structure is at maximum. Damping treatment usually aims to suppress vibration and sound radiation at middle frequencies (200–500 Hz), especially at resonant frequencies. Usually, the damping treatment is also associated with the sound package to improve the sound transmission loss (STL). In some cases, damping is also effective at high frequencies (above 1000 Hz).

Mass can also be used to adjust the panel modal frequency. A mass block (also called a mass damper) can be regarded as a special dynamic damper. After being placed on a panel surface, the mass block can not only change the panel's frequency, but also suppresses its vibration magnitude. Mass dampers are widely used to attenuate interior booming by placing them on a problem panel. First, the panel that has the same frequency as the interior booming frequency is identified, then its mode is analyzed or measured, and finally the optimal location for the damper is determined. However, compared with the stiffness control method, the dynamic tuning range of the mass damper is narrow.

A dynamic damper is an additional mass-spring system that is added onto the panel structure to suppress its vibration at a certain frequency. After identifying the panel structure radiating noise and its frequency, an additional mass-spring system with the same frequency of the panel structure is designed. The mass-spring system suppresses the panel vibration and sound radiation, and reduces booming.

1.3.2.2 Acoustic Cavity Mode

The air inside the body forms an enclosed cavity. The enclosed air is similar to a solid structure and has its own mode. The mode formed by the enclosed air is called the acoustic cavity mode. The mode distribution of the structure is characterized by displacement, whereas the mode distribution of the enclosed air is described by pressure. Figure 1.21 shows the first acoustic cavity mode of a vehicle body.

The modal shapes and frequencies of the acoustic cavity modes are determined by the interior space and the medium, whereas the space depends on the styling and interior design. When the styling is finalized, it is almost impossible to change the cavity modes and frequencies. The frequencies of the cavity modes are relatively low: for example, the first modal frequency for a sedan is between 40 and 60 Hz, and the modal shape (sound pressure) varies along the longitudinal direction of the vehicle body. The pressures are large in some locations, but small in others. The shape looks like an accordion, as shown in Figure 1.21. Similar to the first modal shape, the second modal shape also changes along the longitudinal direction. The third modal shape changes along the lateral direction of the vehicle body, and the high-order modal shapes are more complicated.

The acoustic cavity modes usually bring two kinds of noise problems. The first problem is the modal coupling motion between the acoustic cavity mode and the body panel mode. Excited by an external excitation, the body panel pushes against the cavity. The panel acts like loudspeaker membrane, generating sound. This tiny sound is amplified inside the acoustic cavity mode, inducing a booming when the modal frequencies of the panel and the cavity are the same. The second problem occurs when the frequency of an external noise source, such as an exhaust orifice noise, is coupled with the frequency of the acoustic cavity mode, generating interior booming.

Acoustic cavity mode studies involve three aspects. The first is studying the characteristics of the cavity modal frequencies and shapes, including the measurement and analysis of acoustic cavity modes. The second is studying the coupling relationship between the panel structural modes and acoustic cavity modes, and finding methods to decouple them. The third is studying the influence of the acoustic modes on the sound–sound transfer function: i.e. the relationship between the external sound excitations and the acoustic cavity modes. The mode that induces interior booming is often the first

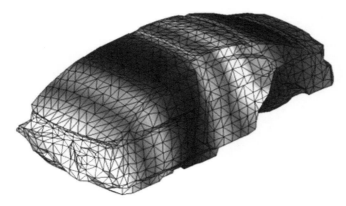

Figure 1.21 First acoustic cavity mode of a vehicle body.

one, so the key focus of these studies is controlling the first cavity mode and finding methods to avoid its excitement.

1.3.2.3 Control of Accessories

Many components, such as the steering system, mirrors, and so on, are installed on the body. If they are inappropriately designed, they can produce noise and vibration problems, such as a steering wheel "nibble" (oscillation at cruising speeds) or shaking mirrors. From the NVH point of view, the accessory structures are divided into three categories: bracket, steering system, and seat.

A bracket, such as a bracket connecting the engine to the body or a hanger connecting the exhaust system to the body, is a bridge between the body and a system. Figure 1.22 shows a bracket (a), and a bracket connecting with an engine mounting (b). The bracket is the transfer path of the external noise and vibration sources to the body, and often brings noise and vibration problems. A bracket with a low modal frequency is easily excited, and a vibration source easily passes along the transfer path and reaches to the body; in addition, the bracket easily radiates sound. For example, a mounting bracket with 300 Hz modal frequency is subjected to excitation from a four-cylinder engine at 4500 rpm; because the fourth-order excitation frequency corresponding to the speed is 300 Hz, the bracket resonates and transmits the engine vibration to the vehicle body, and could induce interior sound resonance. A bracket and its connected system constitute a mass-spring system; when the system frequency is consistent with the external excitation frequency, it resonates. For example, the mass-spring system of a battery and its holder has a relatively low frequency, so the holder is easily excited by the road excitation and the vibration is transmitted to the vehicle body via the bracket.

The bracket can be regarded as an extension of the body. The body structural characteristics should be included in the bracket design, and one principle of bracket design is that no resonance is induced for the bracket. The bracket should be as short as possible

(a)   (b)

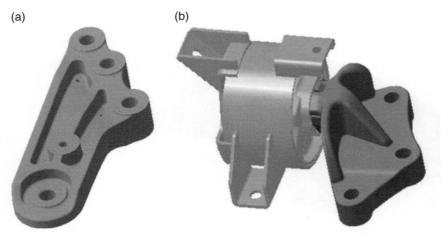

Figure 1.22 Engine mounting brackets. (a) A bracket. (b) A bracket connected with an engine mounting.

so that its frequency is high enough to avoid the excitation frequency, and it must be installed at a body location with high dynamic stiffness.

The steering system is composed of the steering shaft system and CCB system, as shown in Figure 1.23, and the steering wheel is mounted on the steering shaft. Because the steering system is connected to the body, the vibration transmitted to the vehicle body is likely to be transmitted to the steering wheel. There are three typical vibrations on the steering wheel that the driver can perceive: vibration at idling, shake during acceleration, and nibble at cruising. The principle way to avoid these problems is to keep the frequency of the steering system away from the external excitation frequency; therefore, the frequency of the steering system is usually designed to be as high as possible. The general rule is that the frequency should be at least 3 Hz higher than the excitation frequency. For example, the speed of a four-cylinder engine at idle and with the air conditioning on is 900 rpm, and the corresponding second-order frequency is 30 Hz, while the cooling fan speed is 1900 rpm and the corresponding first-order dynamic unbalance frequency is 31.7 Hz, so the steering system frequency must be designed to be 35 Hz or higher.

The frequency of the steering system depends on the frequency of the steering shaft system and the frequency of the CCB system. The frequency of the steering shaft system is determined by the stiffness of the shaft, the positions of the supporting points, the stiffness of the supporting brackets, and the weight of the steering wheel and the airbag. The stiffness of the CCB system is determined by the beam stiffness, the connection between the beam and the A-pillars, the supporting points on the beam, and the stiffness of the supporting brackets. The frequency of the steering system can only be increased if the frequencies of both the steering shaft system and the CCB system are enhanced.

The seat is composed of a seat frame, cushion, and backrest. The frame acts as a support, so it must have sufficient stiffness to separate the seat modal frequencies from the excitation frequencies of the engine and road, and also to avoid resonance. The cushion and backrest are in direct contact with the occupants, affecting their ride comfort, so their frequencies must be kept away from frequencies that are sensitive to the human body in both the vertical and lateral directions.

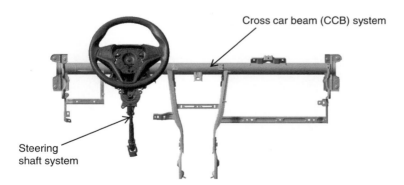

Figure 1.23 Diagram of a steering system.

1.3.3   Sound Package for Vehicle Body

Interior trimming, including sound absorption and insulation materials, is installed on the BIW not only to decorate the interior but also to absorb and insulate noise. Conventionally, non-metallic materials, structures, and techniques associated with acoustical treatment in the vehicle body are referred to as the sound package.

An index to evaluate the effect of the sound package is noise attenuation (NR). It is defined as the difference between the sound pressure level (SPL_{out}) of the outside sound source and the SPL_{in} inside the vehicle, and is expressed as

$$NR = SPL_{out} - SPL_{in}. \tag{1.3}$$

A sound simulator is placed outside the vehicle, and then sound pressures in vicinity of the simulator and inside the body are measured simultaneously. Finally, the NR is obtained by subtracting the sound pressures. For example, to measure the attenuation of engine noise transmitted into the interior, an engine noise simulator is placed in the engine compartment, whereas to measure the attenuation of exhaust orifice noise transmitted into the interior, an exhaust noise simulator is placed near to the exhaust orifice.

In general, the NR increases with frequency increase at a 6 dB/octave slope, as shown in Figure 1.24. The greater the NR, the better the sound insulation effect. For vehicles with a good acoustic package, the NR should be 35–40 dB at 1000 Hz.

Under normal circumstances, the sound package refers to the "special" sound package, including body sealing, sound absorptive materials and structures, and sound insulation materials and structures. The "general" sound package represents a combination of the "special" sound package with damping materials and structures and reinforcement materials and structures. Usually, the damping materials and structures are described in the study of panel vibration.

1.3.3.1   Body Sealing

No matter how good the body is, and no matter how many sound insulation and sound absorptive materials are used, if there are apertures on the body, outside noise will

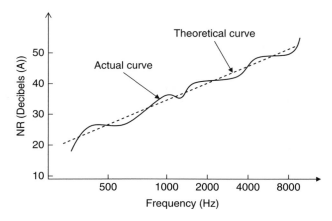

Figure 1.24 Noise reduction (NR) increases with frequency increase.

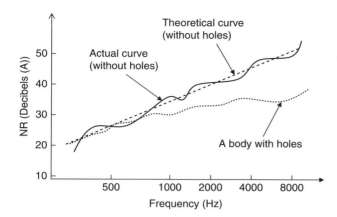

Figure 1.25 Noise reduction (NR) comparison of an ideal body without holes and a body with holes.

directly pass through these holes and enter the interior. Figure 1.25 shows an NR comparison between an ideal body without holes and a body with holes. The apertures significantly reduce the body NR, especially at high frequencies. Therefore, the most basic work of the sound package is to achieve a good body sealing.

Body sealing is divided into static sealing and dynamic sealing. Static sealing refers to body sealing when the vehicle is at a standstill; dynamic sealing refers to body sealing when the vehicle is running. When the vehicle moves, some components could have relative movement (such as the doors and body frame). Good static sealing does not guarantee good dynamic sealing.

The apertures, or holes, on the body are divided into three categories: function holes, manufacturing process holes, and error-state holes. Function holes are those that are deliberately opened on the body panels to realize some designed functions. For example, there are many holes on a dash panel (as shown in Figure 1.26) to allow the steering shaft, air-conditioning tube, shift cable, and so on to pass through it. Manufacturing process holes are opened for certain manufacturing processes and then sealed after the processes are finished. For example, during electrophoresis, the BIW is immersed in an electrophoresis fluid tank; after the process is finished, the liquid must be drained out of the body through holes. Error-state holes are generated by design error and/or manufacturing error. They do not have a function and are not needed for any manufacturing process.

Static sealing measurement methods include the smog method, ultrasound method, and air leakage method. In the smog method, a smoke generator is placed inside the vehicle and releases smoke that penetrates the holes on the body and diffuses to the outside. People standing outside the vehicle observe the smoke and judge the locations and sizes of the holes according to the amount of leaked smoke. In the ultrasound method, an ultrasonic leak detector system is used to send and receive ultrasonic waves. The waves are emitted by a generator inside a vehicle and are detected by a receiver outside the vehicle if there are holes in the body. The characteristics of the ultrasound waves allow the quantity and location of leakage to be easily read on the receiver screen. In the air leakage method, air is blown into the body. When the blower stops, air will flow out of the body if it contains holes, and the pressure inside the body will drop.

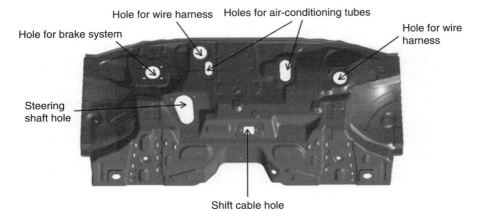

Hole for wire harness   Holes for air-conditioning tubes

Hole for brake system

Hole for wire harness

Steering shaft hole

Shift cable hole

Figure 1.26 Apertures/holes on a dash panel.

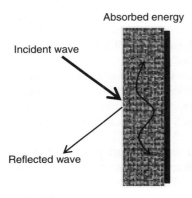

Absorbed energy

Incident wave

Reflected wave

Figure 1.27 Reflection and absorption of a sound wave.

The air leakage method uses the principle of the flow pressure difference inside and outside the vehicle to measure the body leakage.

1.3.3.2 Sound Absorptive Material

When sound waves are transmitted to the surface of a sound absorptive material, some of the energy is reflected back and some is absorbed and converted into heat, as shown in Figure 1.27. It is assumed that sound waves are only reflected and absorbed, and are not transmitted. The sound absorption capacity of the material is represented by the sound absorption coefficient. Figure 1.28 shows the sound absorption coefficients versus frequency for three types of sound absorptive materials. The coefficients increase with the increase in frequency. The coefficients are relatively low at a low-frequency range, are high at the high-frequency range, and then trend to becoming stable after a certain frequency. The material's sound absorption coefficient can be measured inside an impedance tube or reverberation chamber.

Sound absorptive material is soft and porous. Factors affecting the material's sound absorption performance include flow resistance, porosity, structural factor, density, and thickness. The flow resistance represents the material's permeability. The higher the flow resistance, the worse the material's permeability. The material's flow resistance should be controlled within an appropriate range. The porosity is the ratio of the volume of air in the material to the total volume of the material, which reflects the density of the material. The sound absorption coefficient increases as the density of the material increases. But after the density increases to a certain value, the porosity decreases and the flow resistance increases, so its sound absorption performance increases at a low-frequency range, but decreases at a high-frequency range. Therefore, each material should have a suitable density. The structural factor is a dimensionless parameter that reflects the internal shape and arrangement of the material. Structural factors have little

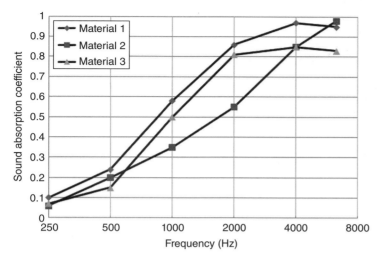

Figure 1.28 Sound absorption coefficients versus frequency for the three types of sound absorptive materials.

impact on the low-frequency sound absorption performance of the material, but have significant influence on the high-frequency absorption performance. The sound absorption coefficient increases as the thickness of the material increases, especially in the low-frequency range, but when the thickness increases to a certain value, the relative increase in the sound absorption coefficient begins to drop.

The porous sound absorptive materials used in vehicles mainly fall under three categories: cotton felt, foam, and glass fiber. Cotton felt is cheap and widely used in economy vehicles. Foam has good sound absorption effects and is widely used in mid-range and luxury vehicles. Glass fiber has good thermal insulation and moisture-proofing effects, so it is often used as a hoodliner and sound insulator inside the engine compartment.

Sound absorptive materials are widely used in automotive interior parts, such as the hoodliner, dash insulator, roof lining, and so on. Sound absorptive materials are also freely placed inside A/B/C-pillars, rockers, door trims, wheelhouses, instrument panels, and so on.

1.3.3.3 Sound Insulation Material and Structure

When sound waves are transmitted to the surface of a sound insulation material, some waves are reflected back, some are absorbed by the material and converted to heat, and some continue to pass through the material, as shown in Figure 1.29. The effect of sound insulation is measured by the amount of STL. There are two ways to measure STL: via an impedance tube or in a special laboratory that combines a reverberation chamber and an anechoic chamber.

Sound insulation structures include single-plates and double-plates. The sound insulation performance of a single-plate is determined by its density (or mass), stiffness, and damping. At a low-frequency range, the sound insulation effect is controlled by the stiffness. Above a certain frequency, the sound insulation is determined by its mass, and increases by 6 dB when the mass is doubled. In this region, the sound insulation increases with frequency, and when the frequency is doubled, the sound insulation

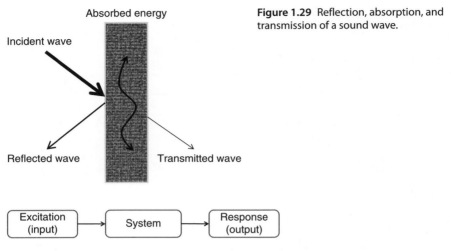

Absorbed energy

Incident wave

Reflected wave

Transmitted wave

Figure 1.29 Reflection, absorption, and transmission of a sound wave.

| Excitation (input) | → | System | → | Response (output) |

Figure 1.30 Relationship between a system and input and output.

increases by 6 dB. At a high-frequency range, coincidence occurs. When the frequency approaches the coincident frequency, the sound insulation decreases rapidly and is also affected by the structural damping. However, double-plate overcomes the shortcomings of single-plates. The double-plate structure has a good sound insulation effect at high frequencies, so it is used in some critical locations on the vehicle, such as the front windshield.

The body panels and glass are sound insulation structures. Sound insulation material is rarely used alone; instead, it is usually combined with sound absorptive materials. For example, the dash insulator is combination of sound insulation and sound absorptive materials.

1.3.4 Body Noise and Vibration Sensitivity

1.3.4.1 Transfer Function and Sensitivity

When an input excitation is applied to a system, an output response is generated, as shown in Figure 1.30. The ratio of the output response to the input excitation is called the transfer function, and is expressed as

$$H(\omega) = \frac{Y(\omega)}{X(\omega)}, \tag{1.4}$$

where $X(\omega)$ and $Y(\omega)$ are the input function and output function, respectively.

The body is a system in which the external noise and vibration sources exerted on the body are inputs, and the interior noise and vibration are responses. For example, when a mounting point between the power plant and body is impacted and exerts a vibration on the body, the generated interior noise and the vibration on the steering wheel are the outputs. The ratio of the interior noise to the vibration on the mounting point is called the sound–vibration transfer function, whereas the ratio of the vibration on the steering wheel to the vibration on the mounting point is called the vibration–vibration transfer function.

In body noise and vibration analysis, the word "sensitivity" is often used. Sensitivity refers to how sensitive the output response is to the input excitation in a system. In fact, sensitivity is the transfer function, but in the automotive NVH world, engineers prefer to use the word sensitivity to emphasize the sensitivity of the output to the input. For example, for vehicle A, a 1 N force is applied to a mounting point and 55 dB (A) of interior sound is generated, so the sound–vibration sensitivity is 55 dB N^{-1}; whereas for vehicle B, a 1 N force is exerted on the same mounting point and 60 dB (A) of interior sound is generated, so the sensitivity is 60 dB N^{-1}. Therefore, we can say that the sound inside vehicle B is more sensitive than that of vehicle A for the same force input.

The sensitivity reflects the characteristics of a system. For a linear system, the sensitivity is independent of the input and output; however, for a nonlinear system, the sensitivity not only relates to the system, but also depends on the input and output. There are many nonlinear systems on a vehicle, such as the seats. The sensitivity between seat cushion vibration (output) and seat rail vibration (input) varies with the magnitude of the external excitation. Strictly speaking, the body is a nonlinear system, but in engineering, the body can be approximately regarded as a linear system.

Body sensitivity is divided into two groups: structural sensitivity and acoustic sensitivity. Structural sensitivity can be further divided into sound–vibration sensitivity, vibration–vibration sensitivity, and driving point dynamic stiffness.

1.3.4.2   Structural Sensitivity

Structural sensitivity reflects the influence of body vibration on interior noise and vibration: i.e. the sensitivity of the interior noise and vibration caused by applying force or vibration excitation on the body. Many points on the body are subjected to external excitations, such as the mounting points where the engine applies force to the body, the connecting points between the exhaust system and the body, the connecting points between the shock absorbers and springs and the body, and the connecting points of the driveshaft bearings.

Sound–vibration sensitivity is the ratio of the interior noise to a unit of excitation force and is expressed as (P/F), where P represents the interior sound pressure and F represents the force exerted on the body.

The severity of vibration on the steering wheel, seat, and floor generated by a unit of force is described by vibration–vibration sensitivity. This is defined as the ratio of vehicle vibration to a unit of excitation force and is expressed by (V/F), where V represents the vibration inside the vehicle and F is the force exerted on the body.

Structural sensitivity represents the transmission of structural-borne sound to the interior. Figure 1.31 shows the transfer processes of sound–vibration sensitivity and vibration–vibration sensitivity.

Control of sound–vibration and vibration–vibration sensitivities is key to controlling structural-borne sound transmission. The two sensitivities should be controlled within certain ranges: for example, generally, the sound–vibration sensitivity should be less than 55 dB N^{-1}.

Dynamic stiffness refers to the ratio of an excitation force to the displacement response and is a function of frequency. If the excitation force and the response are at same point, the dynamic stiffness is called driving point dynamic stiffness, and the point is called the driving point. Driving point dynamic stiffness reflects the strength

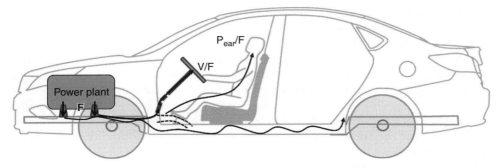

Figure 1.31 Transfer processes of sound–vibration sensitivity and vibration–vibration sensitivity.

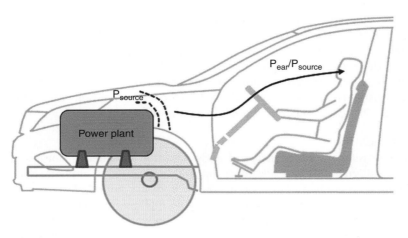

Figure 1.32 Transfer process of acoustic sensitivity.

of the structure at this point. The lower the driving point dynamic stiffness, the higher the structural sensitivity. Therefore, driving point dynamic stiffness is an important factor for controlling structural sensitivity transmission.

1.3.4.3 Acoustic Sensitivity
When a sound source is delivered to the body, some of the sound is reflected back, some is absorbed by the body, and some is transmitted to the interior through the body. Acoustic sensitivity is defined as the ratio of the interior sound pressure (P_{in}) to the outside source sound pressure (P_{out}) and is expressed as P_{in}/P_{out}. The acoustic sensitivity is the sound–sound transfer function, also known as sound–sound sensitivity.

The main sound sources outside the body include engine sound, exhaust tailpipe sound, air intake sound, and sound generated by friction between the tires and the road surface. These sources pass through the body and enter the interior. Figure 1.32 shows the transfer process of acoustic sensitivity. The acoustic sensitivity reflects the impact of the external sound sources on the vehicle interior noise, which depends on the sound insulation and sound absorption performance of the body.

1.3.4.4 Sensitivity Distribution Charts

Sensitivity is a function of frequency. If the sensitivities of different transfer paths are drawn on a sensitivity distribution chart, the noise and vibration contribution of each path at different frequencies can be clearly seen. There are two ways to present the sensitivity distribution.

The first way is to use curves, as shown in Figure 1.33. The abscissa is the frequency, and the vertical axis is the sensitivity value. Figure 1.33 shows the sound–vibration sensitivities of several structural-borne sound transfer paths. The response is the noise perceived by the driver, and the excitation points include the engine mountings, exhaust hangers, and so on.

The second way is to use a color map, as shown in Figure 1.34. The abscissa is the frequency, and each bar on the ordinate represents the sensitivity of one transfer path. The color represents the sound–vibration sensitivity value. In this figure, the excitation points include engine mountings in three directions, x, y, and z, and several exhaust hangers.

From the color map, it is easy to identify the contributions of each transfer path to the interior sound at different frequencies.

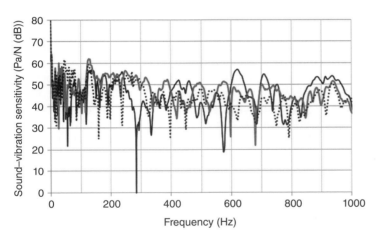

Figure 1.33 Sound–vibration sensitivity curves.

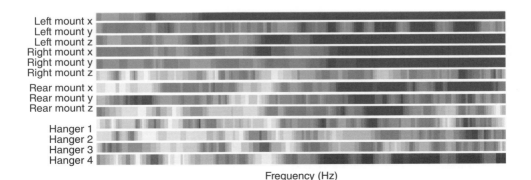

Figure 1.34 A color map of sound–vibration sensitivity.

1.3.5 Wind Noise and Control

The noise induced by the interaction between a moving vehicle and the airflow is called wind noise, also known as aerodynamic noise. When a vehicle is driven at high speed (such as $120\,\mathrm{km\,h^{-1}}$), the wind noise becomes prominent, and could even mask the engine noise and road noise. Customers are usually sensitive to wind noise, and if it is high, they may complain that it sounds as if a window or door is not completely closed, or that it affects their ability to hold a conversation or listen to the radio. Wind noise is dominated by middle- and high-frequency components. Today, people spend more and more time on highways, and their demands for low wind noise have become a prominent requirement for vehicle development.

The research areas of wind noise include:

- classification and mechanism of wind noise
- influence and control of body styling on wind noise
- dynamic sealing
- evaluation, testing, and analysis of wind noise.

1.3.5.1 Classification and Mechanism of Wind Noise

The interaction between the airflow and different body locations generate different noise sources that can be divided into four categories.

The first category is pulsating noise. The airflow acting on the body surface generates a vortex, as shown in Figure 1.35, which forms pressure fluctuations on the body surface. The noise generated by vortex disturbances is called pulsating noise. Beyond the vortex layer, the airflow is stable: this area is called the stable flow layer. In the vortex layer, the airflow creates a lot of small vortices attached to the body surface, which generate the pulsating noise. The noise generated by interactions between the airflow and body protrusions (such as the antenna) is also pulsating noise. Pulsating noise is formed outside the vehicle, but it passes through the body and enters the interior, so the noise transmission is airborne sound transmission.

The second category is aspiration noise. The wind noise outside the vehicle can pass through the body apertures and enter the interior. Even if there are no apertures on the body when it is standstill, small openings could appear between parts such as the doors

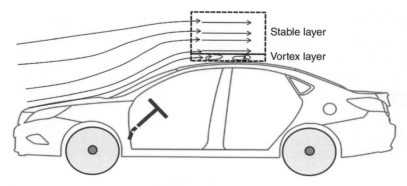

Figure 1.35 Airflow vortex layer and stable layer.

and the body frames when the vehicle moves. Wind noise that penetrates apertures or openings is called aspiration noise.

The third category is buffeting noise. When a sunroof or side window is opened, the body works like a resonator, and a low-frequency booming called buffeting noise will be generated by the interaction between the airflow and the opened body cavity.

The fourth category is cavity noise. Gaps always exist on the body surface, such as gap or margin between an A-pillar and a door. If the gap is large enough, a small cavity will be formed. When the airflow enters the small cavity, the airflow oscillates inside the cavity and generates cavity noise.

Pulsating noise and cavity noise pass through the body and enter the interior. Aspiration noise and buffeting noise directly enter the interior through body openings. If the openings are small, the wind noise is aspiration noise, whereas if the openings are large, the wind noise is buffeting noise. In short, all four categories of noise enter the interior by airborne sound transmission.

The vortices attached to the body surface excite the body panels like numerous small hammers. The excited panels radiate noise to the interior: this is structural-borne sound transmission of wind noise.

1.3.5.2 Influence of Body Styling on Wind Noise

Vehicle styling is one of the main influences on wind noise. The styling includes overall body styling and local structure styling.

In terms of overall body styling, the body should have a streamlined contour, and the transitions should be as smooth as possible. There are many transition surfaces on the body, such as the transition from the engine hood to the front windshield, and these surfaces must have sufficient curvatures and be smooth. Protrusions and large gaps on the body surfaces must be avoided. For example, protrusions underneath the vehicle, such as the exhaust system and subframe, can be hidden by a belly pan cover to achieve smooth surfaces, as shown in Figure 1.36. The small gaps reduce the possibility of generating cavity noise and pulsating noise.

Local structure styling and design involves many components, such as the side mirrors, antenna, door handles, and roof luggage rack. The local structure styling should follow

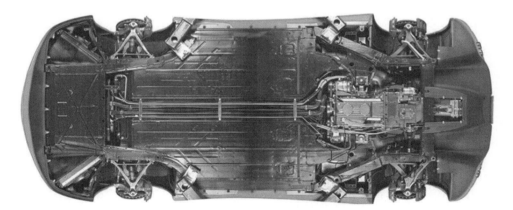

Figure 1.36 Underbody belly pan cover.

several principles. The first is to hide the local components beneath the airflow. For example, the door handle and door should be designed on the same plane. The second principle is that the local structures should have good streamlining so that air flows smoothly along the structures. The third principle is to guide the airflow away from sensitive areas: for example, a deflector can be installed at the front of the roof so that the airflow cannot directly move into the body cavity and buffeting noise is avoided. The fourth principle is to break the single-frequency pulsating noise: for example, a cylindrical antenna can cause an annoying tonal noise, but a spiral antenna can disperse the noise.

1.3.5.3 Control of Dynamic Sealing

The sealing described in the section "Sound Package for Vehicle Body" is static sealing. Static sealing involves sealing holes on the body when the vehicle is at a standstill, and its purpose is to prevent noise, dust, water, etc. from entering the vehicle. Dynamic sealing refers to the sealing between the relatively moving components for a moving vehicle. Gaps or openings between relatively moving components could appear when a vehicle moves even if the static sealing is prefect. A good sealing has the capability to dynamically compensate for gaps that form as a result of the components' movement.

If the body is not dynamically sealed, wind noise will directly pass through the dynamic gaps on the body and enter the interior, generating aspiration noise. Poor dynamic sealing usually generates noise problems at frequencies over 300 Hz.

The main factors affecting dynamic sealing are deformations of the body, doors, and other structures, and deformations of the seals or leakage caused by structural deformations that are not dynamically compensated for. The body and doors must have high enough stiffness and modes so that their structural deformations can be controlled. The deformations should be controlled within the elastic deformations of the seals.

In terms of seals, the main factors affecting dynamic sealing include the amount of elastic deflection and compression load. The design of a seal encompasses material selection, cross-section shape, load analysis, and so on. The variation of compression load with deflection should be as small as possible to meet the requirements of both the dynamic sealing and door closing force. The cross-section of a seal has a great influence on the compression load. There are two typical cross-sections: a single-bulb seal and a double-bulb seal. The curves of the compression load varying with the deflection are relatively flat for double-bulb seals.

1.3.5.4 Evaluation, Testing, and Analysis of Wind Noise

Evaluation and testing of wind noise can be implemented in a wind tunnel and proving ground. The indexes for evaluating wind noise are usually SPL, articulation index, and loudness, and so on. In the proving ground, interior noise can be subjectively evaluated and tested. In the wind tunnel, not only interior noise but also outside wind noise can be tested simultaneously. Beamforming (also called acoustic camera) or acoustic mirror can be used to measure the wind noise outside the vehicle, while a laser vibrometer can be used to measure body panel vibration.

In analysis of interior wind noise, the source–transfer path–receiver model in Figure 1.13 can be used. The source is the wind excitation exerted on the body: i.e. surface pressure fluctuation. The distribution of surface pressure can be measured in the wind tunnel via many small pressure sensors or microphones placed on the body surface.

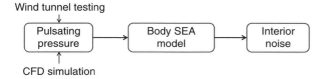

Wind tunnel testing

CFD simulation

Figure 1.37 Source–transfer path–receiver model used in wind noise analysis.

The surface pressure can be also calculated using the computational fluid dynamics (CFD) method. The components of wind noise are in the middle- to high-frequency range, and a statistical energy analysis (SEA) is used to analyze the model, as shown in Figure 1.37.

1.3.6 Door Closing Sound Quality and Control

Many components on the body can be opened and closed, such as the doors, trunk lid, hood, glove box, and sunroof. When they are closed and opened, sound is generated, and excessive sound can have a negative effect on the customer's perception. Among these components, the doors are the most frequently opened and closed: therefore, the sound quality of a closing door is used as an example to illustrate the sound quality in this section.

The study of closing door sound quality includes subjective evaluation and objective evaluation, and methods for controlling the sound quality.

1.3.6.1 Door Closing Sound Quality

In fields such as ship and aircraft engineering, studies of noise and vibration do not include harshness: i.e. the subjective sensation on the human body. But harshness is given special attention on automotive engineering: this is because people have a much closer relationship with automotive vehicles than with aircrafts or ships. In modern society, the automotive vehicle is not only a transport tool, but also an entertainment "toy," so it has a significant impact on people's daily lives. Customers expect comfortable, delightful, and economic vehicles with good sound quality.

Noise and vibration mainly relates to frequency, but it can also relate to engine speed and firing order, which creates the unique characteristics of the vehicle's sound.

A quality is a feature that makes an object unique. Sound is the reception of audible mechanical waves and their perception by the brain: i.e. the human auditory impression. But sound quality is the combination of "sound" and "quality": that is, the unique sensations of the sound. Today, sound quality has become an important attribute of automotive NVH. The work of NVH engineers is not only to reduce noise and vibration, but also, and perhaps even more importantly, to enhance the sound quality. Sound quality has become an important part of automotive DNA.

Door closing sound quality refers to customers' perception of the sound generated at the moment of closing the door, and it can be an important component of a customer's decision to purchase a vehicle. When customers evaluate vehicles at stores or showrooms, they usually pull and push the doors. Along with the vehicle's overall styling, closing the door provides the customer's first impression of the vehicle. If they heard a low-level, single-impact, solid, and clean sound, they might think it is a good vehicle. But if they heard a loud, multi-impact, sharp, fragmented, metal-percussion sound, they might judge it as a bad vehicle, and even have doubts about the vehicle's quality and safety.

1.3.6.2 Evaluation of Door Closing Sound Quality

When customers close a vehicle door and hear a single-impact, clear, vigorous, solid sound, they will perceive it to have good sound quality. But if they hear a sharp, fragmented, multi-impact, "ring-down" sound, they will perceive the sound quality as poor.

Door closing sound quality involves both subjective evaluation and objective evaluation. Subjective evaluation is when a group of people subjectively score the door closing sound quality. Usually, a 10-point scoring system is used, and the higher the score, the better. Objective evaluation is when the sound data are recorded and processed, and

(a)

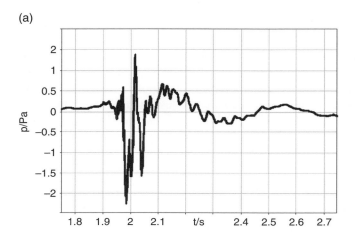

(b)

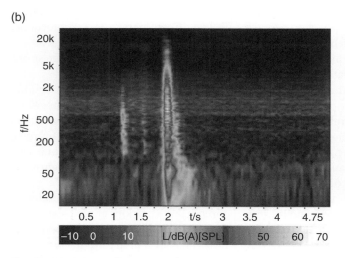

Figure 1.38 Curve and spectrum of a door closing sound. (a) Time-domain curve. (b) Time-frequency color spectrum.

some specific physical indexes are used to evaluate the sound quality. There are many indexes for describing sound quality, such as loudness, sharpness, articulation index, tonality, modulation (fluctuation and roughness), and order. Vehicle sound quality include the power train sound quality and electrical sound quality, and different indexes are used for evaluation of the different systems.

The indexes for objectively describing the door closing sound quality are loudness, sharpness, and ring-down. The time-frequency color spectrum should be analyzed together with the above indexes. Loudness is an auditory sensation index based on sound magnitude that is dominated by low-frequency components. Sharpness is a psychoacoustic index to measure the high-frequency components of a given sound. The sharpness can be interpreted as the ratio of the high-frequency components to the overall sound level. Ring-down refers to the lingering of a sound generated by the collision of two objects, and can be represented by a time-domain curve or a time-frequency color spectrum, as shown in Figure 1.38. From the curve or color spectrum, the frequency composition, number of impacts, and ring-down (or sound attenuation) of the door closing sound can be discerned.

1.3.6.3 Control of Door Closing Sound Quality

During the door closing process, the door and body are subjected to three impacts: the impact between the door and the body frame, the impact inside the door lock, and the impact of the seals during the process of being compressed. When the door is subjected the impact force, the door and accessories (such as glass, internal panels, etc.) generate sound that radiates to the air and can be heard inside and outside the vehicle, and the jiggle on the door panels and the shake of the trimmed parts and glass can be observed.

Door closing sound quality is affected by the stiffness of the door, the structure of the outer and inner panels, the lock structure, the structures of the latch and paw, the seals, the driving point dynamic stiffness of the striker, and so on. These factors must be analyzed to achieve a good door closing sound quality. The influence of the door, lock, and seal on the sound quality is briefly described below.

The outer panel and inner panel of the door must have sufficient stiffness, otherwise they will generate a number of "twitter" sounds. The stiffness of the outer panel can be increased by adding supporting beams and/or pasting reinforcement adhesives on its internal side. The stiffness of the inner panel can be enhanced by punching ribs on its surface. The locations at which door locks and speakers are installed must have sufficient local stiffness. Damping pasted on the panels can suppress their vibration and sound radiation.

The striker is mounted on the body and is subjected to impact from the latch, so the driving point dynamic stiffness of the striker must be sufficiently high. If the lock body is wrapped by soft materials (such as rubber, etc.), it will be well isolated and the radiation sound induced by the internal impact can be reduced.

The seals also have a significant impact on door closing sound quality. The seals prevent direct metal impact between the door and the body, and also reduce the sound generated by the impacts between the latch and striker and by the lock's internal parts. The door closing force should be evenly distributed across the seals to enhance the sound quality and hand sensation of the door closing action. Therefore, the compression load distribution and the elastic deflection of the seals are important factors for controlling door closing sound quality.

1.3.7 Squeak and Rattle of Vehicle Body

S&R refers to the abnormal and irregular sounds generated when a vehicle moves, which can be random and inconsistent. S&R is different from engine noise, road noise, and wind noise, which tend to be regular and last for a relatively long time or exist all the time. Engine noise, for example, is consistent and directly relates to the firing order and resonance of the structures. The research scope of S&R includes:

- mechanism of S&R
- identification of S&R
- control of S&R.

1.3.7.1 Mechanism of S&R

In S&R, squeak refers to a transient noise generated by the friction between two objects, as shown in Figure 1.39a. All forms of friction, including friction between metals, between metal and rubber, and between rubbers, have the potential to generate squeak. Squeaks can occur in many locations of the body: for example, friction between a front door and A-pillar could generate a squeak.

Rattle is a transient noise induced by the impact between two objects, as shown in Figure 1.39b. Rattle can be generated by impacts between metal and rubber or between metals, such as the impact between the latch and striker of a groove box. In another example, the connection between the bolts and internal parts inside the instrument panel (IP) could become loose, so that when the vehicle is driven on rough roads, the internal parts impact one another and cause a rattle.

1.3.7.2 Identification of S&R

S&R can be identified by testing, and can also be predicted by CAE analysis.

Testing can be done on the road or on a four-poster simulator. Testing on the road involves subjectively and objectively evaluating S&R problems by driving a vehicle on different roads and at different speeds. Some S&R can be heard on all roads, but some can only perceived by certain roads; thus, S&R should be tested on all possible road surfaces, including asphalt, cement, corrugated roads, random shock roads, cobblestones, bricks, bumper roads, gravel, and so on. S&R also relates to the vehicle's speed: S&R problems can appear at some speeds, but disappear at others.

For road testing, S&R problems are usually identified by subjective scoring. The evaluators drive the vehicles and score the severities of S&R on different roads. The vehicle S&R scores and road conditions are considered simultaneously; a comprehensive index, the S&R index (SRI), is used to represent the vehicle's overall S&R severity. The SRI is the summation of the vehicle's S&R contributed by all parts and on all tested roads. The

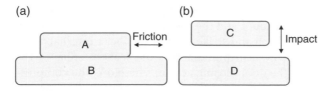

Figure 1.39 (a) Squeak. (b) Rattle.

lower the SRI, the fewer S&R problems there are. After the locations generating S&R have been identified, the S&R magnitudes can be determined by the tested accelerations and SPLs.

S&R problems are also commonly identified on a four-poster simulator. The simulator consists of four individual exciters, and the vehicle's four tires are placed on the corresponding exciters. Each exciter can generate a range of vibration signals, including random signals, sine sweeps, single-frequency inputs, and the collected road spectrum, and the four exciters can move in phase or out of phase. Compared with road testing, one advantage of simulator testing is that the evaluators can carefully identify S&R both inside and outside the vehicle, and can also identify the exact locations generating the S&R by using an NVH stethoscope.

S&R CAE analysis can be used to identify potential S&R-generating locations, and also to predict their probability. The S&R FE analysis includes: analysis of body stiffness and mode, modal analysis of the subsystem, body S&R sensitivity analysis, and overall vehicle response analysis. Insufficient body stiffness is the most important cause of S&R problems, so an analysis of body stiffness and the deflection of the door diagonal can be used to predict impacts between the door and body.

1.3.7.3 Control of S&R

The main causes of S&R problems include inappropriate gaps in body design, poor tolerance control, imprecise assembly or weak installation, poor compatibility of contacted materials, and low structure stiffness and mode.

S&R control involves not only testing the prototype vehicle to identify S&R problems, but also implementing S&R control across the whole vehicle development process. In the design phase, the clearance between adjacent components, the compatibility of materials, and the distribution of fasteners and bolts should be determined. Digital mock up (DMU) checking means identifying NVH and S&R problems in digital vehicle designs and using the data to check gaps between adjacent components, the connection status of wires and pipes with the body and interference between them, the material compatibility characteristics of parts in contact, the connection methods of the connected parts, the spans of the connected points, potential resonance, and so on. FE analysis is used to determine body stiffness, S&R sensitivities, and so on. In the prototype phase, the locations generating S&R are identified by road testing and four-poster simulator testing. In the mass production phase, the main work is to check the quality of parts, the quality control of the entire production process, and the performance of the produced vehicles.

1.4 Noise and Vibration Control During Vehicle Development

It usually takes around 3 years to develop a brand new vehicle, and NVH control takes place throughout the entire development cycle. At the beginning of development, the main work involves benchmark testing and analysis: i.e. analyzing the body NVH characteristics of competitor vehicles and setting up targets for the body's NVH. In the middle phase of development, the NVH work involves optimizing the body structure by CAE analysis, checking the digital prototype, testing the BIW, and analyzing and designing the sound package. During the late phase of the development, the NVH work involves testing the leakage of the BIW and trimmed body, the sound absorption and sound insulation performance, the body mode and sensitivity, and so on.

The NVH modal distribution and the NVH targets are used throughout the development process to guide NVH development. From target setting at the pre-development stage to target checking during the late phase, NVH work is deployed on the basis of the NVH targets, so developing suitable targets and effectively executing them is the key to ensuring good NVH performance of the developed vehicle.

1.4.1 Modal Frequency Distribution for Vehicle Body

The body carries all of the vehicle's systems and components, such as the engine, exhaust, and suspension. These interconnected systems have their own modes, and if their modal frequencies are the same or close, resonance could occur. Therefore, it is necessary to make a modal frequency table, listing the modal frequencies of all the relevant systems, so that it is clear where resonance could occur. In addition, after the frequency of each system has been determined, these systems can be developed independently. Vehicle development is a collaboration between an original equipment manufacturer and various suppliers, and many systems are developed in isolation.

During the early development phase, three modal frequency tables must be prepared: a vehicle modal frequency table, a body modal frequency table, and an excitation source frequency table. Figure 1.40 shows a vehicle modal frequency table.

The vehicle modal frequency table is the cornerstone of vehicle NVH development – it guides the development of the body and connected systems and components. The table includes the modal frequencies of the body and other main systems, along with the engine idling frequency, and it also shows the relationship between the modal

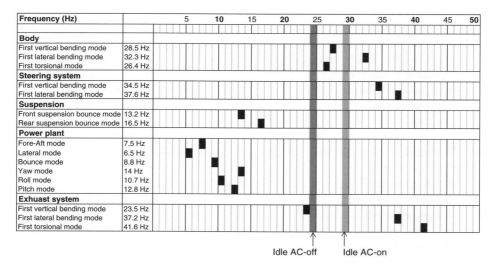

Figure 1.40 A vehicle modal frequency table.

frequencies of the connected systems: i.e. it provides information on the coupling between the body and the connected systems. The body modal frequency must be separated from the modal frequencies of the power plant, exhaust system, suspension system, and so on, as well as being separated from the engine idling frequency.

The body modal frequency table lists the frequencies of the overall body mode and local modes, and the main excitation frequencies. The main body local structures include the front panels, front floor, rear floor, front side frame, roof, steering wheel, mirror, and acoustic cavity, each of which has its own local mode. The local modes generate three kinds of noise and vibration problems: first, a local mode can resonate with the overall body mode; second, a local mode can resonate with a cavity mode; and third, a local mode can be excited by an outside source. The purpose of the body modal frequency table is to separate the local modal frequencies from the modal frequency of the whole vehicle, to avoid coupling between the local modes and the acoustic cavity mode, and to separate the local modal frequencies from the excitation frequencies.

The excitation source frequency table shows the frequencies of all possible sources of body excitation on a table or graph. These include stable excitation frequencies, such as the firing frequency during engine idling, as well as unstable excitation frequencies, such as frequencies during acceleration. There are many excitation sources, such as the engine, transmission, fan, tires, and so on. The purpose of making this table/graph is to compare it with the body modal frequency table to find frequencies at which the body may vibrate, and also to find methods to avoid such vibration.

Chapter 2 describes the body mode frequency table and excitation frequency table in detail.

1.4.2   Body NVH Target System

The body modal frequency distribution is a part of the body NVH target system. The body NVH target system consists of four levels: vehicle-level body NVH targets, trimmed body NVH targets, BIW NVH targets, and component NVH targets.

Vehicle-level body targets influence the whole vehicle's NVH performance, and include the body vibration targets and noise targets. The vibration targets specify bending and torsional modal frequencies for the body structure, and modal frequencies for panel structures (such as the dash panel, etc.) and accessories (such as side mirrors, instrument panels, etc.). The noise targets specify acceptable levels of vehicle air leakage and sound insulation, a range for the acoustic cavity mode, and parameters for door closing sound quality.

The vibration targets for the trimmed body include its bending and torsional modal frequencies, and vibration–vibration sensitivity of external vibration to interior vibration. External vibration refers to excitation applied to the body, such as the excitation applied by the power plant to mounting points on the body. Interior vibration refers to the perceived vibration of the steering wheel, seats, and floor. The noise targets for the trimmed body include the sound–sound sensitivity of external sound excitation transmitted as interior sound, and the sound–vibration sensitivity of external excitation transmitted as interior sound. External sound excitation includes engine radiation noise, exhaust orifice noise, and so on, and interior sound refers to sound heard by the occupants.

The BIW NVH targets include its bending and torsional modal frequencies, the modal frequencies of the panel structures, and the driving point dynamic stiffness. All of the function and manufacturing process holes on the BIW are blocked up in order to check its air leakage; the air leakage of the BIW is set as a noise control target.

The component targets refer to accessories connected to the body and include the modal frequencies of brackets, the engine hood, trunk lid, and so on; the sound absorption coefficient and sound insulation coefficient of the sound package; noise reduction; and targets relating to the door, such as the modal frequencies of the internal and outer door panels, and the driving point dynamic stiffness of the striker.

1.4.3   Execution of Body NVH Targets

The body modal frequency table and the NVH target system provide the guidelines for body NVH development. Strict implementation methods and procedures are necessary to achieve the NVH targets. The vehicle development process is divided into many milestones, and a target implementation plan and execution results should be clear for each milestone. The first goal is to achieve the component targets and BIW targets, then to achieve the trimmed body targets, and finally is to achieve the overall vehicle targets.

In the implementation process, milestone checking is very important. The vehicle program's chief engineer and the NVH chief engineer organize a special meeting to check the input goals and output requirements for each milestone. If the milestone requirements are not met, a risk analysis must be conducted and the work for the next milestone needs to be determined.

1.5   Structure of This Book

This book comprehensively expounds the mechanism and control methods of vehicle body noise and vibration. This book is divided into nine chapters, and covers the overall body vibration and mode, local structure vibration and noise, the sound package, sensitivity, wind noise, sound quality, S&R, and the NVH target system. The contents of each chapter are briefly described below.

Chapter 1, "Introduction" introduces the body structure and the noise and vibration problems induced by the body. It describes structural-borne sound and airborne sound, and their transfer paths to the vehicle interior. Based on analysis of the transfer paths, the key technologies of NVH control, such as body structure vibration control, the sound package, wind noise control, and sound quality control, are explained concisely. Body NVH control during vehicle development is briefly introduced, including establishment of the modal frequency tables and target system, and milestone checking.

Chapter 2, "Vibration Control of Overall Body Structure," describes NVH problems relating to the overall body structure. The overall body stiffness and mode are the cornerstones of vehicle NVH. The overall body stiffness is determined by the overall layout of the body structure, frame stiffness, joint stiffness, adhesive stiffness, and so on. This chapter describes the testing and analysis of body stiffness and modes. The principles of decoupling between a system and an excitation and decoupling between adjacent systems are illustrated.

Chapter 3, "Noise and Vibration Control for Local Body Structures," describes mechanisms of and control methods for body panel vibration and sound radiation. The stiffness and mode of the overall body frame structure affect the whole vehicle's NVH performance and S&R, while the modes of the local panel structures create noise problems at specific frequencies. The body panels include the dash panel, floor, roof, trunk lid, and so on. This chapter introduces coupling between the panel structural modes and the acoustic cavity mode, mechanisms of panel vibration and sound radiation, and the potential interior booming induced by the panels. This chapter also describes methods for changing panel stiffness and suppressing vibration, such as stiffness control, damping treatment, and so on. Finally, the chapter describes NVH control of several key components, such as brackets, the steering system, and the seats.

Chapter 4, "Sound Package," describes the mechanisms of sound absorption and sound insulation materials and structures. Static sealing is the most basic technology for the sound package. This chapter describes the measurement of and control methods for body sealing and cavity baffling. Sound absorptive materials and structures provide the main method for eliminating high-frequency noise, whereas sound insulation materials and structures provide the main means of eliminating low- and middle-frequency noise. This chapter describes in detail the application of sound absorption and sound insulation to the body.

Chapter 5, "Vehicle Body Sensitivity Analysis and Control," explains body NVH sensitivity from the perspective of the source–transfer path–receiver model. This chapter describes the characteristics of noise and vibration sources that are applied to the body, which are closely related to the body's sensitivities. Structural sensitivity represents the transmission of body structural vibration as interior noise and vibration, whereas sound sensitivity describes the transmission of exterior sound sources to the interior. This chapter also introduces driving point dynamic stiffness, which is closely related to structural sensitivity, and methods for controlling it.

Chapter 6, "Wind Noise," describes mechanisms of and control methods for wind noise when the vehicle is driven at high speed. This book classifies wind noise as pulsating noise, aspiration noise, buffeting noise, and cavity noise, and this chapter describes the mechanism of each in detail. In this chapter, the problems of wind noise caused by improper overall styling and local design are expounded, and the methods for controlling wind noise from a design perspective are given. Whereas static sealing is the basis of the sound package, dynamic sealing is the cornerstone of body wind noise control; this chapter describes the mechanisms and control methods for dynamic sealing, along with methods of wind noise testing in wind tunnels and on the road, and the numerical calculation method for wind noise.

Chapter 7, "Door Closing Sound Quality," introduces the concept of sound quality and evaluation indexes, and three kinds of sound quality problems including power train sound quality, electrical sound quality, and door closing sound quality. This chapter describes in detail the characteristics of door closing sound quality, the relationship between the sound quality color map and door components, and control methods for improving door closing sound quality.

Chapter 8, "Squeak and Rattle Control in Vehicle Body," describes the S&R phenomenon. The mechanisms of S&R are very complex, involving a lot of nonlinear problems. The locations and severity of S&R can be identified by testing on the road and/or by using a simulator. This chapter also emphasizes that S&R control can begin during the pre-development phase through body stiffness and modal analysis, subsystem modal analysis, body sensitivity analysis, and vehicle response analysis. Finally, this chapter details S&R control across the entire vehicle development process, including structural integration design, DMU inspection, matching of material friction pairs, and manufacturing process control.

Chapter 9, "Targets for Body Noise and Vibration," describes the principles of body NVH target setting, cascading, and control. Starting with the body structure, the chapter introduces NVH targets for the vehicle-level body, trimmed body, BIW, and components.

A few topics repeatedly appear in different chapters because they are associated with each other, which should help readers to deeply understand some particular inter-related problems.

2

Vibration Control of Overall Body Structure

2.1 Introduction

Body stiffness is not only one of the most important indicators in vehicle body design, but also an important indicator to determine the vehicle's driving quality and performance. This chapter introduces overall body stiffness and modal analysis, whereas Chapter 3 focuses on local body stiffness and modes.

2.1.1 Overall Body Stiffness

Structure stiffness refers to the structure's capability to resist deformation under the action of an external force: i.e. its capability of elastic deformation to restore its original shape. A vehicle body is a structure, so it has certain stiffness.

2.1.1.1 Classification of Vehicle Body Stiffness

A vehicle body undergoes two major forms of deformation: bending deformation and torsional deformation. Accordingly, the body stiffness can be categorized by these deformations: i.e. the body stiffness is divided into bending stiffness and torsional stiffness. Bending stiffness refers to the ratio of an external applied force to its corresponding displacement deformation, which represents the structure's capability to resist bending deformation, and its unit is N mm^{-1}. Torsional stiffness is the ratio of an external applied torque to its corresponding angle deformation, which refers to the structure's ability to resist torsional deformation, and its unit is KN-m rad^{-1}.

The body is divided into the body-in-white (BIW), trimmed body or body-in-prime (BIP), and full vehicle body. Accordingly, body stiffness can be divided into BIW stiffness, trimmed body stiffness and full vehicle body stiffness. All three can be further divided into bending stiffness and torsional stiffness: for example, BIW stiffness includes BIW bending stiffness and BIW torsional stiffness.

The body consists of many panels and parts, so body stiffness can be divided into overall stiffness and local stiffness. Overall stiffness refers to the capability to resist deformation of the full body, which influences the low-frequency noise and vibration performance of the vehicle. Local stiffness refers to the stiffness of local structures such as panels, which affects the local vibration and noise radiated into the passenger compartment. The frequency range for local vibration and sound radiation is relatively wide. Local stiffness includes panel stiffness (the dash panel, roof, floor, side walls, etc.),

Noise and Vibration Control in Automotive Bodies, First Edition. Jian Pang.
© 2019 China Machine Press. All rights reserved. Published 2019 by John Wiley & Sons Ltd.

component stiffness (the seats, steering wheel, etc.), and the stiffness of connecting points between the body and other systems (the engine, exhaust, etc.).

2.1.1.2 Problems Caused by Insufficient Body Stiffness

The body is foundation of the vehicle, and its stiffness determines the vehicle's driving quality and dynamic performance, including factors such as squeak and rattle (S&R), noise, vibration, and harshness (NVH), handling, collision safety, and reliability. Insufficient body stiffness can induce the following problems.

1) S&R. When a vehicle is driven on a rough or coarse road, or subjected to shock input, insufficient body stiffness will induce a large deformation of the body that could force it to impact or slide against the surrounding parts, resulting in S&R. For example, when deformations between a body closure opening and a door are inconsistent, impact between them could generate abnormal sounds. Another example is the squeal sound generated by friction between the body and an instrument panel due to low body stiffness.

2) NVH problems. The body modes are determined by the body's stiffness. Insufficient body stiffness reduces the vehicle's modal frequencies, which could then fall into the excitation frequency range. For example, if a vehicle has a first bending modal frequency of only 22 Hz and the idling speed for its four cylinder engine is 650 rpm, with a corresponding firing order frequency of 21.7 Hz, the body will resonate while the engine idles. Another example is that low body stiffness could make the deformations between the body closure opening and the doors different, which could cause excessive gaps among the body, the doors, and seals, generating wind noise.

3) Handling problems. Turning and changing lines requires high body torsional stiffness. Insufficient torsional stiffness will force the body to have a large torsional angle, which will reduce the handling capability.

4) Reliability problems. Body stiffness is one of the most important factors in determining vehicle fatigue performance and reliability. For example, when the tires of one side are subjected to an impact, if the body torsional stiffness is low, the deformation of that side will be large, which could cause the doors to become stuck in the body frame, and even stop them from being opened. Low stiffness of the local body structures could increase the local stress, so cracks and breakage at these locations could appear after high mileage running.

5) Collision and safety problems. Body stiffness affects the absorption of collision energy and the protection of pedestrians. The requirements for body stiffness in terms of collisions and NVH are in some cases are consistent, but in other cases are the opposite.

In conclusion, the importance of the body stiffness is very clear. The body should be stiff enough to avoid S&R, to achieve good performances for NVH, handling, reliability, and collision safety, and to ensure a high-quality vehicle overall. Body stiffness can even have a direct impact on potential customers, providing first impressions such as, "a non-safe vehicle," "a comfortable car," "a reliable vehicle," "a likable vehicle," etc.

2.1.1.3 Influence of Body Stiffness on NVH Performance

Body stiffness has a significant influence on NVH performance. Insufficient body stiffness directly reduces NVH performance and induces many other noise and vibration problems, such as booming, resonance, and dynamic sealing problems. Figure 2.1 shows the relationship between interior noise and bending stiffness and torsional stiffness for a group of vehicles driven at $40\,km\,h^{-1}$ on rough roads. The overall trend is that higher the body stiffness, the lower the interior noise.

The relationships between interior noise and body stiffness for vehicles under different running conditions, such as idling, accelerating, decelerating, and cruising, are similar to the relationship shown in Figure 2.1. Generally speaking, a vehicle's NVH performance increases with increases in body stiffness. Figure 2.2 shows the relationship between subjective rating and body stiffness.

Of course, the stiffness of the body should not be excessively high. Extremely high stiffness will increase vehicle's weight and cost, and decrease some attributes, such as ride comfort. Therefore, the body stiffness should be controlled in a reasonable range. Reasonable stiffness refers to a good balance among performance, cost, and weight.

2.1.1.4 Importance of BIW Stiffness

The BIW is the core of a full vehicle body, and the BIW's stiffness is the base of the vehicle's stiffness. Many other systems and trim parts are installed onto the BIW, so it can be regarded as a carrier. BIW stiffness is the most important factor in determining the trimmed body stiffness and full body stiffness.

BIW stiffness determines the overall modes of the body. There are many accessories inside the trimmed body or full body, so the overall and local modes are mixed together;

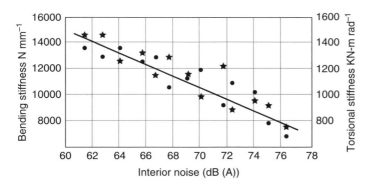

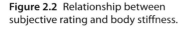

Figure 2.1 Relationship between interior noise and body stiffness.

Figure 2.2 Relationship between subjective rating and body stiffness.

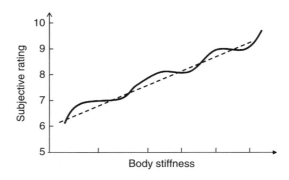

this makes it hard to focus on analyzing the overall modes. The overall modal shapes of a full body and a BIW are similar, but their frequencies are different. Therefore, The BIW modes provide a foundation for understanding the full body modes.

The BIW is relatively simple, and it is easy to analyze the main factors affecting the body stiffness. Therefore, research on BIW stiffness is a very important step for controlling the full vehicle's stiffness and modes.

2.1.2 Overall Body Modes

2.1.2.1 Modes and Modal Analysis

The mode is the inherent characteristics of a structure. For linear systems, the structure's modal characteristics are fixed: i.e. they are independent of external inputs and outputs. Modal parameters of structures include modal shapes, modal frequencies, and modal damping.

The structural modal parameters can be obtained by calculations or experiments. The body modes can be also identified by these two methods.

Calculated modal analysis refers to the method for obtaining the structural modal parameters by calculation. By transferring a system's physical coordinates into modal coordinates, the coupling differential equations in the physical coordinates are transferred into a series of independent equations in the modal coordinates, so the system modal frequencies and modal shapes are acquired in the decoupled modal coordinates. A modal shape means a virtual "vibration shape" for a certain frequency and is constituted by the relative displacements at different locations. Body modal analysis is usually implemented by finite element (FE) methods. A continuous body structure is decomposed into many discrete units by meshing the structure, and then the mass matrix, stiffness matrix, and damping matrix are constructed. After a series of iterative computations, each modal frequency and the corresponding modal shape are obtained. Because the modal parameters are obtained by calculation, the modes are called calculated modes.

Experimental modal analysis refers to a method of acquiring the modal parameters by testing. In body modal testing, the body is lifted by soft ropes or supported on air cushions that form the required "free–free" boundary conditions. Exciters are used to stimulate the strong parts of the body, and the accelerations at different locations and the excited forces at the excitation points are simultaneously measured. After the tested data have been processed and the parameters have been identified, the system's modal shapes and frequencies can be obtained. Because the modal parameters are obtained by testing, the modes are called experiment modes.

Modal analysis is an important method for studying structural dynamics that is widely used in vehicle noise and vibration control. The body modes and frequencies identified by the modal analysis can be used to predict its dynamic responses under external excitation and to provide clues for solving automotive noise and vibration problems.

2.1.2.2 Types of Modes

Like stiffness, the body modes can be classified in three ways: by modal shapes, by body components, and by overall and local structures.

Body modal shapes include bending, torsional, breathing, and composite forms, and the corresponding modes can be divided into bending modes, torsional modes, breathing modes, and composite modes.

Body modes based on the body structure are divided into BIW modes, trimmed body modes, and full body modes, corresponding to the BIW, trimmed body and full vehicle body.

The overall body and local structures generate the overall body modes and local modes, respectively. The BIW modes, trimmed body modes, and full body modes include bending modes and torsional modes, which are the overall body modes. Local modes include body panel modes, such as dash panel modes, ceiling modes, and floor modes, and component modes, such as steering modes and bracket modes.

The BIW modes are the foundation for analyzing the full body modes. The BIW modes are relatively simple to analyze and control, and represent the body's most essential features. There is a close relationship among BIW modes, trimmed body modes, and full body modes, and the trimmed body modes and the full body modes can be inferred from the BIW modes.

2.1.2.3 Importance of Body Modes

Body modal analysis plays a very important part in automotive NVH analysis. Specifically, body modal analysis can solve the following problems:

1) Understanding the dynamic characteristics of body structure in order to design a competitive vehicle. Insufficient design that does not meet the market's needs will make the vehicle uncompetitive. Conversely, gratuitous design that provides much more than the market demands will also be uncompetitive because of the associated high costs.
2) Providing the foundation for decoupling the body and other systems. The systems (such as the power train and exhaust system) connected to the body must be decoupled from the body, otherwise resonance will happen. After the body modes have been determined, they can be used as references for other system designs.
3) Providing the basis for decoupling overall body modes and local modes. In addition to the overall body modes, the local modes, such as dash panel modes, roof modes, and steering modes, must be determined in order to avoid resonance from the overall body modes.
4) Predicting noise and vibration responses. The noise or vibration responses are superimposed by the responses of a number of different modes. After the modes and modal responses have been acquired, the overall dynamic responses can be obtained and major contributed modes can be identified.
5) Providing the basis for noise and vibration control. Modal shapes provide clues as to where to locate other systems or components: for example, excitation sources and concentrated masses should be placed on the modal nodes. Many vehicle noise and vibration problems are related to the body modes, such as booming and body resonance. Only after the body modes have been acquired can the vehicle's noise and vibration sources and transfer paths be identified, which provide clues for analyzing the root causes of noise and vibration.

Body stiffness is acquired in a static condition, whereas the body mode is an important parameter for representing the dynamic properties of the body structure.

The modal parameters depend on the body's stiffness, mass distribution, damping, and so on. The above five aspects illustrate the influence of the body structural modal frequencies and modal shapes on the vehicle's NVH performance, so modal analysis is one of the core methods of body NVH research.

2.1.3 Scopes of Overall Body Vibration Research

The purpose of overall body structural vibration analysis is to analyze the factors influencing the vehicle's noise and vibration, and to find control methods. There are three main research scopes: analyzing and controlling the overall body stiffness, studying the characteristics of the body modes and control methods, and studying the overall body modes and mode distribution for each system.

2.1.3.1 Analysis and Control of Overall Body Stiffness

Body stiffness is one of the most important factors affecting the vehicle's NVH performance, so reasonable overall body stiffness is the key to achieving good NVH performance. The purpose of studying overall body stiffness is to find the major parameters that influence the overall stiffness and methods for controlling them.

Four factors influence overall body stiffness: the overall layout of body frame structures, the cross-sections of each frame or pillar, joint stiffness at intersections, and the stiffness of bonding adhesive between components.

The overall layout of the body frame structures refers to the distribution of frames, cross members, and pillars, as shown in Figure 1.4. In a similar way to a building, the frames must form a closed structure to provide enough stiffness for the whole structure. If they don't form a closed structure, the overall body stiffness will significantly decrease.

The cross-sections of the frames, cross members, and pillars include closed-loop cross-sections, open-loop cross-sections, rectangular cross-sections, circular cross-sections, and so on. Each different cross-section has a different capability to resist bending and torsion deformations, resulting in a range of body bending stiffness and torsional stiffness.

The intersections at frames, cross members, and pillars should be specially designed. Intersection stiffness is a very important factor for overall body stiffness. If the intersection stiffness is weak, the overall body stiffness will also be weak, even if the stiffness of each frame, cross member, and pillar is sufficient.

Some panels, such as windshields, are connected to the body frame not by welding, but by adhesive. The bonding forces of these adhesive connections affect the overall body stiffness.

2.1.3.2 Analysis and Control of Body Modes

Owing to the huge impact of body modes on the vehicle's NVH, it is extremely important to acquire and control the modes. The body modes can be acquired by experiments or calculations. In the experiment method, the body is excited, then the excitation signals and responses are simultaneously measured, and the modal parameters are extracted from the tested data. The calculation method involves obtaining the modal parameters by FE analysis.

Another task for modal analysis is finding the relationship among the full body, trimmed body, and BIW. If the relationship can be found, engineers can control the full vehicle's body modes by controlling the BIW modes in the early stages of product development.

The purpose of modal control studies is to decouple the overall body modes, local modes, and external excitations by designing an appropriate body structure. The body modes can be adjusted by controlling the body's stiffness and mass distribution. Damping has little impact on the modal frequencies; instead, its major influence is on the peaks of the responses. In addition, a dynamic damper can be used to attenuate special resonant peaks.

2.1.3.3 Planning of Body Modes

In order to develop a body modal plan, we need to make three tables or charts: a full vehicle mode distribution table, a body mode distribution table, and a frequency distribution chart for excitation sources. With these three tables or charts, it is easy to describe the relationship among the full vehicle modes, local modes, and excitation frequencies.

The modal frequency distribution for a full vehicle was described in Chapter 1, as shown in Figure 1.40. The modal frequencies of the body, power train, suspension, etc. are placed in the table, making it easy to see whether the body is coupled with other adjacent systems.

The body modal table provides more information for understanding the overall body modes and local modes. The overall body bending modes, torsional modes, and local modes are listed in the table. The local modes include the panels' modes and components' modes. The panels include the dash panel, floor, ceiling, door panels, tanks, side panels, and so on. The components include the steering wheel, mirrors, seats, brackets, and so on. The table focuses on the local modes, and it has two purposes: one is to identify whether the local modes are coupled with the overall body modes, and the other is to find out whether the local modes are excited by external sources.

To understand whether the body modes are coupled with the excitations, the excitation frequencies and orders for all sources are plotted on a chart to generate the third table or chart for body modal control.

2.2 Overall Body Stiffness

When subjected to external forces, the body produces bending and torsional deformations. Therefore, the body needs to have sufficient bending stiffness and torsional stiffness to resist such deformations. The body's bending stiffness and torsional stiffness are closely related to its modal frequencies and shapes.

We can assume that the body is a bilaterally symmetrical structure, and viewed from the side, it can be simplified as a beam. A beam is a simple structure, and its stiffness can be easily obtained. Therefore, this section first analyzes the beam bending stiffness and torsional stiffness, and then introduces the concepts of body bending stiffness and torsional stiffness, testing and analysis methods, and targets.

2.2.1 Body Bending Stiffness

2.2.1.1 Beam Bending Stiffness

When subjected to an external load, a beam deforms. A segment of a beam is shown in Figure 2.3.

The central line of the beam is assumed as origin. The strain (ε) at a location (y) away from the origin is

$$\varepsilon = \frac{x_1 x_2 - dx}{dx} = \frac{(\rho + y)d\theta - \rho d\theta}{\rho d\theta} = \frac{y}{\rho}, \tag{2.1}$$

where ρ and $d\theta$ are the radius and angle of the arc dx, respectively.

The x–y plane is cut off. In the y–z plane, as shown in Figure 2.3b, the bending moment around the z-axis is

$$M = \int y\sigma dA, \tag{2.2}$$

where dA is the area of a micro-element on the cross-section and y is the distance between the micro-element and the z-axis.

Substitute the relationship between the stress and strain $\sigma = E\varepsilon$ and Eq. (2.2) into Eq. (2.1), obtaining

$$\frac{1}{\rho} = \frac{M}{EI_z}, \tag{2.3}$$

where I_z is the moment of inertia of the cross-section to the z-axis, and is expressed as

$$I_z = \int y^2 dA. \tag{2.4}$$

$\frac{1}{\rho}$ represents the curvature of the beam axis after its deformation. According to Eq. (2.3), the larger the EI, the smaller the curvature. EI represents the capacity to resist bending deformation, so EI is called the beam anti-bending stiffness.

The body is supported by the front and the rear suspensions. To analyze its bending stiffness, the body can be regarded as a simply supported beam, as shown in Figure 2.4. A concentrated load, F, is applied on the beam.

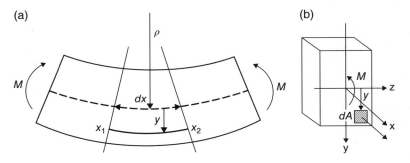

Figure 2.3 (a) A segment of a beam; (b) cross-section of a beam.

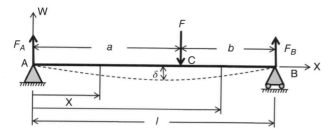

Figure 2.4 A simply supported beam with a concentrated load.

By establishing the equilibrium equations, the reacted force at supporting points A and B can be obtained as follows:

$$F_A = \frac{Fb}{l} \tag{2.5}$$

$$F_B = \frac{Fa}{l}, \tag{2.6}$$

where l is the beam length, and a and b are the distances between the concentrated force and point A and point B, respectively.

In cross-section AC ($0 \leq x \leq a$), the equation of bending moment is

$$\frac{d^2w}{dx^2} = \frac{Fb}{EIl} x, \tag{2.7}$$

where w is the deflection, and E and I are the Young modulus and area moment of inertia of the cross-section, respectively.

Solving Eq. (2.7) and substituting the boundary conditions, the deflection is obtained as follows:

$$w = \frac{Fbx}{6EIl} \left(x^2 - l^2 + b^2 \right). \tag{2.8}$$

In cross-section CB ($a \leq x \leq l$), the equation of bending moment is

$$\frac{d^2w}{dx^2} = \frac{Fb}{EIl} x - \frac{F}{EIl}(x - a). \tag{2.9}$$

Solving Eq. (2.9) and substituting the boundary conditions, the deflection is obtained as follows:

$$w = \frac{Fbx}{6EIl} \left(x^2 - l^2 + b^2 \right) - \frac{F}{6EI}(x - a)^3. \tag{2.10}$$

Assume the concentrated load is applied at the middle point, so the deflection is

$$\delta = -\frac{Fl^3}{48EI}. \tag{2.11}$$

In this case, the beam's bending stiffness at the middle point is

$$k = \frac{48EI}{l^3}.$$

(2.12)

According to Eq. (2.12), the higher the EI, the greater the beam stiffness; the larger the span, the lower the beam stiffness. EI represents the anti-bending stiffness, and its unit is $N\,mm^2$, but it is not real stiffness. The k in Eq. (2.12) is the beam stiffness, and its unit is $N\,mm^{-1}$.

2.2.1.2 Measurement and Analysis of Body Bending Stiffness

The bending stiffness of the body or the full vehicle can be measured on a static deformation test platform or calculated using computer aided engineering (CAE) methods. The supporting boundary conditions and the applied concentrated loads are similar to those for beam testing, as shown in Figure 2.5.

To test the body's bending stiffness, the first step is to place the body on a test platform and adjust the rear axle to make it parallel to the ground and perpendicular to the longitudinal axis of the platform. The supporting brackets of the rear axle are adjusted and then fixed on the platform. Then, the front axle is adjusted so that the body floor is parallel to the ground and the body is symmetrically positioned on the left and right side of the platform. The connecting points between the front and rear shock absorbers and the body should be fixed.

The second step is to install the displacement sensors. The sensors are symmetrically installed on the front and rear side frames, rocker frames, and tunnel beam. The sensors should be perpendicular to the ground.

The third step is to apply load. The concentrated force is applied at the location close to the rear mounting points of the front seats. First, a relatively small load (for example, 1000 N) is applied in order to "pre-load" the body. Then, a relatively large load is applied. The concentrated load is generally between 2000 and 4000 N.

The fourth step is to record the test data from each displacement sensor and force sensor. The stiffness curve will be drawn on the basis of the test data.

Figure 2.6 shows the tested deflection curves. The abscissa is the longitudinal position of the body, the zero point represents the position of the front shock absorber, and the ordinate is the body deflection.

Because the body is a not a completely laterally symmetrical structure, and because the installation of the body on the test platform and the installation of the sensors also cannot be completely symmetrical, the deflections of the body differ slightly on either side.

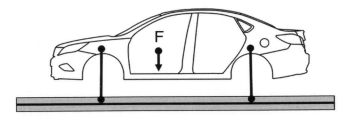

Figure 2.5 Testing diagram of body bending stiffness.

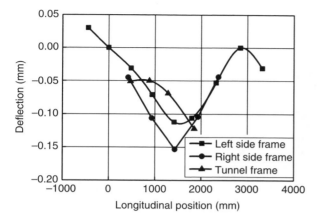

Figure 2.6 Tested deflection curves of a vehicle body.

When the stiffness is analyzed, the average deflection of both sides is used as the body deflection, and the deflections at the front and rear suspensions should be taken into account as well. The body bending stiffness is obtained after dividing the maximum deflection by the applied load. The maximum deflection is

$$\delta_{max} = \frac{\delta_{l\,max} + \delta_{r\,max}}{2} - \frac{\delta_{fl} + \delta_{fr} + \delta_{rl} + \delta_{rr}}{4}, \tag{2.13}$$

where $\delta_{r\,max}$ and $\delta_{r\,max}$ are the maximum deflections of the left and right sides, respectively; δ_{fl} and δ_{fr} are the deflections of the left and right sides of the front suspension, respectively; and δ_{rl} and δ_{rr} are the deflections of the left and right sides of the rear suspension, respectively.

The body bending stiffness is

$$k_{bending} = \frac{F}{\delta_{max}}, \tag{2.14}$$

where F is the applied load on the body.

2.2.1.3 Body Bending Stiffness for Different Vehicle Platforms

The body bending stiffness varies for different vehicles. Figure 2.7 shows a group of BIW bending stiffness values for economy sedans with windshields, and the major distribution ranges are from 8000 to 12 000 N mm^{-1}. Figure 2.8 shows a group of BIW bending stiffness values for mid-sized sedans with windshields, and the major distribution ranges are from 10 000 to 16 000 N mm^{-1}. Figure 2.9 shows a group of BIW bending stiffness values for sports utility vehicles (SUVs) with windshields, and the major distribution ranges are from 8000 to 14 000 N mm^{-1}. Generally speaking, the larger the vehicle, the higher the body bending stiffness; and the more expensive the vehicle, the higher the body bending stiffness. For a sedan and SUV of the same price range, the body bending stiffness of the sedan is usually higher than that of the SUV.

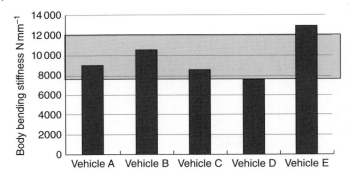

Figure 2.7 A group of BIW bending stiffness values for economy sedans with windshields.

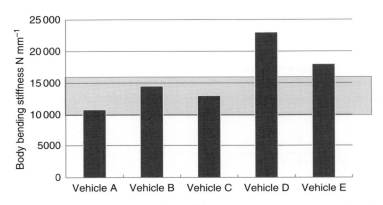

Figure 2.8 A group of BIW bending stiffness values for mid-sized sedans with windshields.

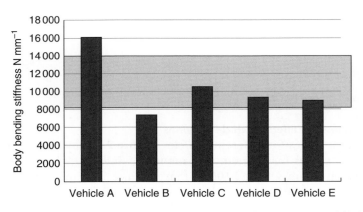

Figure 2.9 A group of BIW bending stiffness values for SUVs with windshields.

Taking other statistical data into account, the BIW bending stiffness for economy sedans, mid-sized sedans, and SUVs should generally be around 10 000, 13 000, and 12 000 N mm^{-1}, respectively, to achieve good NVH performance. The higher the bending stiffness, the better the NVH performance. The bending stiffness of some vehicles can be extremely high, even more than 20 000 N mm^{-1}.

2.2.2 Body Torsional Stiffness

When the loads applied to both sides of a vehicle are not the same, the body will be twisted, which generates torsional deformation. For example, when driven on an uneven road, a vehicle will be subjected to different excitations on the tires of each side. The excitations are transmitted to the body through the suspension, which causes the body to be twisted. Only when the body torsional stiffness is high enough can it resist torsional vibration and meet NVH requirements.

Similar to analysis of the bending stiffness, in the analysis of the body torsional stiffness, we can assume that the body is a beam. First, the torsional stiffness of the beam is analyzed, followed by the body torsional stiffness.

2.2.2.1 Beam Torsional Stiffness

A twisted beam and a small segment dx are shown in Figure 2.10. On a cross-section perpendicular to the x-axis, the torsional deformation at the perimeter where the radius is ρ is

$$s = \rho d\theta = \gamma_\rho dx, \tag{2.15}$$

where $d\theta$ and γ_ρ are the angular deformation on the beam and at the segment dx, respectively.

The relationship between shear stress (τ_ρ) and strain at the perimeter ρ is

$$\tau_\rho = G\gamma_\rho. \tag{2.16}$$

where G is the shear modulus

In this cross-section, the torque is the moment of the internal force to the center, and expressed as

$$T = \int \rho \tau_\rho dA \tag{2.17}$$

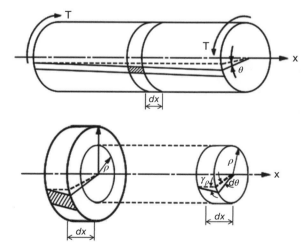

Figure 2.10 A twisted beam and a small segment dx.

Substitute Eqs. (2.15) and (2.16) into Eq. (2.17), obtaining:

$$T = GJ\frac{d\theta}{dx},\tag{2.18}$$

where J is the polar moment of inertia of the cross-section to the center, and is expressed as

$$J = \int \rho^2 dA.\tag{2.19}$$

The angular deformation represents the relative twisted angles of two cross-sections along the rotated axis. According to Eq. (2.18), the twisted angle is

$$d\theta = \frac{T}{GJ}dx.\tag{2.20}$$

For a beam with length l subjected to torque T, its twisted angle is

$$\theta = \int_0^l \frac{T}{GJ}dx = \frac{Tl}{GJ}.\tag{2.21}$$

Eq. (2.21) shows that the bigger the GJ, the smaller the twisted angle. GJ represents the capability to resist torsional deformation; however, GJ is not the beam's torsional stiffness. The beam's torsional stiffness is expressed as follows:

$$k_\varphi = \frac{T}{\theta} = \frac{GJ}{l}.\tag{2.22}$$

The unit for torsional stiffness is N·m rad^{-1}.

2.2.2.2 Testing and Analysis of Body Torsional Stiffness

The torsional stiffness of a vehicle or a body can be obtained by measurements on a static deformation test platform or by CAE analysis. The supporting boundary conditions and applying torque are similar to those for beam testing, as shown in Figure 2.11.

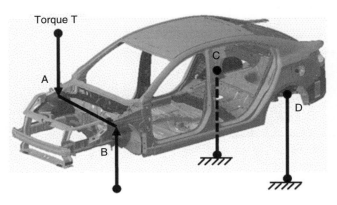

Figure 2.11 Testing diagram of body torsional stiffness.

In the testing of body torsional stiffness, the body is installed on the test platform in the same way as for bending stiffness testing, but the constraint boundary is different. The connecting points between the rear suspensions and the body are constrained, and the torque is applied at the connecting points between the front shock absorbers and the body. The magnitude of the torque is generally 3000 N·mm.

The installation of displacement sensors is the same as for bending stiffness testing. The displacements at the sensor locations are measured, and the displacements at the constrained points and applied torque points can be calculated by interpolation methods. Because the body is a not completely symmetrical structure in the transverse direction, the displacements on the right and left sides of the same cross-section are different. Based on the displacement difference of the two sides, the torsional angle can be calculated and expressed as follows:

$$\theta = \arctan\frac{\delta_A - \delta_B}{l_1} - \arctan\frac{\delta_C - \delta_D}{l_2}, \tag{2.23}$$

where δ_A and δ_B are the displacements on the left and right sides of the front axle, respectively; δ_C and δ_D are the displacements of the left and right rear suspensions; l_1 is the distance between point A and point B; and l_2 is the distance between point C and point D.

Figure 2.12 shows the measured twisted angles of a body when subjected to a torsional torque. The abscissa is the longitudinal position of the body, and the ordinate is the twisted angle at different body positions.

The torsional stiffness can be obtained by dividing the twisted angle by the torque, and is expressed as follows:

$$k_{torsion} = \frac{T}{\theta}, \tag{2.24}$$

where T is the applied torque.

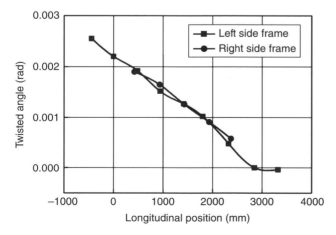

Figure 2.12 Measured twisted angles of a body.

2.2.2.3 Torsional Stiffness for Different Vehicle Platforms

Body torsional stiffness varies for different vehicles. Figure 2.13 shows a group of BIW torsional stiffness values for economy sedans with windshields, and the major distribution ranges are from 700 to 900 KN·m rad^{-1}. Figure 2.14 shows a group of BIW torsional stiffness values for mid-sized sedans with windshields, and the major distribution ranges are from 900 to 1300 KN·m rad^{-1}. Figure 2.15 shows a group of BIW torsional stiffness values for SUVs with windshields, and the major distribution ranges are from 1000 to 1300 KN·m rad^{-1}. Generally speaking, the larger the vehicle, the higher the body torsional stiffness. SUVs have higher torsional stiffness than sedans.

Generally, the BIW torsional stiffness for economy sedans, mid-sized sedans, and SUVs should be higher than 800, 1000, and 1100 KN·m rad^{-1}, respectively. The higher

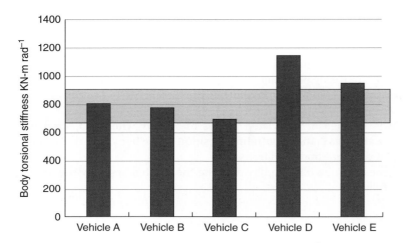

Figure 2.13 A group of BIW torsional stiffness values for economy sedans with windshields.

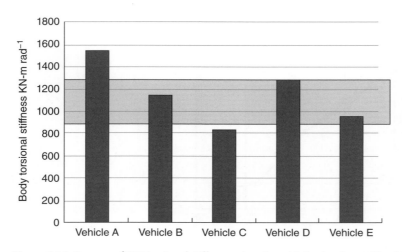

Figure 2.14 A group of BIW torsional stiffness values for mid-sized sedans with windshields.

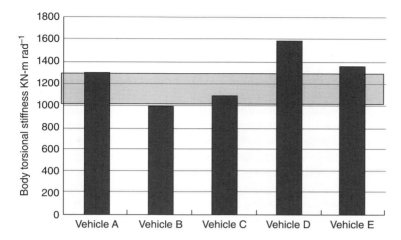

Figure 2.15 A group of BIW torsional stiffness values for SUVs with windshields.

the torsional stiffness, the better the NVH performance. Some vehicles' torsional stiffness is very high, even more than 1600 KN·m rad^{-1}.

2.3 Control of Overall Body Stiffness

The main factors influencing overall body stiffness include the overall layout of the frame and pillar structures, cross-sections, joint stiffness, bonding stiffness of adhesives, and panel stiffness. These factors determine the body overall stiffness and modes, which determine the interior low-frequency booming and body vibration.

The overall layout of the frame structure refers to the arrangement of the frames, cross members, and pillars, and the corresponding joint structures. The body stiffness is significantly influenced by the layout: for example, the stiffness of a closed-loop frame layout is much higher than that of an open-loop layout.

The cross-sections of the frames, cross members, and pillars include rectangular shapes, circular shapes, and irregular shapes. The perimeters of the cross-sections include closed-loops and open-loops. The cross-sectional shapes and loops affect not only the local stiffness of the frames and pillars, but also the overall body stiffness.

Joint stiffness refers to stiffness at the intersections where the frames, cross members, and pillars are joined together. For example, the top joint of an A-pillar consists of the A-pillar, front roof bow, and roof side rail. The joint stiffness should be high enough to guarantee that the A-pillar has sufficient local stiffness, which has a significant influence on the overall body stiffness.

Bonding stiffness refers to the local stiffness of the adhesives used to bond metal sheets or to bond the windshield to the body frame. Bonding stiffness influences the overall body stiffness, especially the torsional stiffness.

Panel stiffness refers to the stiffness of metal sheets such as the door panels. The main influence of panel stiffness is on local modes and sound radiation. Its impact on overall body stiffness is mainly in torsional stiffness. Chapter 3, "Noise and Vibration Control for Local Body Structures," describes panel vibration in detail. The following sections the influence of the overall layout of the frame structures, cross-sections, joints, and bonding stiffness of windshield adhesive on overall body stiffness.

2.3.1 Overall Layout of a Body Structure

The overall layout is the foundation of a vehicle's design. Factors such as vehicle styling, the arrangement of trim parts, manufacturing processes (stamping, welding, and assembly), the choice of materials, and customers' requirements should be included during the layout design. These factors will affect the body stiffness and modes.

From the perspective of body stiffness, the overall layout refers to the structural constitution and distribution of the frames, cross members, and pillars, including the peripheral and geometrical distribution of the frames, the connections among frames, members, and pillars, and the connections and locations of joints.

The frame layout is one of the most important factors determining the overall body stiffness. The body layout should follow several principles. First, the frame structures should be closed loops. Second, the number of frames and the distances between them must meet the body stiffness requirements. Third, resonances between the frame structure and external excitations should be avoided.

2.3.1.1 Closed-Loop Structure and Open-Loop Structure

In a closed-loop structure, the beams constituting the structure form a closed loop, as shown in Figure 2.16. Figure 2.17 displays several open-loop frame structures: Figure 2.17a shows two longitudinal beams that are not connected together at the ends; Figure 2.17b shows a longitudinal beam that is disconnected in the middle; and Figure 2.17c shows a cross member that is disconnected. The stiffness of the closed-loop structure is higher than that of the open-loop structure.

The body frame structure must form a closed loop. The main body frame, including the side frames, front and rear cross members, roof side rails, front and rear roof frames, and so on, is usually designed as a closed-loop structure. However, the main frames for some vehicles, such as pick-up trucks and flat-bed trucks, cannot form a closed loop because of the lack of C-pillars, which reduces the body's torsional stiffness. For a sedan

Figure 2.16 A closed-loop structure.

Figure 2.17 Open-loop structures: (a) disconnected at the end; (b) disconnected at a longitudinal beam; (c) disconnected at a cross member.

and a pick-up truck developed on the same platform, the sedan has higher body torsional stiffness than the pick-up truck. In addition to the main frame, the sub-beams and cross members should form closed loops. Figure 2.18a shows that a tunnel beam, two minor beams, and a cross member are not connected; instead, an open-loop substructure is formed, so the body's torsional stiffness will be influenced. After the sub-beams and cross members are connected, as shown in Figure 2.18b, the substructure becomes a closed loop, so its torsional stiffness is greatly increased.

2.3.1.2 Influence of Number of Beams and Their Distances on Overall Body Stiffness

The number of beams has a great influence on the overall body stiffness. For a frame structure, the more the frames and cross members there are, the higher the stiffness. The sizes of two frame structures in Figure 2.19a and b are the same. The frame structure in Figure 2.19b has higher stiffness than that in Figure 2.19a because it has one more cross member. However, during the design of the body, the number of beams is constrained by the body structure's layout, weight, cost, and other factors. The number of intermediate beams affects not only the overall body stiffness, but also the stiffness of the local panels. For example, by adding a cross member to the roof, the roof panel's modal frequency is increased, and its sound radiation is reduced.

The distance between the main frames will affect the body stiffness. Today, the tendency in vehicle development is to develop several different types of vehicle on the same platform. For example, widening or lengthening an existing body frame could result in a new vehicle. Figure 2.20a shows a base frame structure, and Figure 2.20b

(a)   (b)

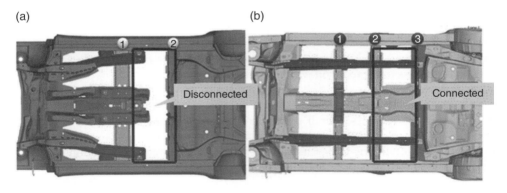

Figure 2.18 A floor structure: (a) open loop; and (b) closed loop.

(a)   (b)

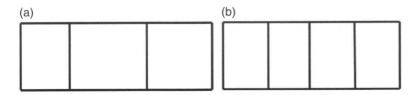

Figure 2.19 Influence of number of beams on body stiffness. (a) Two cross members. (b) Three cross members.

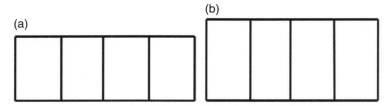

Figure 2.20 (a) Base frame structure; (b) widened frame structure.

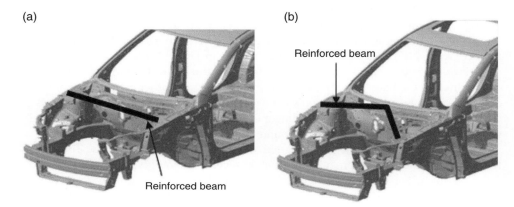

Figure 2.21 Reinforced beams to support shotguns: (a) one cross frame; (b) a V-beam.

provides a widened frame structure. If the same torque is applied to the base structure and widened structure, the widened structure has a higher twisted angle and lower torsional stiffness in comparison with the base structure.

2.3.1.3 Beam Resonance

Beam resonance includes overall resonance and local resonance. If the body frame structure is not well designed and its overall modal frequencies are consistent with the external excitation frequencies, the entire vehicle will be vibrated. For example, during engine idling, customers will not only perceive vibration from the steering wheel, seats and floor, but also observe shaking of the opened doors, indicating both local vibration and full body vibration. In this case, the entire vehicle is excited, so the body frame structure must be analyzed and controlled.

Local beam resonance refers to the resonance of a local beam caused by an external excitation. For example, if the modal frequency of the shotguns of a vehicle is consistent with the engine excitation frequency, the engine compartment will resonate and the vibration will be transmitted to the body, then to the steering wheel, floor, and seats. In the design of the body's overall layout, the local stiffness of each frame must be fully considered. In the late phase of vehicle development, NVH engineers usually face many local noise and vibration problems. Unfortunately, it is too late at this point to modify the overall body structure. To solve such problems, the local frames must be modified. For example, reinforced beams (as shown in Figure 2.21) can be added to support the shotguns in order to increase the local stiffness and modal frequencies, and avoid resonance.

(a)   (b)   (c)   (d)   (e)

Figure 2.22 Several cross-section shapes of body frames and pillars: (a) cross-section of A-pillar; (b) cross-section of rocker frame; (c) cross-section of front roof cross member; (d) cross-section of rear roof cross member; (e) cross-section of floor cross member.

2.3.2 Body Frame Cross-Section and Stiffness Analysis

2.3.2.1 Section Types

The frames, cross members, and pillars have two main functions: one is to support weight, and the second is to construct a body frame structure with sufficient stiffness. Many systems, such as the power plant, are directly mounted on the body: therefore, the structure has to support their weight. A frame structure must have sufficient stiffness to meet performance requirements, such as NVH targets, but must also be relatively light. The frames and pillars are made of thin-walled metal sheets: i.e. they consist of several thin, stacked sheets that have been rolled together. A thin-walled beam has a thickness much smaller than its other geometrical dimensions, such as width and height. Because of their high ratio of stiffness to weight, thin-walled beams are widely used in body structures.

Thin-walled frames and pillars have a variety of cross-section shapes, such as rectangular, circular, and irregular, as shown in Figure 2.22. Their cross-sections are usually designed as irregular shapes to achieve optimal loading capacity and maximum stiffness.

The perimeter of the frame cross-section is usually a closed loop: i.e. a complete ring. Some frames consisting of several metal sheets have several rings. Figure 2.23 shows a cross-section of three metal sheets that includes two closed loops. The loop shape is one of the most important factors in determining frame stiffness. However, the shape is usually constrained by the spatial layout of the design. Therefore, the cross-section design must be balanced between stiffness and spatial layout.

Cross-sections are divided into open cross-sections and closed cross-sections. If the center line of a cross-section is a closed curve or folded line, it is called a closed cross-section. Otherwise, if the center line is a non-closed curve or folded line, it is called an open cross-section, as shown in Figure 2.24.

2.3.2.2 Cross-Section Shape and Stiffness

In beam bending-resistance stiffness (EI) and torsion-resistance stiffness (GJ), E and G are the elastic modulus and shear modulus of the material, respectively, which depend on the material's properties. I and J are the cross-section moment of inertia and the polar moment of inertia, respectively. Eqs. (2.4) and (2.19) show that I and J are only related to the cross-section shape and the axis position. Therefore, in the case of the chosen material, the beam bending-resistance stiffness and torsion-resistance stiffness only depend on the cross-section's shape and size.

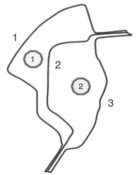

Figure 2.23 A cross-section with two closed loops.

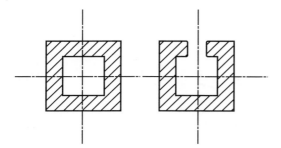

Figure 2.24 A closed cross-section and open cross-section.

(a)   (b)

Figure 2.25 (a) A closed circular cross-section; (b) an open circular cross-section.

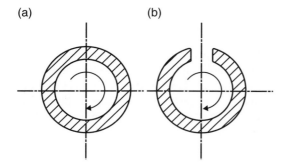

A beam should be designed as a closed cross-section, which will have a much higher stiffness than an open-section beam. Let us compare an open, circular, thin-walled beam and a closed, circular, thin-walled beam to illustrate the difference in stiffness, as shown in Figure 2.25. Suppose two thin-walled beams have an identical circular cross-section radius and thickness, and are made of the same material. Taking torsional stiffness as an example, the difference between an open- and closed-section beam is illustrated as follows.

The torsional stiffness of the open circular cross-section is

$$(GJ)_{open} = \frac{2}{3}G\pi r t^3, \tag{2.25}$$

where r is the circular radius and t is the wall thickness. The torsional stiffness of the closed circular cross-section is

$$(GJ)_{closed} = 2G\pi r^3 t. \tag{2.26}$$

The ratio of the stiffness of the open circular cross-section to that of the closed circular cross-section is

$$\frac{(GJ)_{open}}{(GJ)_{closed}} = \frac{1}{3}\left(\frac{t}{r}\right)^2. \tag{2.27}$$

Because the wall thickness is much smaller than the radius, i.e. $t \ll r$, the torsional stiffness of the open-section beam $(GJ)_{open}$ is much lower than that of the closed-section beam $(GJ)_{closed}$. The same applies for the bending stiffness. Open-section beams should be avoided as far as possible in the vehicle's body design. Most cross-sections in body frames and pillars are closed ones.

The body beam cross-sections are usually very complex and irregular. When a beam is subjected to torque, its deformation forces the cross-section to change to a spatial-curved cross-section, rather than a plane – i.e. a twisted deformation is generated – so calculating the body cross-section stiffness is very complicated. However, the cross-section can be discretized, and then its stiffness can be calculated by geometrical analysis and analytical analysis methods.

Because a closed-section beam has much higher stiffness than an open-section beam, beams should be designed as a closed sections wherever possible. However, for some beams, such as the tunnel beam, open cross-sections cannot be avoided. In some locations, the beams have partial open cross-sections, such as the B-pillars where seat belts are mounted. In locations where the beams have open or partial open cross-sections, the influence of the cross-sections on overall body stiffness must be carefully analyzed.

2.3.3 Joint Stiffness

2.3.3.1 Joints

A joint refers to a juncture point where two or more frames, cross members, and/or pillars intersect. Figure 2.26 shows the joints of a sedan at the A-pillar, B-pillar, and C-pillar. A body can be approximately regarded as a symmetrical structure, so only one side of the body is analyzed. There are seven joints for the sedan, which are defined as follows:

- A1 joint: intersection of the A-pillar, front side frame, and front floor cross member.
- A2 joint: intersection of the A-pillar, cross member under windshield, and shotgun.
- A3 joint: intersection of the A-pillar, side roof frame, and front roof frame.
- B1 joint: intersection of the B-pillar, rocker frame, and middle floor cross member.
- B2 joint: intersection of the B-pillar, side roof frame, and roof cross member.
- C1 joint: intersection of the C-pillar, rear side frame, and rear floor cross member.
- C2 joint: intersection of the C-pillar, rear roof frame, and rear roof cross member.

For an SUV, there are another two joints in addition to the above seven joints, as shown in Figure 2.27:

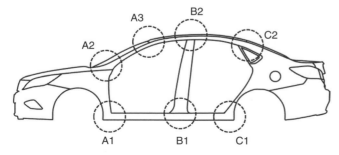

Figure 2.26 Joints of a sedan at A-pillar, B-pillar, and C-pillar.

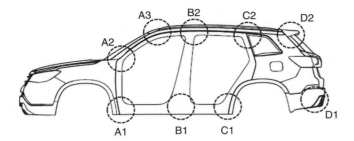

Figure 2.27 Joints of an SUV at A-pillar, B-pillar, C-pillar, and D-pillar.

Figure 2.28 A system consisting of two beams and a joint.

- D1 joint: intersection of the D-pillar, rear side frame, and rear floor cross member.
- D2 joint: intersection of the D-pillar, rear roof frame, and rear roof cross member.

2.3.3.2 Joint Stiffness

The stiffness of a structure consisting of frames and joints depends not only on the frames' stiffness, but also on the joints' stiffness. Suppose a system consists of two beams and a joint, as shown in Figure 2.28. The stiffness of the two beams is given as k_1 and k_2, and the joint stiffness is k_3. The system is regarded as a series connecting system, so its stiffness is

$$\frac{1}{k} = \frac{1}{k_1} + \frac{1}{k_2} + \frac{1}{k_3}. \tag{2.28}$$

The system will only have sufficient stiffness if both the beams and the joint have a suitably high stiffness.

Joint stiffness is one of the most important factors in determining the overall body stiffness. If the joint stiffness is insufficient, many problems will be induced, such as S&R, NVH, and poor reliability. First, improper installation of the joints will decrease the local stiffness, which can easily cause large deformations of the connected parts and increase the diagonal deformations of the door frames. Second, excessive deformations of the door frames will induce S&R and affect the sealing between the door and the body. Third, low joint stiffness will decrease the overall body stiffness, which results in a reduction of the body's bending and torsional modal frequencies, and can cause interior booming and resonances. Fourth, improperly connected joints will cause stress concentration problems, which can cause cracks in the body and fatigue fractures.

Joint stiffness depends largely on the cross-section shapes of the connecting frames and the supporting conditions. If the frames' cross-sections are changed, the joint stiffness will change. Joint stiffness can be regarded as a function of the connecting cross members' parameters. For example, in Figure 2.26, the B2 joint consists of the B-pillar, front side roof frame, middle roof cross member, and rear side roof frame, and the corresponding parameters include:

- B-pillar: I_{x1}, I_{y1}, J_1, A_1
- Front side roof frame: I_{x2}, I_{y2}, J_2, A_2
- Roof cross member: I_{x3}, I_{y3}, J_3, A_3
- Rear side roof frame: I_{x4}, I_{y4}, J_4, A_4
- The joint: I_{x5}, I_{y5}, J_5, A_5.

The most important stiffness components of the joint are the fore-aft bending stiffness (K_{xx}), inward–outward bending stiffness (K_{yy}), and torsional stiffness (K_{zz}). The joint stiffness is a function of the above parameters, and is expressed as follows,

$$K_{xx} = f_1\left(I_{x1}, I_{y1}, J_1, A_1, \ldots, I_{x5}, I_{y5}, J_5, A_5\right) \tag{2.29a}$$

$$K_{yy} = f_2\left(I_{x1}, I_{y1}, J_1, A_1, \ldots, I_{x5}, I_{y5}, J_5, A_5\right) \tag{2.29b}$$

$$K_{yy} = f_3\left(I_{x1}, I_{y1}, J_1, A_1, \ldots, I_{x5}, I_{y5}, J_5, A_5\right). \tag{2.29c}$$

It is difficult to acquire measurements of joint stiffness by testing or calculations; instead, CAE analysis is usually employed, such as the FE method. When joint stiffness is analyzed by the FE method, a portion of the joint is selected, as shown in Figure 2.29. Using the joint center as an origin, a part of the FE model with a certain length (such as 100 or 180 mm) in *x*, *y*, and *z* directions is cut out. Figure 2.29 shows a selected portion of the B2 joint. The strongest frame of the joint is fixed, and torques in three directions are applied at the weakest frame cross-section, then the fore-aft bending stiffness, inward-outward bending stiffness, and torsional stiffness can be obtained.

Figure 2.29 A joint finite element model.

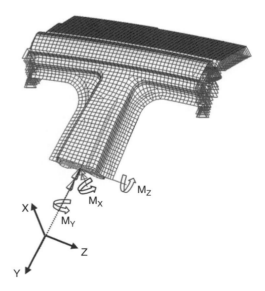

2.3.3.3 Methods to Control Joint Stiffness

There are two methods to control joint stiffness. One is to design a high-stiffness joint and the other is to add additional material at the joint to enhance its stiffness.

The first method is the most important one. Figure 2.30 shows a diagram of the B2 joint that includes the body roof, roof cross member, inner side panel, reinforced plate of the side panel, internal reinforced plate for the mounting holes, B-pillar internal panel, B-pillar reinforced plate, and body side panel. The stiffness of these parts has a significant influence on the joint's stiffness. In a case in which the body panels, frames, and pillars cannot be changed, the design of the inner plate and the reinforced plate is very important. Furthermore, the bending radius of these components, the hole, the welding locations, and gaps also affect the joint stiffness.

The second method is to install a structural foam at the joint, as shown in Figure 2.31.

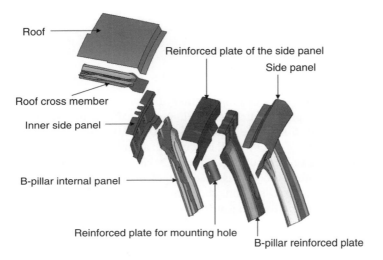

Figure 2.30 Anatomy diagram of a B2 joint.

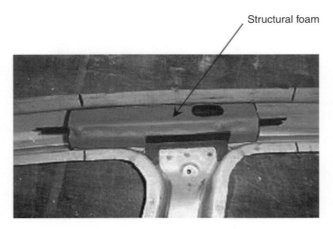

Figure 2.31 Structural foam installed at a joint.

As previously mentioned, joint stiffness makes a great contribution to the overall body stiffness. A convertible has greatly reduced torsional stiffness owing to the lack of a roof structure, so to enhance its stiffness, the floor should be designed to have much higher stiffness than a sedan, especially in terms of joint stiffness.

2.3.4 Influence of Adhesive Bonding Stiffness on Overall Body Stiffness

Structural adhesive consists of thermosetting resin, rubber, and polymer alloy, and its function is to firmly connect two parts to achieve the required strength, stiffness, collision safety, reliability, and so on. For example, the body's metal sheets are connected by structural adhesives. Glass bonding adhesive is used to firmly bond the windshield to the body frame in order to enhance the body stiffness.

Structural adhesives can replace some of the welding and riveting processes that are widely used in the automotive industry. Structural adhesives can be used on some areas that cannot be reached for welding, as well as on the outer surfaces, where welding would affect the vehicle's smooth and beautiful appearance. Structural adhesives can also be used between different materials, such as a metal and non-metal. There are thousands of welding points on a typical vehicle body, but if structural adhesives are used, the number of welding points can be reduced by 10–20%, reducing the body's weight and cost. In addition, the structural strength can be increased, stress concentration and welding cracks can be avoided, the vehicle's capacity to resist corrosion, fatigue, and shock can be improved, and the sealing can be improved, too.

Owing to the light weight of structural adhesives, their application also increases the body's stiffness, resulting in an increase in the body's modal frequencies. Figure 2.32 shows locations where structural adhesives can be used, and Figure 2.33 shows a comparison of frequency response functions (FRF) for two BIWs with the same welding spots and distribution, but where one uses structural adhesives and the other does not. The modal frequencies for the body using structural adhesives are higher than those for the body without structural adhesives. Usually, the body stiffness increases by 20–40% and the modal frequencies increase by 1–4 Hz after application of the adhesives.

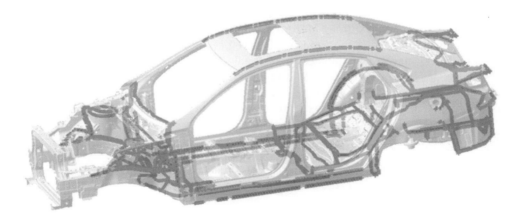

Figure 2.32 Structural adhesives used on a body.

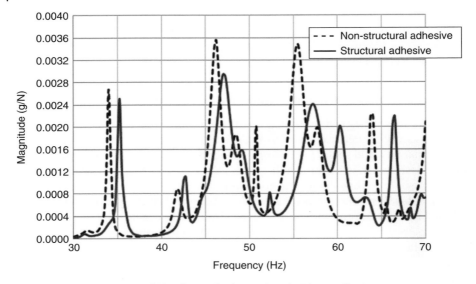

Figure 2.33 A comparison of FRFs for two bodies with and without adhesives.

Of course, structural adhesives are not perfect; there is room for improvement in terms of their strength, durability, and capacity to resist high temperatures. In the automotive assembly process, the application of structural adhesives could also be improved: for example, after structural adhesives have been added, they can be damaged by subsequent processes, such as painting.

Glass bonding adhesives, which are similar to structural adhesives, are used to connect windshields and metal sheets. The front and rear windshield frames are weak structures and are easily twisted under the action of external torque, which accordingly reduces the body's torsional stiffness. After the windshields have been installed, the weak frames are "supported", and their capacity to resist torque increases greatly. Figure 2.34 shows a comparison of bending stiffness and torsional stiffness for two BIWs without and with windshields. Figure 2.34a shows that the BIW's bending stiffness increases only a little after the windshields are added, which demonstrates that the windshield has little impact on bending stiffness. Figure 2.34b shows that the BIW's torsional stiffness increases by 15–35% after the windshields have been added, which demonstrates that the windshields have a great impact on torsional stiffness.

Windshields can only provide good support if the connection between the windshield and the metal sheets is strong enough. For the same body and the same windshields, the body's torsional stiffness could be different for different glass bonding adhesives.

2.3.5 Contribution Analysis of Beams and Joints on Overall Body Stiffness

Stiffness represents the capacity to resist deformation. Deformation appears when a system is subjected to an external force. During the deformation process, the work done by the external force is stored in the system. The more energy that is stored in the system, the larger the deformation, and the lower the stiffness. Therefore, the deformation energy can be used to analyze the system's stiffness and its distribution.

(a)

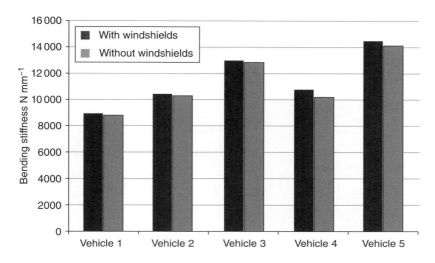

(b)

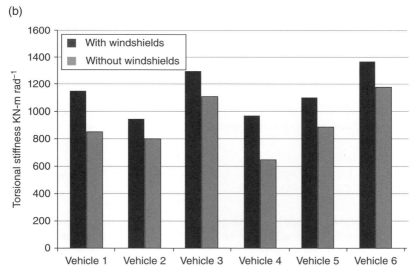

Figure 2.34 Comparison of BIW stiffness with and without windshields: (a) bending stiffness; (b) torsional stiffness.

When the external force reduces, the stored energy is released slowly, which makes deformation gradually recover. The energy stored during the deformation process is called strain energy. Strain energy is a very important parameter for measuring deformation of the system, and is often used to evaluate the system's bending and torsional deformations. The areas with large strain energy represent the weakest areas of the system.

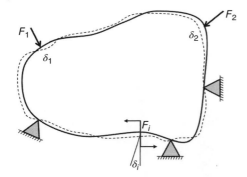

Figure 2.35 A system subjected to general forces and induced general deformation.

An arbitrary system is acted on by several different loads, including force, torque, and moment, as shown in Figure 2.35. Under the action of the external loads, the system is deformed, including displacement deformation and angle deformation. For an elastic object, the stored strain energy depends on the loads and final deformation values, irrespective of the sequential applied loads. All the loads can be represented by general forces expressed by F_1, F_2, ..., F_n, and the corresponding final deformations are called general deformations expressed by δ_1, δ_2, ..., δ_n. Thus, the system's strain energy is

$$W = \frac{1}{2}F_1\delta_1 + \frac{1}{2}F_2\delta_2 + \cdots + \frac{1}{2}F_n\delta_n. \tag{2.30}$$

A beam with length L is subjected a moment M, so the bending deformation $\theta = \dfrac{ML}{EI}$ is generated and the corresponding strain energy is

$$W_M = \frac{1}{2}\frac{M^2 L}{EI}. \tag{2.31}$$

When a beam is subjected to the action of an external torque T, the twisted deformation is $\varphi = \dfrac{TL}{GJ}$, and the corresponding strain energy of the beam is

$$W_T = \frac{1}{2}\frac{T^2 L}{GJ}. \tag{2.32}$$

The areas with large strain energy have a large capacity to store energy and a correspondingly low stiffness. If these areas are reinforced, the structure's stiffness will increase. It is important for a body analysis to identify the locations with highest bending and torsional strain energies. After the locations are reinforced, the overall body bending and torsional stiffness will be significantly increased.

The strain energy distribution on a body can be calculated by FE analysis. Figure 2.36 shows the bending strain energy distribution of a BIW on the frames and joints when it is subjected a static bending load. The strain energy in the frames, at the joints, and in the remaining areas are 44, 26, and 30%, respectively, which shows that the frames are the most important contributors to bending stiffness, with the joints coming in second.

Figure 2.37 shows the torsional strain energy distribution on the frames and joints of the same BIW when it is subjected to a static torque. The strain energy in the frames, at the joints, and in the remaining areas are 42, 39, and 19%, respectively, which shows that the frames and joints have almost the same impact on torsional stiffness.

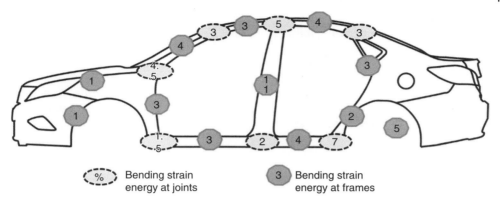

Figure 2.36 Bending strain energy distribution on the frames and joints of a BIW when it is subjected to a static bending load.

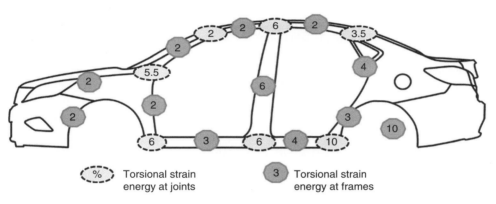

Figure 2.37 Torsional strain energy distribution on the frames and joints of a BIW when it is subjected to a static torque.

The body's stiffness depends not only on the stiffness of the frames, but also on the stiffness of the joints. Thus, the design of the joints is very important, and the stiffness of the joints should be as uniformly distributed as possible.

2.4 Identification of Overall Body Modes

2.4.1 Foundation of Modal Analysis

Modal analysis involves transferring a coupling physical coordinate of a linear and time-invariant system into a decoupled modal coordinate and changing the coupling equations into a series of decoupled equations. The independent equations are described by a series of modal coordinates, and the modal parameters for each equation can be easily obtained. Then, using coordinate transformation from the modal coordinate into the physical coordinate, the solutions in the physical coordinates can be acquired.

For a linear and time-invariant system with N degrees of freedom (DOFs), its physical matrix equation is expressed as follows:

$$[M]\ddot{X}+[C]\dot{X}+[K]X=F, \tag{2.33}$$

where $[M]$, $[C]$, and $[K]$ are the mass matrix, damping matrix, and stiffness matrix, respectively. X and F are the displacement vector and excitation force vector, respectively, and are expressed as follows:

$$X=\left[\, x_1 \ x_2 \cdots x_r \cdots x_N \,\right]^T \tag{2.34}$$

$$F=\left[\, f_1 \ f_2 \cdots f_r \cdots f_N \,\right]^T. \tag{2.35}$$

Expression (2.33) provides the dynamic equations of the system at physical coordinates. Each equation contains the physical coordinate of each point, so the expression is a set of coupled equations. When the system's DOFs are large, it is very difficult, even impossible, to solve the equations. Therefore, the equations must be decoupled in order to make each equation independent. To reach this objective, coordinate transformation is needed.

Laplace transform is applied on both sides of Eq. (2.33), obtaining

$$\left(s^2[M]+s[C]+[K]\right)X(s)=F(s). \tag{2.36}$$

For the time-invariant system, $s = j\omega$. Substitute this into Eq. (2.36), obtaining

$$\left([K]-\omega^2[M]+j\omega[C]\right)X(\omega)=F(\omega). \tag{2.37}$$

The response of any point can be expressed as a linear combination of each mode's response: for example, the response at point l can be expressed as

$$x_l(\omega)=\sum_{r=1}^{N}\varphi_{lr}q_r(\omega), \tag{2.38}$$

where φ_{lr} is the modal shape coefficient at point l for r^{th} mode, and q_r is the modal coordinate for r^{th} mode. For N DOFs, the array of rth modal shape coefficients is

$$\phi_r=\left[\, \varphi_1 \ \varphi_2 \cdots \varphi_r \cdots \varphi_N \,\right]^T, \tag{2.39}$$

where φ_r is the modal shape of rth order, called the rth mode vector.

By putting all of the orders' modal vectors together and forming a matrix, the system's modal matrix can be obtained, which is expressed as follows:

$$\Phi=\left[\, \phi_1 \ \phi_2 \cdots \phi_r \cdots \phi_N \,\right]. \tag{2.40}$$

The modal coordinate q_r represents the contribution of the rth mode to the response. By putting all of the orders' modal coordinates together, the modal coordinate matrix can be obtained, which is expressed as follows:

$$Q_r = \begin{bmatrix} q_1 & q_2 \cdots q_r \cdots q_N \end{bmatrix}^T \tag{2.41}$$

The relationship between the physical coordinate and modal coordinate can be established by the modal matrix,

$$X = \Phi Q. \tag{2.42}$$

Substitute Eq. (2.42) into Eq. (2.37), obtaining

$$\left([K] - \omega^2 [M] + j\omega [C] \right) \Phi Q = F(\omega). \tag{2.43}$$

Due to low damping, the modes for a vehicle body can be regarded as real modes, and the damping is assumed to be proportional damping. By multiplying both sides of Eq. (2.43) by Φ^T, the coupled system in the physical coordinate is transferred into a decoupled system in the modal coordinate, and the corresponding dynamic equations in modal coordinates are

$$\left([K_r] - \omega^2 [M_r] + j\omega [C_r] \right) Q = F_r(\omega), \tag{2.44}$$

where $[M_r]$, $[K_r]$, and $[C_r]$ are the modal mass matrix, modal stiffness matrix, and modal damping matrix in modal coordinates, respectively. All of them are diagonal matrices:

$$[M_r] = \Phi^T [M] \Phi = \begin{bmatrix} M_1 & & & & \\ & \ddots & & & \\ & & M_r & & \\ & & & \ddots & \\ & & & & M_N \end{bmatrix} \tag{2.45}$$

$$[K_r] = \Phi^T [K] \Phi = \begin{bmatrix} K_1 & & & & \\ & \ddots & & & \\ & & K_r & & \\ & & & \ddots & \\ & & & & K_N \end{bmatrix} \tag{2.46}$$

$$[C_r] = \Phi^T [C] \Phi = \begin{bmatrix} C_1 & & & & \\ & \ddots & & & \\ & & C_r & & \\ & & & \ddots & \\ & & & & C_N \end{bmatrix}. \tag{2.47}$$

$F_r(\omega)$ is the excitation vector in the modal coordinate, and is expressed as follows:

$$F_r(\omega) = \Phi^T F(\omega) = \begin{bmatrix} f_1 & \cdots f_r & \cdots f_N \end{bmatrix}^T. \tag{2.48}$$

For the rth mode, Eq. (2.44) can be expressed as

$$\left(K_r - \omega^2 M_r + j\omega C_r\right) q_r = f_r, \tag{2.49}$$

where M_r, K_r, and C_r are the mass, stiffness, and damping of the rth mode. f_r is the rth modal force:

$$f_r = \phi_r^T F(\omega). \tag{2.50}$$

For a special case of a single point input – for example, an excitation at point p – Eq. (2.35) becomes

$$F = \begin{bmatrix} 0 & 0 \cdots f_p \cdots 0 \end{bmatrix}^T. \tag{2.51}$$

The modal force in Eq. (2.48) becomes

$$f_r = \phi_{pr} f_p(\omega). \tag{2.52}$$

Substitute Eq. (2.52) into Eq. (2.49), and the response for the rth modal coordinate is

$$q_r = \frac{\phi_{pr} f_p(\omega)}{\left(K_r - \omega^2 M_r + j\omega C_r\right)}. \tag{2.53}$$

Substitute Eq. (2.53) into Eq. (2.38), and the response at point l is

$$x_l(\omega) = \sum_{r=1}^{N} \frac{\phi_{lr} \phi_{pr} f_p(\omega)}{\left(K_r - \omega^2 M_r + j\omega C_r\right)}. \tag{2.54}$$

The transfer function between the response point (l) and the excitation point (p) is

$$H_{lp}(\omega) = \frac{x_l(\omega)}{f_p(\omega)} = \sum_{r=1}^{N} \frac{\phi_{lr} \phi_{pr}}{\left(K_r - \omega^2 M_r + j\omega C_r\right)}. \tag{2.55}$$

2.4.2 Modal Shape and Frequency of Vehicle Body

2.4.2.1 Description of Body Modes

According to the modal analysis theory, a response at any point is a superposition of each modal response at this point. Each mode includes the modal shape and modal frequency. The modal shape is an imagined vibration "shape" for a certain frequency – of course, the shape is invisible.

A cantilever beam shown in Figure 2.38a is used as an example to illustrate modal shape. Figure 2.38b and c show the first and second bending modes. For the first bending mode, the displacement at a fixed supporting location is zero, and the maximum displacement is at the free end. For the second bending mode, the displacement at the fixed point is zero, then the displacement increases gradually and reaches the maximum value, before decreasing until it reaches zero. After zero, the displacement increases in the opposite direction.

A point where the modal displacement is zero is called a node, such as point A in Figure 2.38c. The point where the modal displacement is at maximum is called an anti-node, such as point B in Figure 2.38c. At a node, the corresponding mode has no contribution to the overall response. All bending, torsional, and other complex modes include nodes and anti-nodes. Concentrated mass and external excitation forces should be placed on or as close as possible to the nodes to reduce excitation inputs and system responses.

The most important modes for a vehicle body are the first bending mode and the first torsional mode. The full vehicle modes and trimmed body modes are difficult to identify and observe because their structures are complex, and also because the local modes and overall modes are mixed together. To clearly observe the body overall modes, the BIW is the best choice. Figure 2.39 shows the first bending mode of a BIW that has the following characteristics:

- The main motion and deformation are in the vertical direction.
- The front end and rear end move vertically in phase.
- The middle portion moves out of phase with the two ends.
- B-pillars move laterally in a breathing motion.
- The roof and the floor move mostly in phase, but occasionally out of phase.

(a)   (b)   (c)

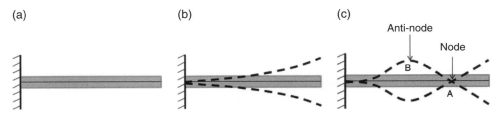

Figure 2.38 (a) Cantilever beam; (b) first bending mode; (c) second bending mode.

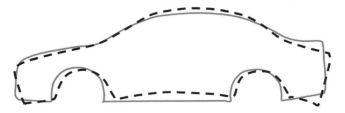

Figure 2.39 First bending mode of a BIW.

Figure 2.40 First torsional mode of a BIW.

Figure 2.40 shows the first torsional mode of a BIW. Its diagonal is twisted. The body torsional mode has the following characteristics:

- The left side and right side of the front end move out of phase.
- The left side and right side of the rear end move out of phase.
- The diagonal points of the front end and the rear end move in phase.

The body's first bending mode and first torsional mode have significant influence on the vibration of the full vehicle, on vibration of the steering wheel, floor, and seats, and on interior booming. If the modal frequencies of these modes are consistent with the frequencies of the power plant and road excitations, the noise and vibrations mentioned above will be easily induced.

For a full vehicle or trimmed body, there exist many other local modes in addition to the overall bending modes and torsional modes, such as dash panel bending modes, door breathing modes, steering wheel bending modes, and so on. The body modes can be divided into four categories.

1) Overall body mode: this refers to the motion shown in an overall body structure, including:
   - Vertical bending mode: the first bending mode is the motion of the front and rear ends moving vertically in phase, while the center moves vertically out of phase with the ends; the second bending mode is the motion of the front and rear end moving vertically out of phase.
   - Lateral bending mode: this motion is the same as for the vertical bending modes, with the difference that the movement is in a lateral direction.
   - Torsional mode: the torsional mode is the twisting motion of the front end and rear end.
2) Body panel mode: this is the local motion of a panel, such as dash panel, roof, or floor, and mainly includes:
   - Dash panel mode.
   - Door breathing mode: lateral motions of the B-pillars induce the door panels to move out of phase.
   - Side panel mode: bending motion perpendicular to a panel.

- Floor mode: bending motion perpendicular to the floor.
- Spare tire well mode: rigid bouncing of the spare tire well.
- Gas tank mode: rigid bouncing of the gas tank drives the floor to move.
3) Secondary beam modes: local motion of the secondary beams, mainly including:
   - Shotgun mode: horizontal or vertical motion of shotguns.
   - Cooling modulus supporter mode: vertical motion of the supporter.
   - Front bumper mode: fore and aft motion of the bumper.
4) Accessory mode: local motions of the accessories installed on the vehicle body, mainly including:
   - Steering system mode: vertical, horizontal, and twisting motion of the steering wheel.
   - Mirror mode: local motion of the mirror.
   - Seat mode: vertical motion of the seat bottom and horizontal motion of the back frame.

This following section introduces the overall body modes. Chapter 3, "Noise and Vibration Control for Body Local Structures," introduces the body panel local modes, local vibration, and accessory modes.

2.4.2.2 Body Modal Frequency

The main target for body mode study is the BIW, because it is the core of a vehicle and the basis of the trimmed body and full vehicle body. There is a certain relationship for modes and frequencies among the three kinds of bodies. Generally, the higher the BIW modal frequency, the higher the modal frequencies of the trimmed body and the full vehicle body. It is also relatively simple to analyze the BIW's modes, and its overall bending and torsional modes, and local panel modes can be clearly identified. Furthermore, it is easy to analyze and test the BIW. Finally, in the early stages of product development, the main task is to design the body's overall structure and to decide the BIW's structural parameters, stiffness, and modes, making the BIW a focus for modal analysis.

Figure 2.41 shows the modal frequencies of the BIW's first bending modes for economy sedans, mid-sized sedans, and SUVs; their frequencies are in the ranges of 40–50 Hz, 45–55 Hz, and 45–55 Hz, respectively.

After the doors, trimmed parts, and seats are installed on a BIW, the BIW becomes a trimmed body. The modal frequencies of the trimmed body drop because of the weight increase. The first bending frequency of the trimmed body is about 20 Hz lower than BIW's. The full vehicle body's first bending frequency is about 1–3 Hz lower than that of the trimmed body. The first bending frequency of the full vehicle body is usually between 20 and 30 Hz. The first bending modal frequency of the BIW is about 1.5–2.2 times that of the full vehicle body.

Figure 2.42 shows the modal frequencies of the BIW's first torsional modes for economy sedans, mid-sized sedans, and SUVs, which are about 40 Hz, 40–50 Hz, and 35–40 Hz, respectively.

The first torsional modal frequency of the trimmed body is about 5–10 Hz lower than the BIW's. The full vehicle body's first torsional frequency is about 1–3 Hz lower than that of the trimmed body. The first torsional modal frequency of the full vehicle body is usually between 20 and 30 Hz. The first torsional modal frequency of the BIW is about 1.1–1.6 times that of the full vehicle body.

(a)

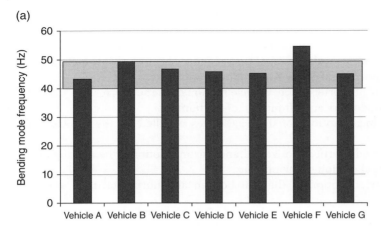

(b)

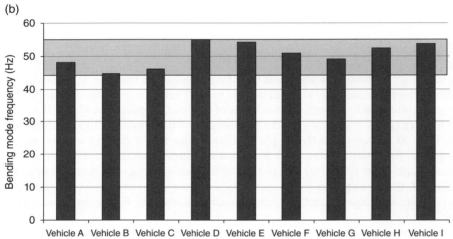

(c)

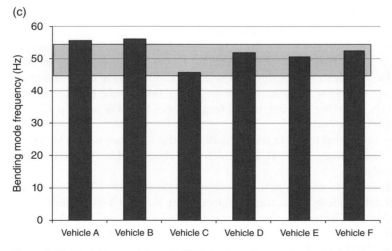

Figure 2.41 Modal frequencies of a BIW's first bending modes for (a) economy sedans, (b) mid-sized sedans, and (c) SUVs.

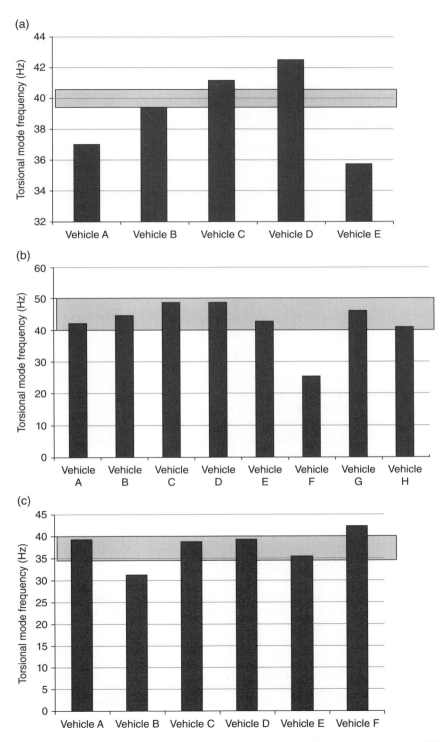

Figure 2.42 Modal frequencies of a BIW's first torsional modes for (a) economy sedans, (b) mid-sized sedans, and (c) SUVs.

Mid-sized sedans and SUVs have much higher bending and torsional stiffness than economy sedans; however, there are no significant differences among the modal frequencies of the three vehicle types. That said, the modal frequencies of some mid-sized sedans and SUVs can be somewhat lower than that of some economy sedans due to their heavier weight and longer body.

The body modes are determined by the body structures. The underbody structures, such as the side frames, tunnel beam, and so on, have a significant influence on the bending stiffness, whereas the upper body structures, such as the roof frames, pillars, windshield, and so on, have a great impact on the torsional stiffness. The first torsional modal frequency of a BIW with windshield is about 3–10 Hz higher than that of a BIW without windshield. But the windshield has just a small influence on the bending modal frequencies.

2.4.3   Modal Testing for Vehicle Body

Body modal testing is based on several assumptions. First, it is assumed that the body is a linear system in which the principle of superposition works. Second, it is assumed that the system is a time-invariant system, which means its dynamic characteristics do not change with time. Third, the system is assumed to be observed, which indicates that all dynamic analysis data can be measured. Fourth, it is assumed that the reciprocal principle works for the system: i.e. a response at point B from the input at point A is equal to a response at point A from the input at point B.

2.4.3.1   Testing System

The modal testing system consists of an excitation system, a response system, and a data processing system, as shown in Figure 2.43.

A body system is too large to be excited by a hammer, so usually exciters are applied to excite the body for modal testing. The excitation system consists of a signal generator, power amplifier, and exciter. The signal generator transmits signals to the amplifiers, and the amplified signals drive the exciters to work. A rod connects an exciter to the body, and a sensor or impedance head is mounted at one end of the rod. The vibration generated by the exciter is transmitted to the body through the rod. The input force should be along the axis of the rod.

One or more exciters can be used for body modal testing. Two principles are used to choose the number of exciters. The first principle is that all tested points have clear signals when the body is excited. For example, if only one exciter is used at the front of the body, an initial input is chosen in order to excite the front end, and then the input force is gradually increased until the rear body is excited. All the tested points should have clear signals and a sufficiently high signal-to-noise ratio. The second principle is that the input energy shouldn't be too large, otherwise the body transfer function could include nonlinear signals that distort the test.

The signals used for body excitation include sine signals, sine sweep signals, random signals, and random triggers.

The response system comprises accelerometers, force sensors, and a data acquisition system. The force sensors are used to measure the excitation forces applied on the body. The accelerometers are distributed on different parts of the body to acquire acceleration responses. The excitation forces and acceleration signals are recorded by the data acquisition system, which is used for subsequent signal processing.

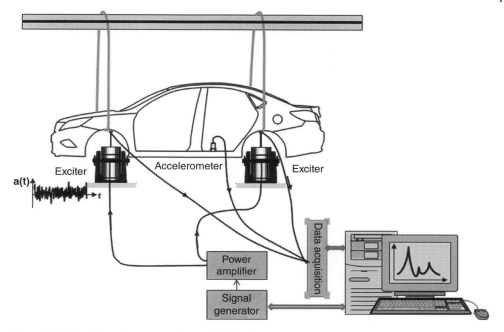

Figure 2.43 Modal testing system for a vehicle body.

The data processing system is a system to process the measured signals such as accelerations and excitation forces. For modal testing and analysis, the purpose of data processing is to acquire the transfer functions between the accelerations and the excitations, from which the modal parameters can be abstracted.

2.4.3.2 Boundary Conditions

For body modal testing, the tested body should be set up as a free–free boundary: i.e. the boundary has no impact on the testing results. Usually, a free–free boundary can be achieved by two methods. One is to hang the body using soft rubber ropes, as shown in Figure 2.44. The other is to support the body with soft air cushions, as shown in Figure 2.45. The points by which the body is hung or supported should be the main mode nodes or locations close to the nodes. The highest frequency of the body rigid mode should be 10% lower than the lowest body elastic modal frequency.

2.4.3.3 Excitation Points

The excitation points are where energy is inputted into the body. In the analytical frequency range, the excitation input should be a white noise signal. To guarantee the quality and energy of the input signal, the excitation points should meet the following requirements:

- The excitation points should have sufficiently large stiffness.
- The excitation points should be as close as possible to the actual excitation locations, and the exciters and impedance heads or force sensors should be easy to install.
- The excitation points should avoid mode nodes and/or supporting points.
- The excitation points should not be along the body's symmetry plane.

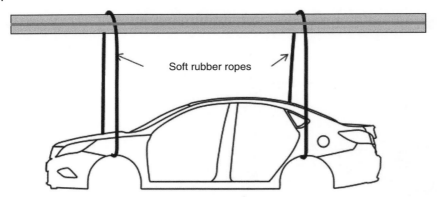

Figure 2.44 A tested body is hung by soft rubber ropes.

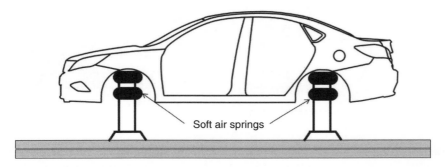

Figure 2.45 A tested body is supported by soft air cushions.

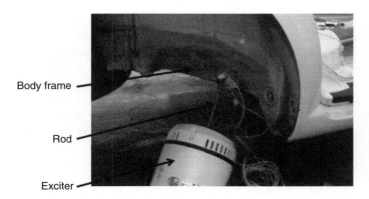

Figure 2.46 An excitation location where an exciter is mounted.

In modal testing, the signals from the force sensors are set up as reference signals. Several locations on the body are excited until an optimal location can be found where the best transfer functions are obtained for all measured points, then this location is used as the excitation point. Figure 2.46 shows an excitation location where the exciter is mounted on a rear side frame and close to the suspension's spring and damper mounting locations. The point satisfies the above requirements.

2.4.3.4 Response Points

The purpose of modal testing is to acquire a body's modal shapes and frequencies. The choice of the response points is very important and should follow several principles. First, these points should reflect the appearance of features of the body structure. They should be uniformly located on the body in order to obtain the contours of the body's modal shapes, especially for some important points such as joints, external excitation points, concentrated mass locations, and boundary points, as shown in Figure 2.47. The small squares in the figure represent locations for installing accelerometers.

Second, the response locations should avoid the nodes of critical modes, such as the first and second bending modes and the torsional modes.

Third, the key body local modes must be included in modal testing, so sufficient test points on local structures must be chosen. The description and analysis of local modes, such as the front panel modes, roof modes, floor modes, and so on, is detailed in Chapter 3. Figure 2.48 shows accelerometer distribution on a dash panel; a sufficiently large number of accelerometers has been used to ensure that the modes will be accurately identified.

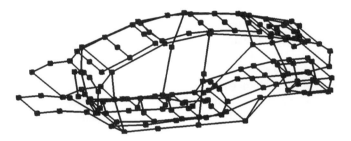

Figure 2.47 Distribution of response points for BIW modal testing.

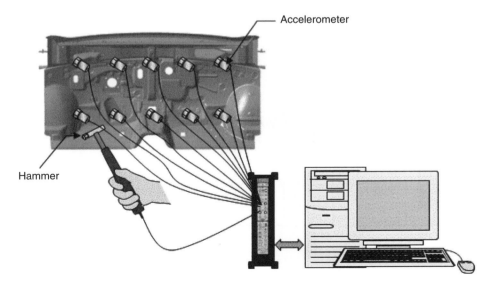

Figure 2.48 Accelerometer distribution on a dash panel.

The transfer function at an excitation point is usually employed as a reference for where an accelerometer should be placed. If an impedance head that includes an accelerometer is used, a separate accelerometer is not needed.

2.4.3.5 Testing and Signal Processing

Before testing, a 3D layout of the points to be tested is plotted using computer software. Because the modal frequencies of interest for a BIW are below 100 Hz, the sampling frequency is set to 256 Hz, and the resolution is set to 0.25 Hz. However, the frequencies of the local modes are relatively higher, usually more than 200 Hz, so the sampling frequency should be increased accordingly.

To ensure the quality of the testing data, the auto power spectrum, linearity, reciprocity, FRF, and coherent function of the tested data must be inspected. The auto power spectra of the exciting forces must be checked to guarantee the characteristics of the white noise: i.e. to ensure the input energy for each frequency is the same. If more exciters are used, the auto spectrum difference for the exciting forces should be less than 2 dB.

The linearity check is to ensure that the body system and the tested data are linear. For a linear system, the transfer functions should be the same for different inputs: i.e. they are independent of inputs. By adjusting the input magnitudes, the transfer functions can be checked.

The reciprocity check involves interchanging the input and output points, and comparing the transfer functions before and after interchanging. For a linear system, the interchanged transfer functions should be the same.

The FRF check involves observing the sharpness of the FRF curve. There should be a clear curve at the resonance and anti-resonance points.

The coherence function check involves examining the relationship between the output signal and the input signal, ensuring that the coherence coefficient between them is high enough. The closer the coherent coefficient is to 1, the better the coherence between the two signals.

From the FRFs, the modal frequencies, modal shapes, and modal parameters can be identified. Figure 2.49 shows a transfer function between an acceleration and an

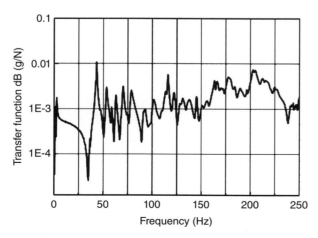

Figure 2.49 A transfer function between an acceleration and an input force.

(a)

(b)

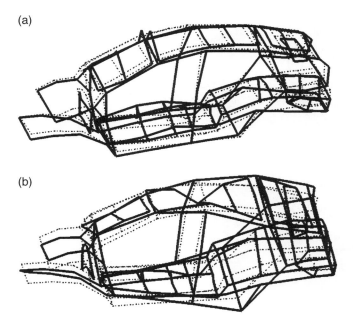

Figure 2.50 (a) The body first bending mode; (b) the body first torsional mode.

input force. There is a resonance at 45 Hz, and the corresponding peak is very sharp. All FRFs at 45 Hz on the body are then extracted, and the corresponding peaks of the imaginary parts are drawn to obtain the modal shape, as shown in Figure 2.50a. The mode shows that the body's front end and rear end move in the same direction, and the middle section moves in the opposite direction, which shows it is the first bending mode. Similarly, at 52 Hz, the body's first torsional mode is acquired, as shown in Figure 2.50b.

2.4.4   Calculation of Vehicle Body Mode

The body is a very complicated elastic continuum. It is impossible to obtain the body modes by analytical methods, so instead the FE method is commonly used. The FE method involves subdividing the elastic continuous body into many smaller and simpler elements called FEs. The system's modal parameters can then be calculated by solving the finite discrete equations.

Because the damping inside the body is very small, and because the damping's contribution to the body's modal frequencies and shapes is insignificant, damping is usually neglected for modal calculations using FE methods. However, when calculating dynamic responses and transfer functions, the damping should be included. Equation (2.33) can be simplified as follows:

$$[M]\ddot{X}+[C]\dot{X}+[K]X = 0. \tag{2.56}$$

Establishment of a body's FE model is the first step of the analysis. Based on computer-aided design (CAD) data, the FE meshing is built using software. Then, the boundary conditions are set up as free–free: i.e. with no constraints. Finally, appropriate methods,

(a)

(b)

Figure 2.51 First bending mode (a) and first torsional mode (b) of a body calculated by finite element analysis.

such as the subspace method, Lanczos method, or modal reduction method, are used to calculate the model.

The major modal frequencies of a body are below 100 Hz, but the modal frequencies of local body modes and other systems attached to the body, as well as excitation frequencies, are usually higher than 100 Hz; therefore, 250 Hz is usually set as the analysis frequency. With the increase of analysis frequency, the calculation load greatly increases, and the accuracy of the calculations decreases. In addition, when a structure's frequencies are high, it is difficult to clearly distinguish its modes.

Figure 2.51 shows the first bending mode and the first torsional mode of a body calculated by FE analysis.

To verify the accuracy of the calculated modes by FE analysis, the calculated and measured modal values are compared. The comparison can be implemented by checking natural frequencies, viewing the modal shapes, using modal confidence factors or FRFs, and so on.

For a natural frequency comparison, the calculated and measured natural frequencies are listed in a table and compared. If the calculated and measured frequencies are within 3% of each other, the calculated mode and measured mode are regarded to be the same.

When comparing modal shapes, the calculated and measured modes are animated, and the modal shapes and motions are observed to confirm their correlation.

When using modal confidence factors, the calculation model must be reduced to the same DOFs as the measured model, because the computational model has much higher DOFs than the measured model. There are many methods for processing this reduction. The characteristics of the reduced model and the original model should be consistent.

Based on comparison between a measured modal vector (ϕ_t) and a calculated modal vector (ϕ_c), the modal confidence factor is defined as

$$MAC = \frac{\left|\phi_t^T \phi_c\right|^2}{\left(\phi_t^T \phi_t\right)\left(\phi_c^T \phi_c\right)}. \tag{2.57}$$

In a case where ϕ_t and ϕ_c are the same mode, $MAC \approx 1$. In a case where ϕ_t and ϕ_c are different modes, $MAC \approx 0$. MAC is between 0 and 1. The higher the MAC value, the more relevant the two modes. Usually, the MAC value for related modes is higher than 0.8. After the MAC values have been calculated for all calculated modes and measured modes, a $MAC(i,j)$ matrix can be formed. If the calculated modes and measured modes are consistent, the diagonal values in the matrix will be close to 1, and the non-diagonal values will be close to 0.

For an FRF comparison, damping must be included in the calculation. The calculated FRF and the measured FRF are plotted together, and their shapes, frequencies, and peaks are compared to judge the credibility of the calculated and measured values.

2.5 Control of Overall Body Modes

The body's structural response is determined by the body's modal characteristics and the external excitations. When the excitation frequency and the structural modal frequency are the same, resonance occurs. Therefore, the design of the body's modal shapes and frequencies is very important, otherwise the body could induce interior vibration and noise, such as booming during engine idling and body shaking during acceleration. The overall body modes can be controlled from three aspects: first, the excitation frequencies can be analyzed and separated from the systems' frequencies; second, a perfect modal table can be developed; and third, control measures can be implemented to achieve modal separation or suppress the noise and vibration.

2.5.1 Separation and Decoupling of Body Modes

Modal decoupling of the overall body includes separation between the overall body modes and excitation frequencies, separation between the overall body modes and the modes of attached systems, and separation between the overall body modes and the local body modes

2.5.1.1 Separation Between Overall Body Modes and Excitation Frequencies

To control body vibration, we must understand the characteristics of vibration sources, including engine excitation, road excitation, wind excitation, fans and other rotating machinery excitation, random excitation, and pulse excitation. Chapter 5, "Vehicle Body Sensitivity Analysis and Control," has a detailed analysis of the features of various kinds of vibration sources. The following sections briefly introduce the frequency characteristics of several of the main body excitations.

- Engine excitation sources: for a four-cylinder engine, the idling speed range is usually from 600 to 1000 rpm, corresponding to a second order firing order frequency of 20–33.3 Hz. The engine speed for vehicle acceleration and cruising is usually from 1000 to 6000 rpm, corresponding to a second order frequency of 33.3–200 Hz.
- Road and tire excitation sources: road and tire excitation depends on the vehicle's speed and tire size. The tire's first order dynamic unbalance excitation frequency is usually 10–20 Hz.
- Rotating machinery excitation sources: rotating machineries such as the fan, generator, compressor, and pump will excite the body. Fan excitation is the most common; the fan speed is generally in the range of 1500–3000 rpm, and the corresponding frequency of the first order dynamic unbalance is 25–50 Hz. There are many blades on the fan, so high-order and high-frequency excitations will be induced as well.
- Random excitation and pulse excitation sources: a vehicle in motion can be subjected to random and pulse excitations from the road. Airflow excitation also has random characteristics. Some accessories installed on the body are subjected to pulse excitations: for example, pulse excitations from the fuel pipe will induce the floor to vibrate, which radiates noise into the passenger compartment.

The body modal frequencies must be separated from the above excitation frequencies to avoid resonance. Table 2.1 lists the modal frequencies of a body and steering wheel system, along with the engine idling speed and fan speed, and their corresponding excitation frequencies.

In Table 2.1, when the air conditioning is on (AC-on condition), the engine's second order excitation frequency is 25 Hz. This is close to the body's first order bending modal frequency, 26 Hz, so resonance of the body could be induced by the engine excitation. To avoid resonance, the body structure must be modified to increase its bending frequency, or the engine speed should be adjusted.

The excitation frequency of the fan is 31.6 Hz, which is very close to the steering system's modal frequency, 31 Hz. The fan's vibration is transmitted to the body, then to the steering system, which causes the steering wheel to resonate. To avoid resonance, the steering wheel must be modified or the fan's speed adjusted. Generally speaking, to avoid resonance, the structural modal frequency and the excitation frequency must be separated by at least 3 Hz.

Table 2.1 Modal frequencies for a vehicle's body, steering wheel system, engine idling speed, and fan speed, and the corresponding excitation frequencies.

	Speed (rpm)	Frequency (Hz)
First body bending mode	—	26
First body torsional mode	—	38
Steering wheel system mode	—	31
Engine idling speed (AC off)	700	23.3
Engine idling speed (AC on)	750	25
Fan speed	1900	31.6

2.5.1.2 Separation between Overall Body Modes and Attached Systems' Modes

The main systems connected to the body include the power plant, suspension system, exhaust system, driveshaft system, and steering system.

The power plant's modes include rigid modes and elastic modes. The rigid modes are mainly used for vibration isolation analysis, and the corresponding frequencies are relatively low, usually between 5 and 15 Hz. The elastic modes' frequencies are relatively high, usually more than 200 Hz, so the power plant modes and body modes are not coupled.

The suspension's bounce modes are generally between 10 and 16 Hz, which is unlikely to couple with the body modes, but these modes are easily coupled with the power plant modes and road excitation frequencies. The fore-aft modes of the suspension system are usually between 15 and 30 Hz, which overlap with some of the body modes.

The exhaust system is connected with the body by hangers, and its vibration is transmitted to the body. The exhaust modes include bending modes and torsional modes, and the frequencies are usually between 20 and 50 Hz – the body bending modes and torsional modes are usually within the same frequency range. Therefore, the exhaust modes must be separated from the body modes.

The driveshaft system is connected with the body by bearings and isolators, and its vibration is transmitted to the body. Generally, the modal frequencies of the driveshaft are between 90 and 250 Hz. These frequencies do not overlap with the overall body modes, but may overlap with the local body modes. Thus, in driveshaft design, its modes must be decoupled from the body modes, including the local panel modes.

In principle, the system modes must be separated from the body modes by at least 3 Hz, but generally it is better to achieve a 5 Hz separation.

2.5.1.3 Separation between Overall Body Modes and Local Body Modes

If the local body modes are coupled with the overall body modes, the overall body vibration will worsen the local structure vibration. For example, if a vehicle with a first body torsional modal frequency of 38 Hz and front floor modal frequency of 39 Hz is subjected to external excitation of 38 Hz, the body will resonate, which will significantly increase the vibration of the floor. Another example is the steering system: this usually has a modal frequency of between 25 and 35 Hz, meaning it has a wide frequency overlap with the body's modes. Thus, the steering system can easily be induced to vibrate by the overall body vibration.

In short, the local panel modes and accessory modes must be well separated from the overall body modes.

2.5.2 Planning Table/Chart of Body Modes

Modal separation and decoupling is one of the most important aspects of vehicle noise and vibration control. To achieve the three decoupling principles mentioned above, we must develop a modal planning table/chart to guide the development of each system. Three modal tables/charts – i.e. a full vehicle modal planning table, a body modal planning table, and a source excitation chart – should be planned in the early stages to guide development of the vehicle. If the modal planning is followed, each system can be worked on independently by different teams, and modifications during later phases can be kept to a minimum, along with the associated costs.

2.5.2.1 Full Vehicle Modal Planning Table

The full vehicle modal table includes all of the modal frequencies of the body and other systems, as shown in Figure 1.40. Frequencies below 100 Hz are listed in the table, because the main body modes are lower than 100 Hz. In some cases, the listed frequencies are below 50 Hz, because the body modes below 50 Hz are the most important. Figure 1.40 provides the modal frequencies of the body, power plant, suspension, exhaust, etc. and the engine idle excitation frequencies. The purpose of the table is to make sure that adjacent systems are decoupled and that the body modes avoid the excitation frequencies. After the table has been finalized, each system will be given its own modal targets, and development work on the systems can be carried out independently.

In Figure 1.40, the body first bending and first torsional modal frequencies are 28.5 and 26.4 Hz, respectively, and the frequencies of the systems connected to the body are as follows.

• Frequencies of the power plant rigid body modes: 7.5, 6.5, 8.8, 10.7, 12.8, and 14 Hz.
• Frequencies of the suspension bounce modes: 13.2 and 16.5 Hz.
• Frequencies of the exhaust modes: 23.5, 37.2, and 41.6 Hz

From the above, it is clear that the body modes are decoupled from the systems' modes.

The engine speeds during AC-on and AC-off conditions are 750 and 900 rpm, respectively. For a four-cylinder engine, the excitation frequencies of the second order are 25 and 30 Hz respectively. Therefore, the body modes are separated from the idle excitation frequencies.

After the full vehicle modal table has been determined, the body can be developed independently. As a result of the particular relationship among the full vehicle modes, trimmed body modes, and BIW modes, the full vehicle's modal targets can be inferred from modal information from the BIW, which becomes a foundation for body development.

2.5.2.2 Body Modal Table

The body modal table puts the frequencies of the overall body modes and all local modes together, as shown in Table 2.2. The purpose of the table is to separate the overall body modes from the local modes, and also to separate them from external excitations. The main local structures include the front panel, front floor, rear floor, roof, side panels, luggage compartment floor, trunk lid, engine hook, steering wheel, mirror, seats, and acoustic cavity.

These local modes generate two kinds of noise and vibration problems. First, they could be resonant with the overall body modes. For example, if the modal frequency of the front floor is 39 Hz and the body's first torsional frequency is 38 Hz, the floor could easily be excited when the body is subjected to an external excitation at this frequency. Second, the local modes could be excited by external excitations directly. The panels are easily excited by engine, wind, and road inputs. For example, when vibration from a compressor is transmitted to the dash panel through the pipes, the panel could vibrate and radiate noise.

After all of the modal data for a vehicle have been collected, they are put into a table, as shown in Table 2.2. A new table similar to Figure 1.40 can be generated, and the coupling status of adjacent structures will be clearly indicated.

Table 2.2 Body modal table: overall body modes and local modes.

Mode category	Mode description
Engine excitations	Idling excitation (AC Off)
	Idling excitation (AC On)
	Firing order excitation
Overall body modes	First bending mode
	First torsional mode
Panel modes	Dash panel bending mode
	Front floor bending mode
	Roof bending mode
	Fender bending mode
	Rear package tray bending mode
	Front door breathing mode
	Rear door breathing mode
	Engine hood bending mode
	Engine hood torsional mode
	Trunk lid/liftgate bending mode
	Trunk lid/liftgate torsional mode
Cavity modes	Passenger cabin cavity mode
	Trunk cabin cavity mode
Accessory modes	Steering wheel bending mode
	Instrument panel bending mode
	Side mirror mode
	Inner mirror mode
	Seat transverse mode
	Seat longitudinal mode
	Gas tank bounce mode
	Spare tire bounce mode
Bracket modes	Power plant left bracket mode
	Power plant right bracket mode
	Power plant rear bracket mode
	Battery tray mode
	...
Secondary frame modes	Body front end mode
	Cooling module mode
	Bumper bending mode
	Bumper torsional mode

2.5.2.3 Body Mode and Excitation Chart

There are many rotary machines in a vehicle, such as the generator, water pump, oil pump, compressor, blower, and fan. They are divided into two categories. The first category is rotary machines with fixed speeds, such as the engine speed at idle and the fan speed. For example, the idling speeds for a vehicle under AC-off and AC-on conditions are 700 and 900 rpm, respectively, with corresponding second order excitation frequencies of 23.3 and 30 Hz, respectively. Suppose the fan has two speeds, 1800 and 2600 rpm, and the fundamental frequencies for the two speeds are 30 and 43.3 Hz, respectively. After all these parameters have been listed, as in Table 2.3, we can clearly see the relationship between speed and frequency.

The second category is rotary machines whose speeds have ratio relationships with the engine speed. The orders of these machines depend on not only the engine order, but also their internal blades, bearings, and so on. Table 2.4 lists the ratios between the machines' speeds and the engine speed, the numbers of blades, bearings, and slots inside the machines, and their corresponding engine orders.

Table 2.2 gives a long list of the body modes. In order to illustrate the relationship between body modes and excitation, the partial body modes of a vehicle are provided in Table 2.5.

All data in Tables 2.3–2.5 are plotted on a speed–frequency–order chart, as shown in Figure 2.52. The horizontal coordinate is frequency, and the longitudinal coordinate is

Table 2.3 Relationship between speed and frequency.

	Speed (rpm)	Frequency (Hz)
Engine: AC-off	700	23.3
Engine: AC-on	900	30
Fan: high speed	1800	30
Fan: low speed	2600	43.3

Table 2.4 Ratios between rotary machines' speeds and engine speed, and their corresponding engine orders.

Rotary machine	Ratio of rotary machines/engine	Number of blades, bearings, or slots	Corresponding engine order
Engine second order	1 : 1	—	2
Engine fourth order	1 : 1	—	4
Compressor	1.25 : 1	No. of blades: 5	6.25
Generator	1.3 : 1	No. of blades: 10	13
		No. of bearings: 12	15.6
		No. of slots: 16	20.8
Water pump	1 : 1	No. of blades: 9	9
...	...	...	...

Table 2.5 Partial body modes.

System or part	Mode description	Frequency (Hz)
Body	First bending mode	26
	First torsional mode	35
Front dash panel	Bending mode	185
Floor	Bending mode of front floor	75
	Bending mode of rear floor	120
Roof	Bending mode of front roof	75
	Bending mode of rear roof	52
...	...	...

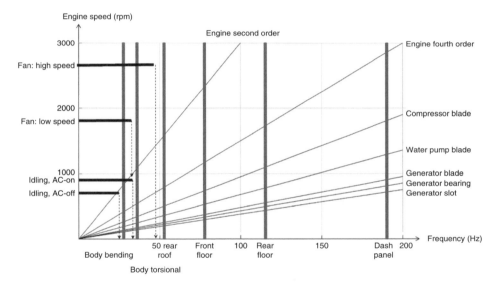

Figure 2.52 Excitation relationship between body modes and rotary machines.

rotational speed. The order excitations of the engine and related rotary machines are plotted as scattering order lines. The frequencies corresponding to the body modes are plotted as vertical lines. The speeds for the fixed-speed rotary machines are plotted as horizontal lines, and the corresponding frequencies are plotted as vertical lines. After these lines have been drawn, a body mode and excitation chart can be built.

From the chart, we can clearly see the relationship between the body modes and excitation sources, and can quickly diagnose problems. For example, an interior booming at 185 Hz is identified during acceleration at an engine speed of 1800 rpm. Draw a horizontal line at 1800 rpm and it meets with lines of the engine, fan, and compressor excitations. The frequencies for the engine second and fourth order excitations are 60 and 120 Hz, respectively. The frequencies for the fan at low speed and the compressor are 60 and 187.5 Hz, respectively. Therefore, the excitation source can be judged to be the compressor. The compressor's excitation is transmitted to the dash panel through the pipelines. The dash panel has a 185 Hz modal frequency, so it is excited and radiates noise to the interior.

Figure 2.52 shows only some of the body modes and excitation sources. A full chart of body modes and excitation sources is much more complicated, but the analysis method is the same.

2.5.3 Control of Overall Body Modes

The frequencies for major body modes, such as the first bending mode and the first torsional mode, should be designed to be as high as possible. The modes that have significant influence on body vibration should be far away from the excitation frequencies. For some special cases where coincidences between the body modal frequency and excitation frequency are inevitable, the excitation should be placed on the body nodes or close to the nodes.

The body modes mainly depend on the body's stiffness and mass distribution, so tuning of the modes can be started from these two aspects. Because of the small effect of structural damping on overall body modes, the damping can be ignored for modal analysis. However, for local panel modes, damping is very important to restrain vibration and decrease noise radiation. In addition, some parts connected with the body can be regarded as dynamic dampers, so appropriate design of these parts can reduce body vibration.

2.5.3.1 Stiffness Control for Body Modes

Body stiffness control can be implemented from four aspects: overall stiffness distribution, frame cross-section stiffness, joint stiffness, and bonding adhesive stiffness. The strain energy distribution of a body subjected to static bending loads and/or torsional loads can be analyzed to reveal weak stiffness points: i.e. locations with high levels of strain energy.

Figures 2.36 and 2.37 show the strain energy distribution of a body subjected to a static load, and weak stiffness locations can be identified from the diagrams. Similar to strain energy analysis, the bending and torsional modal strain energy can be analyzed. From the static strain energy distribution and modal strain energy distribution, the contribution of each frame and joint to the body's stiffness and modes can be identified. The structures that significantly influence the overall body bending stiffness and modes are the A-pillars, B-pillars, and C-pillars, the roof structure close to the A-pillars and B-pillars, the rocker frame, and the side frames. To increase the bending stiffness and modal frequencies, these structures should be modified. The structures that significantly affect the body's overall torsional stiffness and modes are the B-pillars and C-pillars, the rocker frame, the rear side frame, and the windshields. Modification of the structures can effectively increase the body's torsional stiffness and modal frequencies.

In the early stages of vehicle development, the targets for body frame stiffness and joint stiffness, along with the body modes, are determined by benchmarking and experimental data. The major parameters influencing the overall body bending stiffness and bending modes are the cross-section moment of inertia of frames/pillars and the joints' torsional stiffness. Figure 2.53 shows the bending stiffness target values of a body's frames and joints. The basic values of the frames' stiffness and the joints' torsional

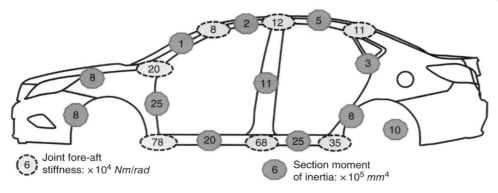

Figure 2.53 Bending stiffness target values for frames and joints of a body.

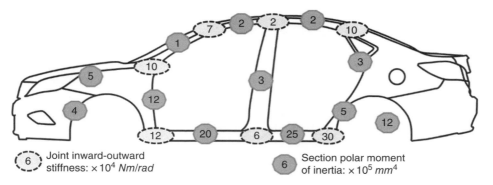

Figure 2.54 Torsional stiffness target values for frames and joints of a body.

stiffness are listed below the figure. The numbers inside the circles and hexagons represent coefficients corresponding to the joints' stiffness and the frames' stiffness, respectively. For example, 11 noted at the B-pillar means that the cross-section's moment of inertia is $11 \times 10^5 \, mm^4$, and 68 at the B1 joint represents that its torsional stiffness is $68 \times 10^4 \, Nm/rad$.

The major parameters affecting the overall body torsional stiffness and torsional modes are the frames' polar moment of inertia (J) and the joints' torsional stiffness. Figure 2.54 provides the torsional stiffness targets for the frames and joints of a body. The meanings of the numbers inside the circles and hexagons are similar to those for the bending stiffness.

2.5.3.2 Body Modal Control by Mass Method

The body modal frequencies are determined by stiffness and mass. If we assume that mass is uniformly distributed along the body, adding mass will reduce the overall body frequencies. When the trim parts are installed on a BIW to form a trimmed body, the body mass increases, so the trimmed body modal frequencies are lower than those of the BIW. The ratio between the bending modal frequency of the trimmed body and the BIW bending modal frequency is approximately equal to the square root of their mass ratio. Compared with the BIW, the weight of the trimmed body is higher, but

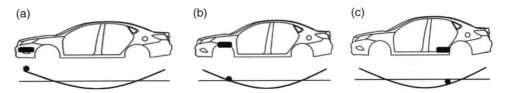

Figure 2.55 Influence of a battery placed at different locations on body bending modal frequency: (a) at front end; (b) on top of front tire; (c) at behind of rear seat.

simultaneously, the closures such as doors increase the body torsional stiffness, so its torsional modal frequencies decrease slightly.

The same mass can have a different influence on the body's modal frequency when it is placed at different locations. When placed at a node, mass has little impact on the body's modes. Figure 2.55 shows a 25 kg battery placed at three locations of a BIW: at the front end, on top of the front tire, and behind the rear seat. The BIW bending modal frequency without the battery is 52 Hz. After the battery is placed at above three locations, the modal frequency changes to 47.5, 51.2, and 49.8 Hz, respectively. The BIW modal frequency reduces after adding the mass, but the amount of reduction differs according to where the battery is placed. When the battery is placed at the front end, the modal frequency is greatly affected and reduced by 4.5 Hz because the front end is the location of the mode's highest magnitude. When the battery is placed on top of the front tire, the location of a node, the modal frequency is little influenced and reduces by only 0.8 Hz. The seat location is not a node, but is close to a node, so the modal frequency changes moderately, reducing by 2.2 Hz.

Adding mass at a location where the mode's amplitude is highest – i.e. at an anti-node – has the greatest influence on modal frequency. For example, the anti-node of the first body bending mode is at front end, so after a front bumper has been installed, the body bending modal shape and magnitude will be substantially changed.

If a concentrated mass is also an excitation source, it must be installed on the body's nodes. For example, a power plant is a concentrated mass, but also an excitation source: therefore, it must be placed on a node. Another example is the suspension system: it should be placed at the body's nodes because excitation from the road will be transmitted to the body through the suspension.

2.5.3.3 Dynamic Damper for Vehicle Body

In addition to tuning the body's modes through varying the body's stiffness and mass, the modes can be altered by designing systems as dynamic dampers. The principle of dynamic dampers is introduced in Chapter 3. A dynamic damper consists of a main system and an additional system, and the resonance of the main system can be eliminated by the additional system's vibration. For example, a power plant can be regarded as an additional mass-spring system, while the body and suspension comprise the main system, as shown in Figure 2.56. The power plant can be used as a body damper to change the body's bending modal frequency and modal shape. Another example is a cooling module being used as a body damper, as shown in Figure 2.57. The isolation parameters of the cooling module can be tuned to change the body's first bending mode.

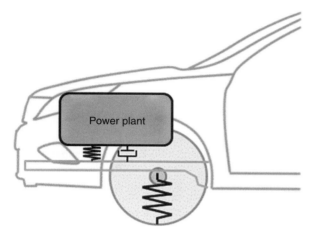

Figure 2.56 A power plant used as a body damper.

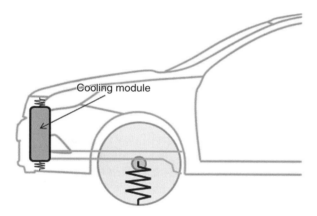

Figure 2.57 A cooling module used as a body damper.

Bibliography

Crocker, M.J. (2007). *Handbook of Noise and Vibration Control*. John Wiley & Sons.

Doyle, J.F. (1997). *Wave Propagation in Structures*. New York: Springer-Verlag.

Ewins, D.J. (1984). *Modal Testing: Theory and Practice*. Research Press.

Griffiths D., Green E.R., Liu K.J. Application of the Modal Compliance Technique to a Vehicle Body in White. SAE Paper 2007-01-2355; 2007.

Heylen, W. and Lammens, S. (1997). *Modal Analysis Theory and Testing*. New York: Prentice Hall.

Isomura Y., Ogawa T., Monna H. New Simulation Method Using Experimental Modal Analysis for Prediction of Body Deformation during Operation. SAE Paper 2001-01-0494; 2001.

Jans J., Wyckaert K., Brughmans M., Kienert M., Van der Auweraer H., Donders S., Hadjit R. Reducing Body Development Time by Integrating NVH and Durability Analysis from the Start. SAE Paper 2006-01-1228; 2006.

Khanse K.R., Pathak S.P. Test Set-Up of BIW (Body in White) Stiffness Measurements. SAE Paper 2013-01-1439; 2013.

Kim K.C., Choi I.H., Kim C.M. A Study on the Development Process of a Body with High Stiffness. SAE Paper 2005-02-2465; 2005.

Kim H.S., Yoon S.H. A Design Process using Body Panel Beads for Structure-Borne Noise. SAE Paper 2007-01-1540; 2007.

Kim K., Choi I. Design Optimization Analysis of Body Attachment for NVH Performance Improvements. SAE Paper 2003-01-1604; 2003.

Koizumi T., Tsujiuchi N., Nakahara S., Oshima H. Mode Classification Analysis using Mutual Relationship between Dynamics of Automobile Whole-Body and Components. SAE Paper 2007-01-3500; 2007.

Lilley K.M., Fasse M.J., Weber P.E. A Comparison of NVH Treatments for Vehicle Floorpan Applications. SAE Paper 2001-01-1464; 2001.

Magalhães M., Ferraz F., Agostinho A. Comparison Between Finite Elements Model and Experimental Results for Static Stiffness and Normal Vibration Modes on a Unibody Vehicle. SAE Paper 2004-01-3351; 2004.

Merlette N., Wojtowicki, J.L. FEA Design of a Vibration Barrier to Reduce Structure Borne Noise. SAE Paper 2007-01-2164; 2007.

Moeller M.J., Thomas R.S., Maruvada H., Chandra N.S., Zebrowski M. An Assessment of a FEA NVH CAE Body Model for Design Capability. SAE Paper 2001-01-1401; 2001.

Shorter P.J., Gardner B.K., Bremner P.G. A Review of Mid-Frequency Methods for Automotive Structure-Borne Noise. SAE Paper 2003-01-1442, 2003

Silva I., Magalhães M. Influence of Spot Weld Modeling on Finite Elements Results for Normal Modes Vibration Modes of a Trimmed Vehicle Body. SAE Paper 2004-01-3358; 2004.

Sung S.H., Nefske, D.J. Assessment of a Vehicle Concept Finite-Element Model for Predicting Structural Vibration. SAE Paper 2001-01-1402; 2001.

Thomas, W.T. and Dahleh, M.D. (1998). *Theory of Vibration with Application.* Prentice Hall.

Ver, I.L. and Beranek, L.L. (2006). *Noise and Vibration Control Engineering: Principles and Applications.* John Wiley & Sons.

Yamamoto T., Maruyama S. Feasibility Study of a New Optimization Technique for the Vehicle Body Structure in the Initial Phase of the Design Process. SAE Paper 2007-01-2344; 2007.

You Y.K., Yim H.J., Kim C.M., Kim K.C. Development of an Optimal Design Program for Vehicle Side Body Considering the B.I.W. Stiffness and Light Weight. SAE Paper 2007-01-2357; 2007.

3

Noise and Vibration Control for Local Body Structures

3.1 Noise and Vibration Problems Caused by Vehicle Local Structures

3.1.1 Classification and Modes of Local Body Structures

The vibration and control of the overall body structure is introduced in Chapter 2. Chapter 3 describes vibration, sound radiation, and control of the local body structures. In this book, the local body structures are divided into four categories: panel structures, acoustic cavity structures, accessory structures, and secondary-beam (or sub-beam) structures, as shown in Figure 3.1. The local structures have their own local modes and can generate vibration and noise. The local modes for the four kinds of structures are briefly described in the following sections.

3.1.1.1 Panel Structures

After the main frames and pillars, most of the rest of the body consists of various panel structures. The panels are connected with the body frames and pillars by welding or other means, forming a closed body, and they include fixed panels and movable panels. Fixed panels include the dash panel, roof, floor, side panels, fenders, wheelhouse, rear package tray, and windshield, whereas movable panels include the doors, engine hood, and trunk lid.

The panels are major contributors of body vibration and noise. Their corresponding modes are called local panel modes.

3.1.1.2 Acoustic Cavity Structures

The acoustic cavity refers to the enclosed air space in the passenger compartment. For a sedan, in addition to a passenger acoustic cavity, there is a trunk acoustic cavity. Similar to a solid structure, the acoustic cavity has its own modes, called acoustic cavity modes. Figure 3.2 shows the acoustic cavity of a sedan.

Excited by external excitations, the enclosed air moves, potentially creating acoustic cavity resonance. The motion of the acoustic cavity is usually related to the local body structures, so acoustic cavity modes are categorized with local body modes.

Noise and Vibration Control in Automotive Bodies, First Edition. Jian Pang.
© 2019 China Machine Press. All rights reserved. Published 2019 by John Wiley & Sons Ltd.

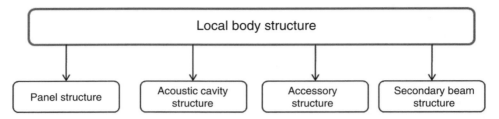

Figure 3.1 Classification of local body structures.

Figure 3.2 An acoustic cavity of a sedan.

3.1.1.3 Accessory Structures

Many accessories are installed on the body, such as the steering system, instrument panel, interior rearview mirror, side mirrors, fuel tank, spare tire, and antenna. Some accessories are made up of many parts: for example, the side mirror consists of the mirror head, mirror glass, base, motor, and so on, while the steering system consists of the steering shaft, cross car beam (CCB), and other components.

These systems have their own individual modes called body accessory modes. Some accessory modes vary with external conditions: for example, the fuel tank mode is related to the quantity of gas, while the spare tire mode is related to the tire's mass and fixtures. Accessory modes are a category of local body mode.

3.1.1.4 Secondary-Beam (or Sub-Beam) Structures

The side frames, cross members, and pillars comprise the body frame structure, which determines the overall body modes. In addition to the main frames, there are some "weak" beam structures on the body, such as the bumpers and roof cross rail, as shown in Figure 3.3. As used here, the word weak indicates that the stiffness of these beams is lower than that of the frames and pillars. The secondary beams are not subjected to the main loads from the engine and road. Compared with the main frames, the secondary beams have a much lower influence on overall body vibration and noise.

Secondary-beam structures have their own modes – the secondary-beam modes – which include the bumper modes, roof cross rail modes, and so on. Secondary-beam modes can each cause noise and vibration problems.

3.1.2 Noise and Vibration Problems Generated by Local Modes

A vehicle is subjected to various dynamic loads, such as engine excitation and road excitation. When (i) a load frequency is consistent with the modal frequency of a local

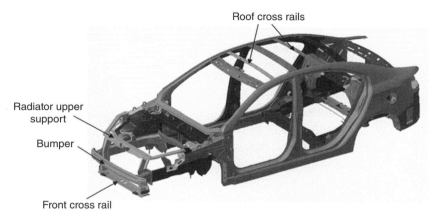

Figure 3.3 Some secondary beams of a body.

structure, (ii) the frequency of a local mode is consistent with the overall body mode frequency, or (iii) the modal frequencies of two adjacent components are consistent, resonance could occur. Owing to their relatively low frequencies, the body panel modes are easily coupled with the body acoustic cavity modes, causing an uncomfortable interior "booming" for the vehicle's occupants. In addition, squeak and rattle (S&R) could occur inside local structures such as body panels, doors, or the instrument panel when they are subjected the external excitations. The noise and vibration problems induced by above four categories of local structures are described in the following sections.

3.1.2.1 Noise and Vibration Problems Induced by Panel Structures

A body panel is similar to a speaker diaphragm. When the panel is excited, it will vibrate and radiate sound. The panel vibrates in its normal direction (i.e. perpendicular to the panel), so the corresponding modes move in the form of a bending motion in the normal direction as well. However, movable panels, such as the doors, engine hood, and trunk lid, have torsional modes in addition to bending modes. The modal frequencies of the body panels vary from dozens of hertz to several hundred hertz, so they are easily excited by engine and road excitations. In addition, when a panel mode is coupled with a body acoustic mode, interior booming could be generated.

The main local panel modes include the dash panel mode, roof mode, floor mode, engine hood mode, trunk lid mode, and door panel modes. Table 3.1 lists the body panels with corresponding mode descriptions. The noise and vibration problems caused by some panel modes are introduced in the following sections.

There are many holes on a dash panel where wires and pipes pass through, as shown in Figure 3.4. The wires and pipes are connected to other systems, so they might transmit vibration from those systems to the dash panel. For example, an air conditioning pipe connected to a compressor on the engine passes through the dash panel, so the pipe could transfer vibration from the compressor, engine, and fluid inside the pipe to the dash panel. If the excitation frequencies are consistent with the modal frequencies of the dash panel, the panel will resonate and radiate sound to the interior.

Table 3.1 Body panels and corresponding mode description.

Body panel	Description of panel modes
Dash panel	Bending mode
Front floor	Bending mode
Rear floor	Bending mode
Roof	Bending mode
Side panel	Transverse bending mode
Rear package tray	Bending mode
Door panel	Bending mode, torsional mode, breath mode
Engine hood	Bending mode, torsional mode
Trunk lid/liftgate	Bending mode, torsional mode

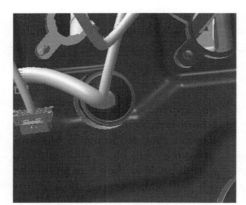

Figure 3.4 Pipes passing through a dash panel.

A number of accessories are installed on the dash panel, such as the vacuum pump and air-conditioning blower. These accessories can excite the dash panel directly and cause it to radiate noise to the interior.

The systems installed on the floor include the exhaust system, driveshaft, and fuel pipe, as shown in Figure 3.5. The exhaust system transfers vibration from the engine and the pulsation of fluid inside the pipe to the body through its hangers. In rear-wheel-drive and four-wheel-drive vehicles, the driveshaft transfers vibration from the transmission and axle to the body. These excitation sources can induce the floor to vibrate, which could be directly perceived by passengers and could radiate noise to the interior.

Even a tiny fuel pipe connected to the floor by fasteners could cause a serious vibration and noise problem. The pulsation of fluid inside the fuel pipe excites its surface, and the pipe's vibration is transmitted to the floor, causing the floor to radiate noise to the interior. Figure 3.6 shows two cases of interior noise during engine idling: in one, there is a rigid connection between the fuel pipe and the floor, and in the other there is

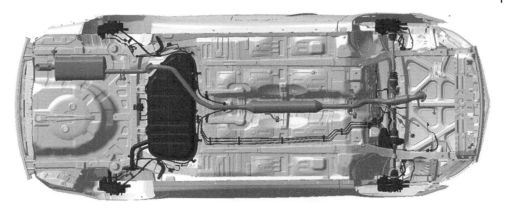

Figure 3.5 Floor and systems installed on the floor.

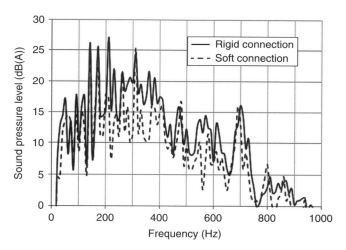

Figure 3.6 Interior idle noise comparison for rigid connection and soft connection between fuel pipe and floor.

a soft pad between the pipe and the floor. After the soft pad is used, the overall interior noise during idling is reduced by about 1.5 dB(A), and there is a significant noise reduction in the middle frequency range (200–600 Hz).

A roof usually consists of a large, thin plate with low stiffness. When driving, wind or road excitations can easily cause the roof to radiate noise to the interior. If the roof's modal frequency is consistent with an acoustic cavity modal frequency, significant air pressure pulsation will occur, resulting in serious booming. Therefore, the roof must be supported by a number of beams, as shown in Figure 3.7. In recent years, more and more beams have been added to roofs to enhance the roof's modal frequency, and the gaps between beams have become shorter and shorter.

The structural modal frequency of a liftgate is relatively low and is usually close to the body's longitudinal acoustic cavity mode frequency. When it is subjected to an external excitation, especially road excitation, loud booming will be perceived by the occupants

Supporting beams

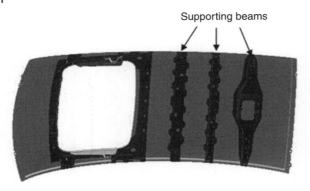

Figure 3.7 A roof and supporting beams.

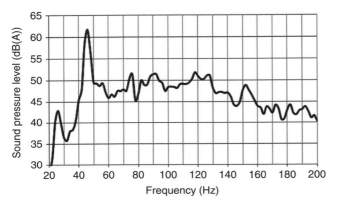

Figure 3.8 A huge interior booming at 46 Hz in a vehicle driven on a rough road.

because of the coupling between the structural mode and the acoustic cavity mode. Figure 3.8 displays a huge interior booming at 46 Hz for a vehicle driven on a rough road at a speed of 60 km h^{-1}. The reason for this is that both the modal frequency of the lift-gate and the first body acoustic cavity modal frequency are 46 Hz, so they are coupled.

3.1.2.2 Noise Problems Induced by Acoustic Cavity Modes

The body acoustic cavity modes include longitudinal modes, transverse modes, and composite modes, as shown in Table 3.2.

The acoustic cavity modes can cause two types of booming problems. The first is induced by coupling between the acoustic cavity mode and a panel's structural mode. Body panels are typically flexible and thin-walled plates that push the acoustic cavity to move when subjected to an external excitation. Figure 3.9 shows the first acoustic cavity mode of a sedan. The modal shape variation is in the longitudinal direction – i.e. along the vehicle's axis (the x-direction) – and there is almost no change in the transverse (y) and vertical (z) directions. The frequency of the first cavity mode for most passenger cars is between 40 and 60 Hz. If a panel's structural mode in the x-direction is within

Table 3.2 Body acoustic cavity modes and description.

Acoustic cavity	Description of cavity modes
Passenger acoustic cavity	Longitudinal mode, transverse modes, and composite mode
Trunk acoustic cavity	Longitudinal mode

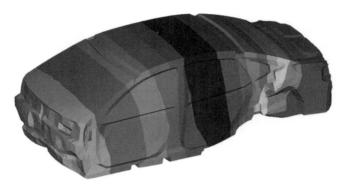

Figure 3.9 First acoustic cavity mode of a sedan.

this range, booming could be induced by coupling between the panel's mode and the acoustic cavity mode. Therefore, the modes of panels perpendicular to the x-direction, such as the dash panel, liftgate or trunk lid, and windshields, must be analyzed. If the frequency of an interior booming is 100 Hz, the higher order acoustic cavity modes should be analyzed. For example, if the second cavity mode frequency is 100 Hz as well and its modal shape is in the transverse direction (y-direction), the modes of the side panels, such as the side wall panels and door panels, must be analyzed.

The second type of booming is generated by coupling between an external noise source and a cavity mode. For example, when the tailpipe noise frequency and a cavity modal frequency are coupled, booming could be generated.

3.1.2.3 Noise and Vibration Problems Caused by Local Accessory Modes

Table 3.3 lists some accessories and their modes, including those for the brackets, steering system, inside rearview mirror, side mirrors, fuel tank, seats, and instrument panel. Poorly designed accessories will cause interior noise and vibration problems. Some examples are used to illustrate the problems.

Brackets, such as the engine brackets and exhaust hangers, are used to connect the body with other systems. Brackets with a low modal frequency are one of the most common causes of noise and vibration problems. For example, if the frequency of an engine bracket is 285 Hz and the first bending mode frequency of the power plant is 280 Hz, the bracket can be easily excited, generating resonance and transmitting vibration to the body.

Steering wheel vibration that is directly perceived by the driver's hands is the most common vehicle vibration problem. Many sources, such as engine vibration, cooling fan vibration, and blower vibration, can excite the steering wheel. Steering wheel vibration

Table 3.3 Body accessories and modes.

Accessory structure	Description of accessory modes
Bracket	Bending mode, torsional mode, complex mode
Steering system	Lateral bending mode, vertical bending mode, torsional mode
Instrument panel	Bending mode, torsional mode
Outside mirror	Bending mode
Inside rearview mirror	Bending mode
Seat	Bending mode, torsional mode
Fuel tank	Rigid bounce mode
Spare tire	Rigid bounce mode

Table 3.4 Secondary-beam structures and modes.

Local structure	Description of local modes
Body front end structure	Vertical bending mode, lateral bending mode, torsional mode
Cooling module supporter	Bending mode
Front bumper	Vertical bending mode, lateral bending mode
Roof cross member	Bending mode
Shotgun	Lateral bending mode, vertical bending mode

can occur in many driving conditions, such as idling, acceleration, and cruising. In certain conditions, "nibble" on the steering wheel can be perceived as a result of road excitation.

Vibration problems can also be caused by side mirrors and internal rearview mirrors with low modal frequencies. When subjected to road and engine excitations, the mirrors will be excited, and the objects reflected in the mirrors could look fuzzy.

The bounce mode of the fuel tank can generate interior booming. The frequencies of the bounce modes are usually between 20 and 30 Hz, depending on the amount of fuel in the tank. If the tank is inappropriately connected to the body, its bouncing motion could couple with a floor mode, which could induce floor vibration and interior booming.

Accessory modes and control methods are described in Section 3.6 of this chapter, which focuses on the brackets, steering system, and seats.

3.1.2.4 Noise and Vibration Problems Caused by Secondary-Beam Structures

Secondary-beam modes include those for the bumpers, roof cross rail, and cooling module supporter; Table 3.4 lists some secondary-beam modes. Two examples are used below to illustrate the noise and vibration problems caused by secondary-beam structures.

The frequency of the transverse bending mode of a shotgun is usually low, which could generate severe transverse vibration of the vehicle. To avoid such vibration, a crossbar can be added between the two shotguns.

The front roof frame and rear roof frame are connected with the front and rear windshields, respectively. The modal frequencies of the frames and windshields are usually low, so the structures are easily excited by road inputs, which can generate interior booming. To reduce the booming, a mass or damper can be added onto the frames.

3.1.3 Control Strategy for Local Modes

Local body modes can generate noise and vibration problems. If the local modes are not effectively controlled, the passengers will perceive steering wheel vibration, floor vibration, and interior booming, which will greatly affect their riding comfort. To avoid body resonance and reduce interior vibration and noise, the overall vehicle modes and local modes must be effectively planned.

There are five methods for controlling the local modes: modal frequency separation, stiffness control, damping control, mass control, and dynamic damper control.

3.1.3.1 Modal Frequency Separation

The primary task of modal frequency separation is to develop a complete mode distribution chart. All of the local body modes, the overall body modes, and the excitation frequencies, such as engine excitation, are listed on the table, as shown in Table 2.2. From the table, it is clear whether the local mode frequencies are consistent with the excitation frequencies.

First, the local modes must be separated from the overall vehicle modes. If the modes are not separated, the vehicle's vibrations can be easily transmitted to the local structures. For example, if the modal frequency of the steering system is consistent with the body's bending modal frequency, the vehicle's vibration will cause the steering wheel to vibrate. Another example is modal separation between the front floor mode and the overall body mode, because the modal frequencies for the floor and the vehicle are usually between 25 and 35 Hz.

Second, the local modes must be separated from the excitation frequencies. For example, the modal frequency of the dash panel must be separated from the excitation frequencies of the components that pass through it, such as the air conditioning pipes. The frequencies of the exhaust system and the frequencies of the floor must be separated, and the exhaust hangers must be placed on the floor's modal nodes. The roof structure can be reinforced by adding roof cross-rails in order to avoid resonance induced by road excitation. The modal frequency of the steering system must be separated from all potential excitation frequencies, such as the engine firing frequencies.

Third, some components can be used as dynamic dampers to attenuate body vibration at corresponding frequencies. For example, a bumper connected to the side frames can resist the body's torsional vibration. If the bumper frequency is consistent with the body's torsional modal frequency, the bumper can work as a dynamic damper. Similarly, a cooling module that is connected to a front cross rail by rubber pads can be regarded as a dynamic damper to suppress vibration of the front body at a certain frequency.

3.1.3.2 Stiffness Control Strategy

Stiffness is a major factor to consider when adjusting the modal frequencies of a structure. The modal frequency of a thin plate is very low, but if it is punched with a few ribs or provided with a few supporting beams, its modal frequency will significantly increase. For example, a dash panel can be easily excited, causing it to radiate noise, but the radiation can be suppressed by strengthening its stiffness. Lateral movement of the B-pillars is the main reason behind the generation of the door breathing mode, so increasing the stiffness of the B-pillars and door panels can attenuate the amplitude of this breathing mode movement. The structural modal frequency of a trunk lid is often consistent with the frequency of the first acoustic cavity mode, so strengthening the trunk lid can avoid coupling between the structural mode and the acoustic cavity mode.

3.1.3.3 Damping Control Strategy
Damping can be used to suppress the vibration and sound radiation of plates (i.e. panels). Large vibrations of a flat plate can be attenuated by placing a damping layer on its surface.

3.1.3.4 Mass Control Strategy
Mass can be also used to tune structural frequencies, but its tuning frequency range is not as wide as the range for stiffness. For example, by adding a mass onto the above-mentioned trunk lid, the coupling between the structural mode and the cavity mode can be avoided.

3.1.3.5 Dynamic Damper Control Strategy
Dynamic dampers can be used to reduce a narrow-band vibration. By adding an additional system that has the same frequency as the main system's resonance frequency, the vibration amplitude of the main system at this frequency will be attenuated.

3.2 Body Plate Vibration and Sound Radiation

Structural vibration and sound radiation can be studied through analytical, numerical, and experimental methods. It is very difficult to use the analytical method to solve a plate's vibration response and sound radiation because this method works only for some basic structures, such as simply supported plates, and is limited by boundary conditions, such as sound radiation analysis of a piston surrounded by infinite baffles (i.e. only the piston moves and radiates sound). The numerical method is the most commonly used one, such as the finite element (FE) method for analyzing the modes and vibration response of complex plate structures, and the boundary element method for calculating the plate's sound radiation. In engineering, the experimental method is also often used to test the body modes and their contribution to sound radiation.

However, the analytical method still can be used to compare changes in vibration and sound radiation as a result of body structure modifications. For this purpose, a body panel can be simplified as an ideal structure: for example, a four-sided, simply supported, thin plate. The plate is analyzed to find locations with severe vibration and strong sound radiation, and then these locations are reinforced or damped. The below section analyzes the vibration and sound radiation of an ideal structure.

3.2.1 Vibration of Plate Structure

The vibration mechanism of a plate is very complex, and analytical solutions can be obtained only for certain basic plate structures, such as a simply supported rectangular plate. The body panels are connected to the body frames and pillars. The frames and pillars have a much higher stiffness than the panels, and the deformations at intersections between the panels and the frames/pillars are much smaller than those on the panels: therefore, the deformations at the intersections can be approximately regarded as zero. It is difficult to determine the boundary conditions of the deflection and moment at intersections. If the deflection and the deflection angle are close to zero, but the moment is not, the boundary can be approximately regarded as a fixed boundary condition, while if the deflection and the moment are close to zero, but the deflection angle is not, the boundary can be roughly regarded as a simply supported boundary condition. Here, a simply supported plate is used to illustrate plate vibration problems.

The transmission of vibrations inside a thin plate is mainly characterized by bending waves, and the characteristics of these vibrations are analyzed below. Figure 3.10 shows a rectangular plate; its length, width, and thickness are a, b, and h, respectively.

The vibration equation for the thin plate is

$$D_0\left(\frac{\partial^4 w}{\partial x^4} + 2\frac{\partial^4 w}{\partial x^2 \partial y^2} + \frac{\partial^4 w}{\partial y^4}\right) + \rho h \frac{\partial^2 w}{\partial t^2} = q(x,y,t),$$

(3.1)

where ρ is density, w is deflection in the z-direction, q is surface load, and D_0 is the bending stiffness of the plate, expressed as

$$D_0 = \frac{Eh^3}{12(1-\mu^2)},$$

(3.2)

where μ is the Poisson ratio and E is the Young modulus.

For a simply supported rectangular plate, its vibration mode can be expressed as

$$w_{i,j}(x,y) = A_{i,j}\sin(i\pi x/a)\sin(j\pi y/b)e^{j\omega t},$$

(3.3)

where $A_{i,j}$ is the vibration amplitude and i,j is the mode number in x- and y-directions, respectively: i.e. $i,j = 1,2,3,\ldots$

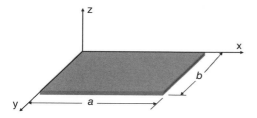

Figure 3.10 A rectangular plate.

The vibration mode function is satisfied with the boundary conditions of the four-sided, simply supported plate, i.e.

$$\left. w \right|_{x=0} = \left. \frac{\partial^2 w}{\partial x^2} \right|_{x=0} = 0 \qquad\qquad (3.4a)$$

$$\left. w \right|_{x=a} = \left. \frac{\partial^2 w}{\partial x^2} \right|_{x=a} = 0 \qquad\qquad (3.4b)$$

$$\left. w \right|_{y=0} = \left. \frac{\partial^2 w}{\partial y^2} \right|_{y=0} = 0 \qquad\qquad (3.4c)$$

$$\left. w \right|_{y=b} = \left. \frac{\partial^2 w}{\partial y^2} \right|_{y=b} = 0. \qquad\qquad (3.4d)$$

By substituting the mode functions and boundary conditions into Eq. (3.1), the system natural frequencies can be obtained, as

$$\omega_{i,j} = \pi^2 \left(\frac{i^2}{a^2} + \frac{j^2}{b^2} \right) \sqrt{\frac{D_0}{\rho h}}. \qquad\qquad (3.5)$$

The body panels consist of many small plates, and the frames can be regarded as the plates' boundaries. The length of the small plates is typically 300–400 mm, and the thickness is about 1 mm, so a simply supported rectangular plate with a side length of 300 and 400 mm, and a thickness of 1 mm is used here as an example. Table 3.5 lists the plate's geometrical dimensions and physical parameters. The vibration can be analyzed by an analytical method or a numerical method (such as the FE method). Table 3.6 lists frequencies for the first 10 modes of the plate calculated by Eq. (3.5) and by the FE method. The results of the FE analysis are lower than the theoretical values, and the calculated difference between the two methods increases with the increase of the analyzed frequency because the plate's stiffness is lowered by FE meshing, but the modal frequencies for the first modes are very close.

Table 3.5 Geometrical dimensions and physical parameters of a rectangular plate.

Parameters	Symbol	Unit	Value
Length	a	m	0.4
Width	b	m	0.3
Thickness	h	mm	1
Density	ρ	$kg\,m^{-3}$	7850
Young's modulus	E	GPa	210
Poisson ratio	μ	—	0.3

Table 3.6 Frequencies for the first 10 modes of a rectangular plate calculated by analytical formula and by the FE method.

Mode	Formula (Hz)	FE method (Hz)	Error (%)
(1, 1)	42.7	42.6	0.23
(2, 1)	88.8	88.5	0.34
(1, 2)	124.6	124.4	0.16
(3, 1)	165.6	165.1	0.30
(2, 2)	170.7	169.8	0.52
(3, 2)	247.6	245.6	0.80
(1, 3)	261.2	260.7	0.19
(4, 1)	273.2	272.3	0.33
(2, 3)	307.3	305.4	0.62
(4, 2)	355.1	351.9	0.90

Figure 3.11 shows the modal shapes, as calculated by the FE method, for the first 10 modes of a simply supported rectangular plate. The modes are spatially distributed and each mode can be represented by two numbers, such as mode (2, 3). The first number, "3", represents the modal shape in the x-direction, and the second number, "2", represents the modal shape in the y-direction.

The most fundamental modal shape for this rectangular plate is mode (1, 1), and its corresponding frequency is 42.6 Hz. The mode is symmetrical in both the x- and y-directions, and moves up and down. The mode amplitude is at the middle point. The modal frequency for mode (2, 1) is 88.5 Hz, and its modal shape is oppositely symmetrical around the center line on the y-axis, where the displacements at the axis are zero. The axis is a node-line. On the left and right sides of the axis, the mode moves up and down, respectively, and their phases show a 180° difference. The modal frequency for mode (1, 2) is 124.4 Hz, and its modal shape is similar to that of mode (2, 1), but the symmetry axis is along the x-axis. As the mode number increases, the modal shapes become more and more complex.

It is useful to understand the plate's modal shapes in order to control its sound radiation. For example, mode (1, 1) is similar to vibration of a single degree of freedom (DOF) particle, so it can be regarded as a monopole sound source. Mode (1, 2) is similar to two oppositely moving objects, so it can be regarded as two sound sources with 180° phase difference, namely a dipole sound source. The radiation sound from the two sources in the far field cancel each other out. From the perspective of far field sound radiation, mode (1, 1) must be controlled, whereas it is not necessary to control mode (1, 2).

The modal frequencies and shapes for simple structures such as the simply supported rectangular plate can be obtained by analytical methods, but these methods do not work for complex structures. To obtain the modal parameters for complex structures, only numerical methods such as the FE method can be used. A vehicle body is a very complex structure, so usually the FE method is used to obtain its modal parameters and dynamic responses. The contribution of each mode to sound radiation can only be

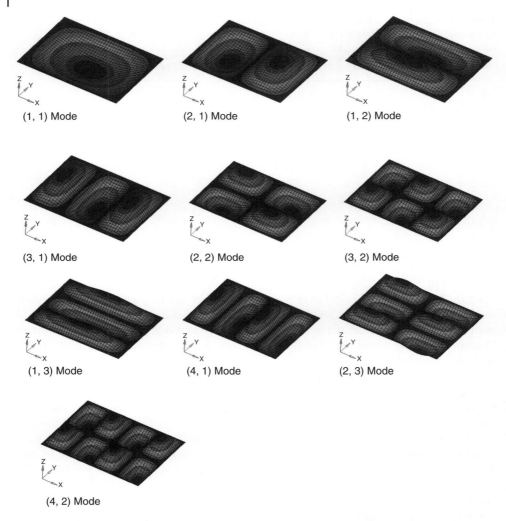

(1, 1) Mode　　　　　(2, 1) Mode　　　　　(1, 2) Mode

(3, 1) Mode　　　　　(2, 2) Mode　　　　　(3, 2) Mode

(1, 3) Mode　　　　　(4, 1) Mode　　　　　(2, 3) Mode

(4, 2) Mode

Figure 3.11 Modal shapes for the first 10 modes of a simply supported rectangular plate.

analyzed after the body modes have been obtained. Then the body structure can be modified by reinforcing it, adding mass, adding dampers, and so on in order to attenuate the sound radiation.

3.2.2 Sound Radiation of Plate Structure

Before analyzing the plate's sound radiation, the sound radiation of a point source should be analyzed. The point source (as shown in Figure 3.12) is a monopole sound source, and its sound pressure is expressed as follows:

$$p(r,t) = j\frac{\omega \rho_0 Q}{4\pi r} e^{j(\omega t - kr)}, \tag{3.6}$$

Figure 3.12 Point source and sound radiation.

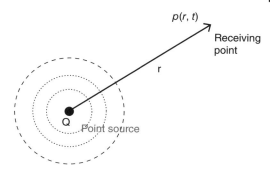

where r is the distance between the radiated point and the source point; ρ_0 is air density; ω is frequency; Q is volume velocity; and k is wave number. $j\omega\rho_0Q$ is the monopole sound source intensity.

A partial expression of Eq. (3.6) is called a free space Green function, $G(r, \omega)$, which is expressed as follows:

$$G(r,\omega) = \frac{e^{-jkr}}{4\pi r}.$$

$$(3.7)$$

Substitute Eq. (3.6) into Eq. (3.7), and another expression for the Green function can be written as

$$G(r,\omega) = \frac{1}{j\omega\rho_0}\frac{p(r,t)}{Q}e^{-j\omega t}.$$

$$(3.8)$$

The Green function can be interpreted as a transfer function between the sound pressure at the far field and the volume velocity of the sound source. It is very important to know this to understand the sound radiation of body panels.

In a similar way to structure vibration analysis, the sound radiation calculation only works for simple structures with special boundary conditions. Compared with the structure vibration calculation, the sound radiation calculation is more complicated. Currently, analytical solutions can only be used for very simple structures, such as rectangular plates and circular plates, and the analysis must assume that the structure is surrounded by hypothetical, infinite baffles.

The sound radiation of a rectangular plate surrounded by infinite baffles (shown in Figure 3.13) can be used to introduce the concepts and issues relating to plate sound radiation, such as radiation efficiency, sound intensity, and so on. The length and width of the rectangular plate are a and b, respectively.

The rectangular plate can be discretized down into many small pieces. Each piece can be regarded as a point source, and its area is δS. Suppose the normal speed of the nth piece is u_n, then its volume velocity is $u_n\delta S$. The vibrated plate radiates sound waves to both sides. If only the sound radiation from one side is considered, the volume velocity will be

$$Q_n/2 = u_n\delta S.$$

$$(3.9)$$

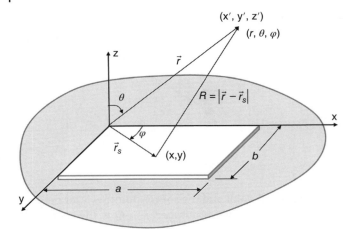

Figure 3.13 Sound radiation of a rectangular plate surrounded by infinite baffles.

Substitute Eq. (3.9) into Eq. (3.6), and the radiated sound pressure at any point (x', y', z') in the space contributed by any point on the rectangular plate (x, y) can be written as

$$p\left(\vec{R},t\right)= j\frac{\omega\rho_0 u_n\left(\vec{r}_s\right)\delta S}{2\pi R}e^{j\left(\omega t-kR\right)}, \tag{3.10}$$

where $u_n\left(\vec{r}_s\right)$ is the velocity of the plate surface, $\vec{r}_s$ is the vector of point (x,y) on the plate, $\vec{r}$ is the vector of the observed point $(x', y', z'$ or $r, \theta, \varphi)$ to the origin, and R is the modulus of $\left|\vec{r}-\vec{r}_s\right|$, i.e. $R=\left|\vec{r}-\vec{r}_s\right|$. If the distance between the observed point and the origin is much larger than the plate's dimensions, i.e. $R\gg a,b$, then $R\approx r-x\sin\theta\cos\varphi-y\sin\theta\sin\varphi$.

The overall sound pressure radiated by the plate to one side is the summation of each point's contribution. According to the Rayleigh formula, the sound pressure in the space radiated by the plate can be expressed as

$$p\left(\vec{r},t\right)= \frac{j\omega\rho_0}{2\pi}e^{j\omega t}\int_S \frac{u_n\left(\vec{r}_s\right)e^{-jkR}}{R}dS. \tag{3.11}$$

Differentiate Eq. (3.3) and the normal velocity distribution of the simply supported rectangular plate can be obtained, expressed as

$$u_n\left(x,z\right)= j\omega A_{ij}\sin\left(i\pi x/a\right)\sin\left(j\pi z/b\right)e^{j\omega t}. \tag{3.12}$$

Substitute Eq. (3.12) into (3.11), and the radiated sound pressure at any observed point (x', y', z') is

$$p\left(x',y',z',t\right)=-\frac{\omega^2\rho_0 A_{ij}e^{j\omega t}}{2\pi}\int_0^a\int_0^b\frac{\sin\left(i\pi x/a\right)\sin\left(j\pi y/b\right)e^{-jkR}}{R}dxdy. \tag{3.13}$$

In the far field, the relationship between particle velocity and sound pressure is

$$v(r,\theta,\varphi,\omega) = \frac{p(r,\theta,\varphi,\omega)}{\rho_0 c}. \tag{3.14}$$

The sound intensity at location R is

$$I(r,\theta,\varphi,\omega) = \frac{1}{2} Re\left[p^*(r,\theta,\varphi,\omega) \cdot v(r,\theta,\varphi,\omega) \right], \tag{3.15}$$

where * represents conjugate.

The simply supported rectangular plate in the infinite baffles moves harmonically. In the far field, $R \gg a,b$, its radiated sound intensity, is expressed as follows:

$$\bar{I}(r,\theta,\varphi,\omega) = 2\rho_0 c\omega^2 A_{ij}^2 \left(\frac{kab}{\pi^3 rij}\right)^2 \left\{ \frac{\cos\left(\frac{\alpha}{2}\right)\cos\left(\frac{\beta}{2}\right)}{\left[(\alpha/i\pi)^2 - 1\right]\left[(\beta/j\pi)^2 - 1\right]} \right\}^2, \tag{3.16}$$

where $\alpha = ka\sin\theta\cos\phi$, and $\beta = kb\sin\theta\sin\phi$.

When i and j are odd numbers, the terms in the brace are $\cos\alpha/2\cos\beta/2$. When i is an odd number and j is even number, the terms are $\cos\alpha/2\sin\beta/2$. When i is an even number and j is an odd number, the terms are $\sin\alpha/2\cos\beta/2$. Finally, when i and j are even numbers, the terms are $\sin\alpha/2\sin\beta/2$.

When the frequencies of the sound waves are very low, the wavelengths are very long (much longer than the plate's dimensions), so the maximum value of the sound intensity in the far field can be approximately expressed by

$$\bar{I}_{max} \approx 2\rho_0 c\omega^2 A_{ij}^2 \left(\frac{kab}{\pi^3 rij}\right)^2. \tag{3.17}$$

The sound radiation power can be obtained by integrating the far-field sound intensity over the hemisphere area of the infinite baffle:

$$\bar{W}_{ij}(\omega) = 2\rho_0 c\omega^2 A_{ij}^2 \left(\frac{kab}{\pi^3 rij}\right)^2 \int_0^{2\pi}\int_0^{\frac{\pi}{2}} \left\{ \frac{\cos\left(\frac{\alpha}{2}\right)\cos\left(\frac{\beta}{2}\right)}{\left[(\alpha/i\pi)^2 - 1\right]\left[(\beta/j\pi)^2 - 1\right]} \right\} \sin\theta d\theta d\varphi. \tag{3.18}$$

Radiation efficiency is defined as the ratio of the radiated sound energy in the space to the plate vibration energy, and is expressed as

$$\sigma_{ij} = \frac{\bar{W}_{ij}}{\rho_0 cab\langle \bar{u}_n^2\rangle_{ij}}, \tag{3.19}$$

where $\langle \bar{u}_n^2 \rangle$ is the mean square velocity of the plate. For a simply supported rectangular plate, the mean square velocity is

$$(\bar{u}_n^2) = \frac{1}{8}\omega^2 |A_{ij}|^2. \tag{3.20}$$

Substitute Eqs. (3.18) and (3.20) into Eq. (3.19), and the modal radiation efficiency of the simply supported rectangular plate is

$$\sigma_{ij} = \frac{16k^2ab}{(\pi^3 ij)^2} \int_0^{2\pi} \int_0^{\frac{\pi}{2}} \left\{ \frac{\cos\left(\frac{\alpha}{2}\right)\cos\left(\frac{\beta}{2}\right)}{\sin\left(\frac{\alpha}{2}\right)\sin\left(\frac{\beta}{2}\right)} \left[(\alpha/i\pi)^2 - 1\right]\left[(\beta/j\pi)^2 - 1\right] \right\} \sin\theta \, d\theta \, d\varphi. \tag{3.21}$$

In the case of very long wavelengths, i.e. $\alpha << i\pi$ and $\beta << j\pi$, and when i and j are odd numbers, the radiation efficiency is

$$\sigma_{ij} = \frac{16k^2ab}{(\pi^3 ij)^2} \int_0^{2\pi} \int_0^{\frac{\pi}{2}} \cos\left(\frac{\alpha}{2}\right)^2 \cos\left(\frac{\beta}{2}\right)^2 \sin\theta \, d\theta \, d\varphi = \frac{32}{\pi^3} \frac{1}{i^2 j^2}\left(\frac{b}{a}\right)\left(\frac{ka}{\pi}\right)^2. \tag{3.22}$$

3.3 Body Acoustic Cavity Mode

3.3.1 Definition and Shapes of Acoustic Cavity Mode

3.3.1.1 Definition of Acoustic Cavity Mode

The body panels surround the air to form an enclosed interior space. A structure (such as a body-in-white) has inherent modal shapes and frequencies, and likewise an enclosed air space (such as a vehicle's interior) also has its own modal shapes and frequencies. The mode formed by the enclosed air is called the acoustic cavity mode. When the structure is vibrated by an external excitation, or when the enclosed air is subjected to outside turbulence, the pressure inside the enclosed space will change, which generates sound. The air inside the enclosed space is subjected to two external disturbances: panel vibration and acoustic waves.

Panel vibration, such as dash panel vibration, roof vibration, and so on, causes tiny changes in the enclosed air volume that generate loud sounds. For example, the interior volume of a passenger car is approximately $3\,\text{m}^3$, and a change in the enclosed air volume of $0.002\,\text{m}^3$ will generate an interior sound pressure of $1.4\,\text{Pa}$, or $94\,\text{dB}$. The vibrated panel disturbs the enclosed air in a similar way to a speaker; as shown in Figure 3.14, a vibrated dash panel pushes the air, so the enclosed air volume changes and generates sound.

The interior air can also be directly pushed by the sound wave. The outside sound (such as an exhaust orifice sound) passes through the body and enters the interior to push the enclosed air. The external sound is also analogous to a speaker because it pushes the enclosed air, which makes the air volume change. If the frequency of the sound source is consistent with the acoustic cavity's modal frequency, interior booming

(a)

(b)

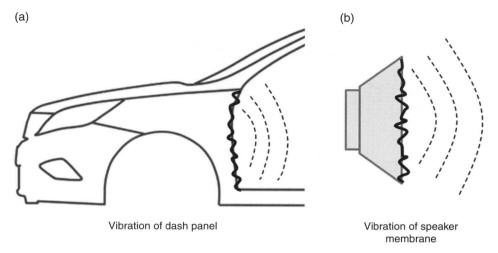

Vibration of dash panel

Vibration of speaker
membrane

Figure 3.14 Analogous comparison between a dash panel and a speaker. (a) Vibration of dash panel. (b) Vibration of speaker membrane.

Table 3.7 Comparison between structure mode and acoustic cavity mode.

	Structural mode	**Acoustic cavity mode**
Physical parameters	Displacement	Pressure
Outside excitations	Force	Pressure disturbance
Modal shapes	Displacement variation	Pressure variation

is generated. In addition, some sound sources inside the interior (such as a blower) directly push the enclosed air.

When the acoustic cavity is acted on by external excitations, the pressure inside the cavity changes, so pressure is a parameter for describing the cavity's mode. The mode distribution of a structure is represented by displacement, whereas the acoustic mode distribution of an enclosed cavity is characterized by pressure. The structure mode and acoustic cavity mode are very similar in many ways. Table 3.7 shows a comparison between the two types of modes.

3.3.1.2 Shapes of Acoustic Cavity Mode

The body cavity modes can be obtained by either testing or computer aided engineering (CAE) analysis. A continuous structure has many modes, such as the first bending mode, second torsional mode, and so on. An enclosed air body also has many modes, which are called the first cavity mode, second cavity mode, and so on.

Figure 3.9 shows the modal shape of first acoustic cavity mode of a vehicle body, and the colors represent pressure magnitudes. The pressures at different locations are different, and pressures at some locations are zero. A line where all the pressures are zero is called a node line, which is similar to a node in a structure mode. In this modal shape, the pressure changes are almost all along the vehicle's longitudinal direction

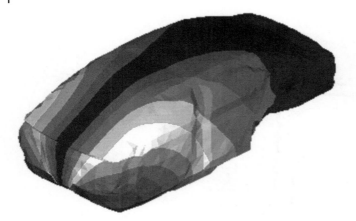

Figure 3.15 Third acoustic cavity mode of a vehicle body.

(i.e. the fore-aft direction or x-direction). There are nearly no changes in pressure in the transverse (y) direction and vertical (z) direction.

Each body cavity mode has its own specific shape, i.e. a specific sound pressure distribution. For example, Figure 3.15 is the third acoustic cavity mode for the same vehicle body, which is different from its first mode in Figure 3.9. The modal shape shows little change in the longitudinal (x) direction, some changes in the vertical (z) direction, and huge changes in the transverse (y) direction. As the mode order increases, the modal shapes become more and more complex. The higher order modes have complex changes in all directions.

3.3.2 Theoretical Analysis and Measurement of Acoustic Cavity Mode

3.3.2.1 Calculation of Acoustic Cavity Mode

Calculation of the acoustic cavity modes is based on classical acoustic theory. The ideal fluid analysis is based on several assumptions. First, the fluid is compressible. Second, the medium is continuous and uniform. Third, the medium is standstill when there is no sound disturbance. Fourth, during sound propagation, the processing of the medium, which is dense and sparse, is adiabatic. Fifth, the wave transmitted in the medium has a small amplitude, i.e. the sound pressure is far lower than the static sound pressure. Under these assumptions, three equations can be established for the medium: a motion equation, continuity equation, and ideal gas equation. By combining the three equations, a three-dimensional acoustic wave equation is obtained:

$$\frac{1}{c^2}\frac{\partial^2 p}{\partial^2 t} - \nabla^2 p = 0, \tag{3.23}$$

where c is sound speed, t is time, p is sound pressure (which is a function of time and space, i.e. $p(x,y,z,t)$), and ∇ is the Laplace operator, which is expressed as follows,

$$\nabla^2 = \frac{\partial^2}{\partial x^2} + \frac{\partial^2}{\partial y^2} + \frac{\partial^2}{\partial z^2}. \tag{3.24}$$

For an enclosed space, the enclosed air has mass, stiffness, and damping. Eq. (3.23) can be discretely processed by the Galerkin method, i.e. the enclosed space is divided into several finite elements. The wave equation in the flow field can be written as an FE matrix equation as follows:

$$M_f \ddot{P} + C_f \dot{P} + K_f P = 0, \tag{3.25}$$

where M_f, C_f, and K_f are the fluid equivalent mass matrix, damping matrix, and stiffness matrix, respectively.

In a similar way to structural modal analysis, Eq. (3.25) can be solved to acquire the sound pressure distribution, i.e. the acoustic cavity modes. There is no external excitation in Eq. (3.25), so it is the cavity free motion equation.

The acoustic cavity is a very complex structure, and it is very hard to achieve theoretical calculations. Therefore, the FE method is widely used to calculate the cavity modes. The cavity FE model is built up using acoustic fluid units, and Figure 3.16 shows a cavity FE model in a vehicle body without seats. The denser the elements, the higher the accuracy of the calculation; but, of course, the longer the calculation will take. Within the analyzed acoustic wavelength corresponding to the maximum frequency, six to eight acoustic elements must be guaranteed.

Figure 3.17 shows the first five acoustic cavity modes for the body in Figure 3.16. Figure 3.17a shows the first mode, and its frequency is 57 Hz. The mode moves along the vehicle's longitudinal (x) direction. The sound pressure varies along the x-axis, and the sound pressure becomes zero at some locations. In the cross-sections perpendicular to the x-axis, the pressures are almost the same: namely, there are no pressure variations in the transverse (y) and vertical (z) directions. The sound pressure variation in the longitudinal direction looks like an accordion: thus, this mode is called a longitudinal mode. The first acoustic cavity mode is the first longitudinal mode. When the cavity is excited by an external excitation, the locations with high variation in sound pressure represent the sensitive areas. The locations where the sound pressure does not change are called nodes, node lines, and node surfaces.

Figure 3.17b shows the second cavity mode, and its frequency is 102.5 Hz. This mode is a longitudinal mode as well, so it is called the second longitudinal mode. Its modal shape is similar to the first mode.

Figure 3.17c shows the third cavity mode, and its frequency is 119.5 Hz. In this mode, the sound pressure changes little in the longitudinal (x) and vertical (z) directions, but there are large changes in the lateral (y) direction: therefore, this mode is a transverse mode. Because this is the first appearance of a transverse mode, it is called the first

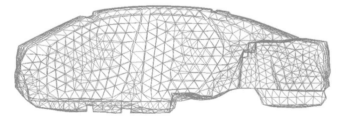

Figure 3.16 An FE model for a body acoustic cavity.

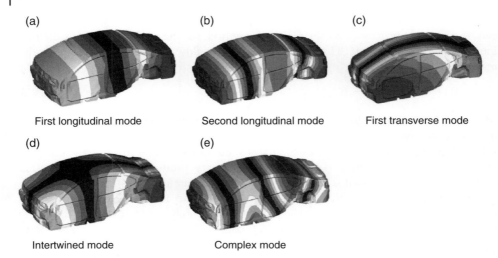

(a)   (b)   (c)

First longitudinal mode   Second longitudinal mode   First transverse mode

(d)   (e)

Intertwined mode   Complex mode

Figure 3.17 First five acoustic cavity modes for a body cavity without seats. (a) First longitudinal mode. (b) Second longitudinal mode. (c) First transverse mode. (d) Intertwined mode. (e) Complex mode.

transverse mode. In a similar way to the longitudinal modes, the sound pressures vary in different locations.

Figure 3.17d is the fourth cavity mode, and its frequency is 141.4 Hz. In this mode the sound pressures change in both the longitudinal (x) and transverse (y) directions, so the modal shape is a complicatedly intertwined shape.

Figure 3.17e is the fifth cavity mode, and its frequency is 153.6 Hz. Typically, the modal shapes after the fifth mode become complicated.

For passenger vehicle bodies, the first mode is always a longitudinal mode, the second mode is also usually a longitudinal mode (except in a few cases), and the third mode is usually a transverse mode. The fourth mode is a vertically and horizontally intertwined one. The fifth and higher order modes are complex, and they are intertwined in all three directions.

3.3.2.2 Testing of Acoustic Cavity Mode

In addition to FE analysis, the acoustic cavity mode can be acquired by testing.

First, the profile of the body cavity is drawn, and the locations for placing microphones are determined. A series of microphones are placed inside the cavity at distances of between 20 and 30 cm in order to ensure a sufficiently high measurement accuracy.

There are two ways to install the microphones. One way involves knotting six to eight microphones on a bar, as shown in Figure 3.18, then moving the bar back and forth, and up and down. First, the bar is placed in a reference position and the test data are recorded; second, the bar is moved back and forth to different positions and the data are recorded; third, the bar is moved up and down to different positions and the data are recorded. After all of the data in all planned positions have been recorded and processed, the body acoustic cavity modes can be obtained.

The second way is to install many microphones inside the cavity, 20–30 cm apart: i.e. a series of microphone arrays is formed, as shown in Figure 3.19. In this way, all of the

Figure 3.18 Microphones knotted onto a bar.

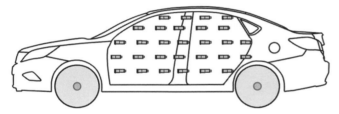

Figure 3.19 Microphone array inside a vehicle.

data needed to process the cavity modes can be acquired simultaneously in one go. This way is much faster than the first way, but it requires many more microphones, so the cost is much higher.

Usually, a volume velocity source is used as an excitation source for acoustic cavity mode testing, as shown in Figure 3.20. Another excitation method is to use a high-quality, low-frequency speaker. The volume velocity source cannot be placed at mode nodes. If the node positions cannot be determined before testing, the sound source can be tried in different locations, and then the correlation function between the responses at the microphones and the source excitation can be used to choose the appropriate excitation location.

The cavity modes and frequencies can be obtained by the tested transfer functions between the sound pressure responses and the excitation source. Figure 3.21 shows a tested first acoustic cavity mode of a body. The pressures vary in the body's longitudinal direction, which is consistent with the calculated results.

3.3.2.3 Influence of Seat on Acoustic Cavity Mode
Figure 3.22 shows an FE model of the cavity including the seats, and Figure 3.23 provides the first five modes. Compared with the cavity modes in Figure 3.17, the cavity modal shapes with seats are little different from those without seats; in fact, their modal shapes are almost the same. The first and second modes are longitudinal modes, the third mode is a transverse mode, the fourth mode is an intertwined mode, and the fifth mode is a complex mode.

Figure 3.20 Volume velocity source.

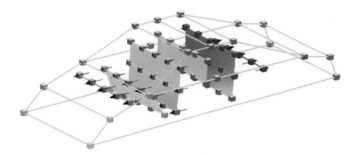

Figure 3.21 A tested first acoustic cavity mode of a body.

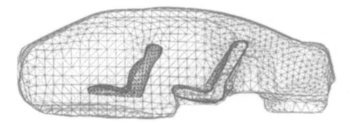

Figure 3.22 An FE model of a cavity including seats.

The modal frequencies of the cavity modes with and without seats are listed in Table 3.8, so the influence of seats on the modal frequencies can be compared. After the seats have been installed, the first modal frequency reduces from 57 to 50.1 Hz, the second modal frequency reduces from 102.5 to 96.2 Hz, the third modal frequency reduces from 119.5 to 118.4 Hz, the fourth modal frequency reduces from 141.4 to 139.8 Hz, and the fifth modal frequency reduces from 153.6 to 150.8 Hz. The results

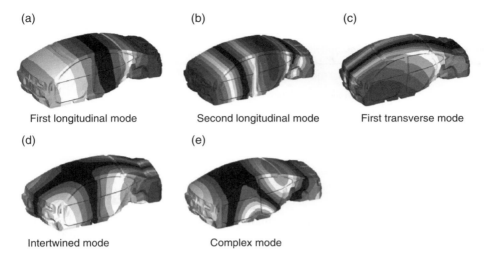

(a)

First longitudinal mode

(b)

Second longitudinal mode

(c)

First transverse mode

(d)

Intertwined mode

(e)

Complex mode

Figure 3.23 First five modes of the body cavity including the seats. (a) First longitudinal mode. (b) Second longitudinal mode. (c) First transverse mode. (d) Intertwined mode. (e) Complex mode.

Table 3.8 Modal frequencies of cavity modes with and without seats.

Mode no.	Modal shape	Frequency without seats (Hz)	Frequency with seats (Hz)	Frequency comparison (%)
1	First longitudinal mode	57.0	50.1	12.1
2	First longitudinal mode	102.5	96.2	6.14
3	First transverse mode	119.5	118.4	0.92
4	First intertwined mode	141.4	139.8	1.13
5	Complex mode	153.6	150.8	1.82

show that the seats reduce the body cavity modal frequency, and they have more impact on the lower modes than the higher modes.

The seat cushion and backrest are made of foam materials. The foams are filled with air to form a special chamber. When the foam cavity mode is coupled with a body cavity mode, the modal frequency of the combined cavity modes will be lowered. Therefore, the influence of the seats must be considered in the body cavity modal analysis. If the occupants are inside the cavity, the cavity modal frequency will further decrease.

Table 3.9 provides a frequency range of the body cavity modes for most passenger cars. The modal shapes and frequency range of the acoustic cavity can be compared with the structural modes to identify the source of interior booming and which panels may have produced the problem.

The frequencies of most interior booming are between 30 and 100 Hz, so the first three cavity modes, especially the longitudinal modes, should be prioritized. If the booming is induced by coupling between a structural mode and an acoustic cavity

Table 3.9 Frequency range of body cavity modes for most passenger cars.

Mode no.	Mode description	Frequency (Hz)
1	Longitudinal (first longitudinal mode)	40–60
2	Longitudinal (second longitudinal mode)	60–100
3	Transverse (first transverse mode)	90–130
4	Intertwined (first intertwined mode)	100–150
5	Complex	>150 Hz

mode, the panels perpendicular to the longitudinal axis (x-direction) should be analyzed in particular.

After a vehicle's interior size has been finalized, the cavity modal shapes and frequencies are set. It is almost impossible to change the cavity modal frequencies, but the seats could be carefully designed to position the occupants' ears as close to the mode nodes as possible.

3.3.2.4 Simple Calculation Formula for Acoustic Cavity Modal Frequency

The cavity modal shapes and frequencies can be calculated by the FE method. However, during the vehicle development process, sometimes the modal frequencies need to be quickly estimated. The body cavity can be simplified as a rectangular box, as shown in Figure 3.24, and its modal frequencies can be quickly estimated by the following expression:

$$f_{ijk} = \frac{c}{2}\sqrt{\left(\frac{i}{L_x}\right)^2 + \left(\frac{j}{L_y}\right)^2 + \left(\frac{k}{L_z}\right)^2}, \tag{3.26}$$

where c is the sound speed, L_x, L_y, and L_z are the lengths of the rectangular cavity in the x-, y-, and z-directions, respectively; and i, j, and k are the mode numbers in the x-, y-, and z-directions, respectively.

The modal shape can be calculated by the following simple formula:

$$\Phi_{air} = \cos\frac{i\pi x}{L_x}\cos\frac{j\pi y}{L_y}\cos\frac{k\pi z}{L_z}. \tag{3.27}$$

The FE model in Figure 3.16 is simplified as a rectangle, the dimensions of which are determined by the interior maximum dimensions, as shown in Figure 3.24. The maximum dimensions in three directions for the model in Figure 3.16 are $L_x = 3248\,mm$, $L_y = 1491\,mm$, and $L_z = 1282\,mm$, respectively. After the FE model of the rectangle has been built, its calculated modal frequencies can be compared with the results of the real cavity model. Table 3.10 lists comparisons of the first three modal frequencies calculated by the real cavity FE model, the rectangular FE model, and the simple calculation formula.

The values from the simple formula and the rectangular FE model are consistent, but they are different from those of the real cavity model because their structures are different. The second longitudinal mode frequency is double that of the first longitudinal

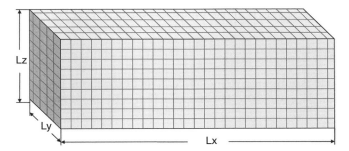

Figure 3.24 A rectangular box.

Table 3.10 Comparison of first three modal frequencies calculated by real cavity FE model, rectangular FE model, and simple calculation formula.

Mode no.	Modal shape	Simple formula (Hz)	Rectangular FE model (Hz)	Real cavity FE model (Hz)
1	First longitudinal mode (1,0,0)	52.3	52.3	57.0
2	Second longitudinal mode (2,0,0)	104.7	104.7	102.5
3	First transverse mode (0,1,0)	114.0	114.1	119.5

mode frequency in the results of the simple formula and the rectangular FE model, but there is no such relationship for the real cavity modes because the real model is complex and not completely symmetrical. Despite the difference, it is still useful to use the simple formula to calculate the body acoustic cavity modal frequency. First, the calculated values are not significantly different among the three methods. Second, the calculated results have the same trend. Third, the simple formula can be used to quickly estimate the cavity frequency, which is helpful for quickly evaluating engineering problems.

3.3.3 Coupling of Acoustic Cavity Mode and Structural Mode

The enclosed air and the body panels are linked to each other. Panel vibration excites the cavity, disturbs the air, and forces the sound pressure to change. At the same time, the air motion pushes against the panel and excites it, causing it to vibrate. Thus, the cavity and the panel interact: i.e. the fluid and the solid structure interact. The coupling relationship between the fluid and the solid structure can be represented by a coupling matrix, R.

The force exerted on the fluid by the panel is expressed as

$$F_1 = -R\ddot{U},$$ (3.28)

where U is the panel vibration displacement and $\ddot{U}$ is the acceleration.

The force exerted on the panel by the fluid is expressed as

$$F_f = RP^T,$$ (3.29)

where P is the sound pressure inside the cavity.

If the vibration input of the body panel is applied on the acoustic cavity, Eq. (3.25) becomes

$$M_f \ddot{P} + C_f \dot{P} + K_f P = -R\ddot{U}. \tag{3.30}$$

The excitations that the panel is subjected to include external excitations from the engine, road, etc. and force from the cavity. The dynamic equation of the panel structure is expressed as

$$M_s \ddot{U} + C_s \dot{U} + K_s U = F_s + RP^T. \tag{3.31}$$

Combine Eqs. (3.30) and (3.31) to form

$$\begin{bmatrix} M_s & 0 \\ R & M_f \end{bmatrix} \begin{Bmatrix} \ddot{U} \\ \ddot{P} \end{Bmatrix} + \begin{bmatrix} C_s & 0 \\ 0 & C_f \end{bmatrix} \begin{Bmatrix} \dot{U} \\ \dot{P} \end{Bmatrix} + \begin{bmatrix} K_s & -R^T \\ 0 & K_f \end{bmatrix} \begin{Bmatrix} U \\ P \end{Bmatrix} = \begin{Bmatrix} F_s \\ 0 \end{Bmatrix}. \tag{3.32}$$

The purpose of analyzing coupling between the cavity and the panels is to calculate the impact of panel vibration on the cavity's sound performance. The application of the coupling analysis involves three aspects. The first aspect is to find the characteristics of sound pressure changes inside the cavity. The panel pushes the cavity to move and causes its sound pressure to change. The second aspect is to identify and analyze the panel that is inducing interior booming, which is generated when a panel's vibration frequency is consistent with the cavity frequency. The third aspect is to analyze each panel's contribution to the overall interior sound pressure. After the frequencies and magnitudes of sound pressure contributed by each panel have been analyzed, the major contributors can be identified, which provides clues on how to modify the corresponding panels to attenuate interior booming.

3.3.4 Control of Acoustic Cavity Mode

When the body styling and interior design have been finished, the acoustic cavity modes are set in place. It is impossible to change their shapes and frequencies. Therefore, any interior booming induced by the cavity modes can be controlled in only three ways.

First, in the early stages of vehicle design, the first acoustic cavity modal frequency is set as a vehicle noise, vibration, and harshness (NVH) control target. The cavity modal frequency is listed in a vehicle modal table in order to separate it from the excitation frequencies of the engine and other power train systems, and the road.

Second, the impact of occupants and seats on the cavity modes and frequencies should be considered. After the seats have been installed, the cavity modes will change, so in seat design, both the ride comfort and ergonomics should be primarily considered. When the interior is occupied by passengers, the cavity mode will further change. The positions of the passengers' ears are determined by the seat's structure and installation, and the sitting posture. The ear positions should be as close as possible to the node line of the cavity mode in order to reduce the perception of interior booming.

Third, the panels' modes in relation to the cavity mode must be controlled. The main reason for interior booming is coupling between a panel's structural mode and the acoustic cavity mode. Owing to the difficulty of changing the cavity mode, careful design of the panels is very important. Separation between panel modes and the cavity mode is an important principle of body panel design.

3.4 Panel Contribution Analysis

3.4.1 Concept of Panel Contribution

A body panel generates bending vibration when subjected to an external excitation, and then radiates sound to the interior, as shown in Figure 3.25. There are many panels on the body, such as the dash panel, floor, roof, side panels, door panels, and trunk lid. The sound radiation contribution of each panel to the interior noise is different. Only after the major contributing panels have been identified can the interior noise be effectively controlled by suppressing the panels' vibration. In addition, each panel consists of many small pieces or zones, and the contribution of each piece to the interior sound is different. The panel vibration can only be effectively suppressed after the pieces with strong radiation capacity have been found. The purpose of panel contribution analysis is to identify the vibration and sound radiation of each panel and each piece, and then to find the panels and pieces that need to be controlled.

Panel vibration induces the surrounding air to move. The air volume velocity is applied on the cavity, and disturbs the enclosed air, which changes the pressures at the occupants' ears. The interior noise is a superposition of the radiated sounds contributed by all panels. The transfer processing of the panel vibration to the sound at the occupants' ears is shown in Figure 3.26. The sound radiation capacity of the panel is determined by its volume velocity, and the interior sound level depends on the transfer function between the sound at the ears and each panel's vibration.

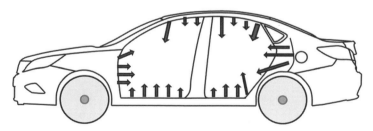

Figure 3.25 Body panel vibration and sound radiation to passenger compartment.

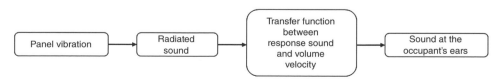

Figure 3.26 Transfer processing of panel vibration to sound at the occupant's ears.

3.4.2 Contribution Analysis of Panel Vibration and Sound Radiation

The body cavity can be regarded as an enclosed space surrounded by panels. Figure 3.27 shows an enclosed space. A volume sound source is placed at point C, and the enclosed panel (S) is an elastic body that vibrates and radiates sound. Point O is set as the coordinate origin, and other points use it as a reference point. A sound response at any point inside the space (point A) is contributed by the volume sound source and the panel radiation. The sound pressure at point A is

$$p = p_v + p_s, \tag{3.33}$$

where p_v and p_s are the radiated sound pressures at point A contributed by the volume sound source and the panel radiation, respectively.

The sound source inside the space can be regarded as a point source that directly radiates sound to a receipt point. According to Eq. (3.10) and integrating the enclosed small volume, the sound pressure at the receipt point (A) is:

$$p_v(r,t) = j\omega\rho_0 \int_V G(r_0,r_c,\omega)Q(r_c)dV, \tag{3.34}$$

where r_c is the distance between the source point (C) and the origin (O), Q_c is the volume sound source, and $G(r_0,r_c,\omega)$ is the Green function between point C and point A.

The radiated sound generated by vibration of the enclosed panel (S) can be integrated by Helmholtz integration:

$$p_s(r_0,\omega) = \oint_S \left(G(r_0,r,\omega)\frac{\partial p_A(r,\omega)}{\partial n} - p_A(r,\omega)\frac{\partial G(r_0,r,\omega)}{\partial n} \right)dS, \tag{3.35}$$

where $G(r_0,r,\omega)$ is the Green function of the volume sound source at point B on the panel surface to the sound pressure at point A, and expressed as

$$G(r_0,r,\omega) = \frac{1}{j\omega\rho_0}\frac{p_A(r_0,\omega)}{Q_B(r,\omega)}, \tag{3.36}$$

where $Q_B(r,\omega)$ is the volume sound source at point B and $p_A(r_0,\omega)$ is the sound pressure at point A.

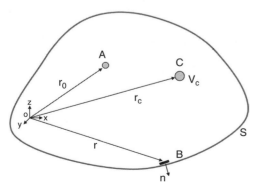

Figure 3.27 Volume sound source and panel radiation sound source inside an enclosed space.

In Green function testing, it is usual to place the volume sound source at a receipt point (such as an occupant's ear) and to test the sound pressure at the enclosed surface (such as the panel). According to the reciprocal principle, another expression of the Green function can be obtained,

$$G(r_0,r,\omega) = \frac{j}{\omega\rho_0} \frac{p_B(r,\omega)}{Q_A(r_0,\omega)}, \tag{3.37}$$

where $Q_A(r_0, \omega)$ is the volume sound source at point A, and $p_B(r, \omega)$ is the sound pressure at point B.

Differentiate above equation, obtaining

$$\frac{\partial G(r_0,r,\omega)}{\partial n} = \frac{j}{\omega\rho_0 Q_A(r_0,\omega)} \frac{\partial p_B(r,\omega)}{\partial n}. \tag{3.38}$$

The relationship between the derivative of the sound pressure at n-direction and sound speed is

$$\frac{\partial p_A}{\partial n} = -j\omega\rho_0 v_A(r,\omega). \tag{3.39}$$

Because the speed is a vector, the relationship between the sound pressure at point B and the sound speed can be expressed as

$$\frac{\partial p_B}{\partial n} = j\omega\rho_0 v_B(r,\omega). \tag{3.40}$$

Substitute Eqs. (3.38–3.40) into Eq. (3.35), obtaining

$$p_s(r_0,\omega) = \oint_S \left(\frac{p_B(r,\omega)}{Q_A} v_A(r,\omega) + \frac{v_B}{Q_A} p_A(r,\omega) \right) dS. \tag{3.41}$$

The sound pressure at point A is contributed by the sound source inside the enclosed space and the radiated sound from the panel's vibration, and is expressed as follows:

$$p(r_0,\omega) = j\omega\rho_0 \int_V G(r_0,r_c,\omega)Q(r_c)dV + \oint_S \left(\frac{p_B(r,\omega)}{Q_A} v_A(r,\omega) + \frac{v_B}{Q_A} p_A(r,\omega) \right) dS. \tag{3.42}$$

In panel contribution analysis, usually there is no volume sound source inside the body, so the first term in Eq. (3.42) disappears. Because the sound speed transmitted to the panel from source A is very low, the corresponding contribution can be ignored; therefore, the sound pressure at point A contributed by the body panels is

$$p(r_0,\omega) = \oint_S \frac{p_B(r,\omega)}{Q_A} v_A(r,\omega) dS. \tag{3.43}$$

Assume point i is a point or small block on the enclosed panel, and its sound radiation to the point A is

$$p_{A_i}(r_0,\omega) = \frac{p_i(r,\omega)}{Q_A} v_{A_i}(r,\omega)\delta S_i. \tag{3.44}$$

Many points or small blocks on the panel radiate sound to point A, so the ratio of the sound pressure contributed by point i to the overall sound is

$$t_i(\omega) = \frac{P_{Ai}(r_0,\omega)}{p(r_0,\omega)}. \tag{3.45}$$

$t_i(\omega)$ represents the contribution of the ith block on the panel to the overall interior sound. If the contributions from all the small blocks are plotted, it will be easy to distinguish their contributions to the overall sound. The contribution from each small block at different frequencies and different engine speeds can be identified as well. Through panel contribution analysis, the main contributing locations will be found, which is helpful in guiding the design of the panel structure in order to reduce its sound radiation.

The sound source can be placed near the panel in order to obtain the transfer function between the interior sound and the panel's structural vibration; however, the analysis or testing work is considerable. According to the reciprocal principle in Eqs. (3.43) or (3.44), the sound source can also be placed on the receipt point and the sound pressures can be measured near the panel, so the transfer function can be acquired in this way as well.

Below, a body panel is used as an example to illustrate panel contribution analysis. First, the body is divided into seven panels: the roof, front floor, rear floor, dash panel, left side panel, right side panel, and trunk lid. Figure 3.28 shows the overall interior sound pressure and each panel's contributions when the vehicle is driven on a rough road. It is clear from the figure that the roof is the biggest contributor to the interior sound, and there is a peak at 75 Hz. The corresponding subjective evaluation shows a booming at 75 Hz. Second, to identify which location on the roof contributes the booming, it is further divided into small blocks, as shown in Figure 3.29. Figure 3.30 plots the contributions to the 75 Hz booming from all of the blocks, including positive, zero, and negative contributions.

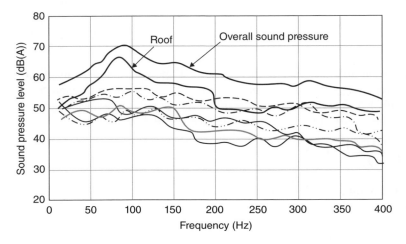

Figure 3.28 Overall interior sound pressure and each panel's contributions.

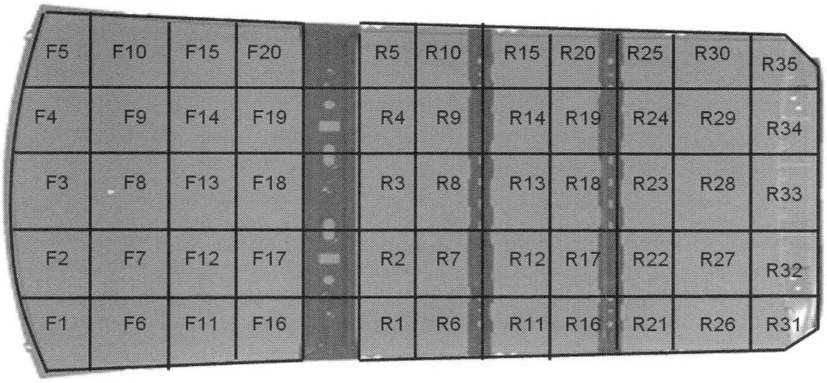

F5 F10 F15 F20 R5 R10 R15 R20 R25 R30 R35
F4 F9 F14 F19 R4 R9 R14 R19 R24 R29 R34
F3 F8 F13 F18 R3 R8 R13 R18 R23 R28 R33
F2 F7 F12 F17 R2 R7 R12 R17 R22 R27 R32
F1 F6 F11 F16 R1 R6 R11 R16 R21 R26 R31

Figure 3.29 A roof is further divided into small blocks.

Sound contribution
F1 F3 F5 F7 F9 F11 F13 F15 F17 F19 R1 R3 R5 R7 R9 R11 R13 R15 R17 R19 R21 R23 R25 R27 R29 R31 R33 R35

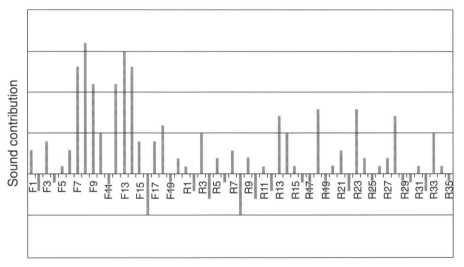

Figure 3.30 Contributions to the 75 Hz booming by all blocks, including positive, zero, and negative contributions.

The panels can be divided into three categories according to how their vibration contributes to the overall sound radiation: positive contribution panels, negative contribution panels, and neutral panels. If the sound pressure contributed by a panel has the same phase as the overall sound pressure, and the panel is also a major contributor, it is called a positive panel; the overall sound pressure increases with the panel's vibration. If the sound pressure contributed by a panel has the opposite phase to the overall sound pressure, and the panel is also a major contributor, it is called a negative panel; the overall sound pressure decreases with the panel's vibration. If the sound pressure contributed by a panel is much lower than the overall sound pressure, it is called a neutral panel.

Therefore, a panel with large vibrations is not necessarily a large contributor to the interior sound. The interior overall noise is a vector summation of the sound radiation contributed by each panel. The interior noise can only be reduced if the positive panels are identified and their vibrations suppressed. In the above example, the major panels contributing to the 75 Hz booming are those in the middle of the roof, so the roof's local design must be modified to reduce the booming.

3.4.3 Testing Methods for Panel Vibration and Sound Radiation

In addition to using the volume source, some other methods can be used to measure panel vibration and sound radiation, such as the window method, sound intensity, laser measurement, near-field acoustic holography (NAH), and beamforming.

3.4.3.1 Window Measurement Method

In order to study the contribution of a panel's vibration to the interior noise, the body panel is divided into several areas (often around 20–30). First, the body panel is covered with sound absorption and insulation layers with a thickness of about 50 mm, as shown in Figure 3.31a, and the interior noise is measured. Second, a piece of insulation layer in one area is removed, leaving a window, as shown in Figure 3.31b; then the interior noise is tested again. By comparing the two measurements, the noise contribution of the area can be obtained. The same technique is repeated for the remaining areas to acquire the contribution of each area to the interior noise. This method is called the window measurement method.

The advantages of this widely used method are that it is simple and the measurement result is more reliable than in the panel sound radiation method. However, because only one window's contribution to the interior noise is tested each time, the phase relationship between panels is lost, which results in inaccuracy at low frequencies. The

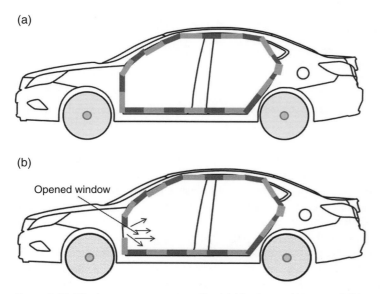

(a)

(b)

Opened window

Figure 3.31 Window measurement method: (a) body panel is covered; (b) one area is opened.

window method is applicable only for frequencies higher than 250 Hz. In addition, owing to the heaviness of the insulation materials, it is hard to place them on the roof, so the method is inconvenient in such cases.

3.4.3.2 Measurement of Sound Intensity

Sound intensity is defined as the average sound energy per unit of time and per unit of area perpendicular to the sound wave's propagation direction, and is expressed as

$$I = \frac{1}{T} \int_0^T \text{Re}(p)\text{Re}(v) dt, \tag{3.46}$$

where p and v are sound pressure and sound velocity, respectively, and Re represents the real part.

Sound intensity can be obtained by two methods. The first method is to directly test the sound pressure and medium velocity, and then to calculate the sound intensity with Eq. (3.46). The second method is to use two adjacent microphones to measure the sound pressures, and then to calculate the medium velocity and the sound intensity. Owing to the difficulty in directly measuring the medium velocity, the first method is rarely used, but the second method is widely employed in engineering. In an ideal fluid medium, for sound waves with small amplitudes, the medium velocity and sound pressure exist in the following relationship

$$v = -\frac{1}{\rho} \int \frac{\partial p}{\partial r} dt. \tag{3.47}$$

Two adjacent microphones are used to measure the sound pressure gradient, so the average sound pressure and medium velocity can be obtained, and expressed as

$$p = \frac{p_1 + p_2}{2} \tag{3.48}$$

$$v = -\frac{1}{\rho} \int \frac{p_1 - p_2}{\Delta r} dt, \tag{3.49}$$

where p_1 and p_2 are the sound pressures measured by the two microphones, and Δr is the distance between the two microphones.

Thus, the sound intensity can be expressed as follows:

$$I = -\frac{1}{2} \text{Re} \left[\frac{p_1 + p_2}{2} \left(-\frac{1}{\rho} \int \frac{p_1 - p_2}{\Delta r} \right)^* dt \right], \tag{3.50}$$

where * represents the complex conjugate.

Both sides of Eq. (3.50) are transferred by the Fourier transform, and the sound intensity can be represented by a cross-spectrum of sound pressures, expressed as follows:

$$I(\omega) = -\frac{\text{Im}(G_{12}(\omega))}{2\rho\omega\Delta r}, \tag{3.51}$$

where $G_{12}(\omega)$ represents the cross-spectrum between sound pressures p_1 and p_2, and Im represents an imaginary number.

Equation (3.50) shows that the sound intensity can be measured by placing two microphones together. Two microphones are placed on a stand to form a sound intensity probe, as shown in Figure 3.32. A sound intensity measurement system consists of a probe, an analyzer, and an indicator.

Sound intensity is a vector, with a magnitude and direction. Due to the vector's characteristics and the probe's sensitivity to sound direction, the measurement of sound intensity can be used to identify the sound sources. Figure 3.33 shows the position relationship between a sound source and the probe.

The angle between the axis of the sound intensity probe and the connection line between the sound source and the probe center is θ. When $\theta = 0°$ or $\theta = 180°$, the sound intensity has its maximum value, $I(0°) = I(180°)$. When $\theta = 90°$, the sound intensity is zero. θ varies between $0°$ and $180°$, and the sound intensity at any angle θ is noted by $I(\theta)$, and is expressed as

$$I(\theta) = I(0°)\cos\theta. \tag{3.52}$$

Figure 3.34 shows a schematic diagram of a probe sweeping through the sound source. From point A to point B, the sound intensity changes from high level to low level, and from point B to point C, the sound intensity varies from low level to high level. Point B is a turning point, and the sound intensity is zero. The location of the sound source can be identified by this sweeping.

Sound intensity is a vector with strong directionality, so the environment has little impact on its measurement. Sound pressure is a scalar, and a particular acoustic environment (such as anechoic chamber or reverberation chamber) is needed to measure it.

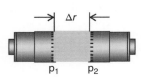

Figure 3.32 A sound intensity probe.

Compared with the sound pressure measurement, the sound intensity measurement has wider applications, and it is routinely used in sound power measurement, sound source localization, sound insulation measurement, and so on. However, calibration of the sound intensity measurement system is more difficult than in sound pressure measurement, because it includes the magnitude and phase. In the automotive NVH field, sound intensity probes are widely

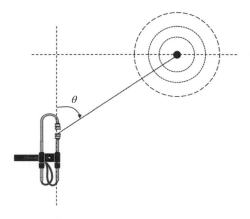

Figure 3.33 Position relationship between a sound source and the probe.

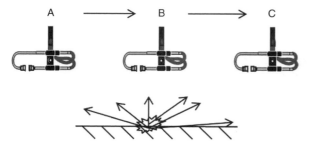

Figure 3.34 A schematic diagram of a probe passing through the sound source.

used in outdoor near-field sound measurement because they are convenient, quick, and relatively unaffected by ambient noise.

3.4.3.3 Laser Measurement Method

Laser measurement is a non-contact vibration measurement method that includes laser Doppler vibration measurement, laser holographic vibration measurement, and electronic speckle interferometry. The laser Doppler vibration measurement method is widely used in the automobile industry.

The wavelengths of the radiated sound vary with the relative motion between the sound source and the observed position. This phenomenon is known as the Doppler effect. In front of a moving sound source, the sound wave is compressed: the wavelengths become shorter and the frequencies become higher. Behind the moving sound source, the sound wave is elongated: the wavelengths become longer and the frequencies become lower.

The Doppler effect applies not only to sound waves, but also to other types of waves, such as electromagnetic waves, light waves, and so on. When a laser is projected onto a moving object, the frequency of the reflected wave changes with the speed of the moving object. A laser Doppler vibrometer uses the Doppler effect to measure the velocity and displacement of moving objects.

Sometimes it is difficult to place accelerometers on a body panel to test its vibration contribution. First, it is costly to place so many accelerometers to obtain each panel's vibration. Second, the weight of the placed accelerometers will change the panel's frequencies and modes. Third, additional excitation sources could be induced in wind tunnel testing due to interaction between the accelerometers and the airflow. However, laser measurement overcomes these difficulties. Because it is non-contact and no sensors are needed, the panel's structural characteristics will not be altered.

Figure 3.35 is a schematic diagram showing measurement of body panel vibration using the laser method. The laser instrument projects a reference light beam and a measuring light beam on the body panel, and the panel's vibration signals are measured. If a vehicle is placed in a wind tunnel or on a dynometer (dyno) and then excited, each panel's vibration under the moving condition can be acquired. If the body is excited by a shaker, its modal shapes can be identified. Figure 3.36 shows the vibration responses of one side of a body obtained by laser measurement.

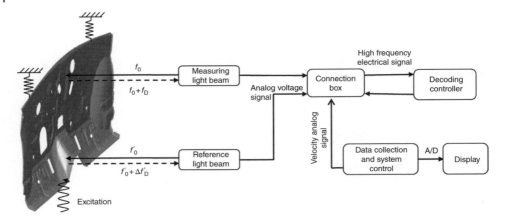

Figure 3.35 Schematic diagram of measuring a body panel's vibration by laser.

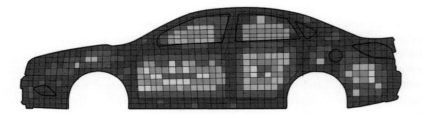

Figure 3.36 Vibration responses of one side of a body, as measured by laser.

In recent years, laser measurement has been widely adopted owing to its non-contact nature and speed. But it has a caveat: that is, the laser beam must be perpendicular to the tested plate, so only the vertical vibration of the plate can be tested.

3.4.3.4 Near-Field Acoustic Holography

NAH is a method for identifying near-field sound source characteristics using acoustic holography. Near field refers to the area that is closer to the sound source than one or two wavelengths of the highest frequency. For example, if the highest analyzed frequency of the sound source is 5000 Hz, the corresponding near field is the area 7–14 cm away from the sound source. In the near field, sound waves are very complex, being neither spherical waves nor plane waves, and there is no fixed relationship between the sound pressure level and the distance.

A microphone array consisting of a series of microphones is needed for NAH testing, and a holographic measuring surface in the near field of the sound source must be constructed, as shown in Figure 3.37. The sound source characteristics can be identified by the relationship between the measuring surface and the source surface. The sound pressure on the measuring surface is a convolution integral of the sound pressure on the sound source surface and the Green function. By using this and applying the two-dimensional Fourier transform on the Helmholtz equation, then performing a transformation calculation of the spatial domain and spatial wave number domain, the NAH method reconstructs the sound field of the sound source

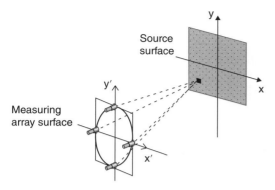

Figure 3.37 A sound source surface and a microphone array surface.

surface and obtains the sound pressure distribution on the tested object's surface. The results are displayed as an image.

Many microphones are orderly arranged on the array. An array design has two aspects: one is the array size, and the other is the distance between the microphones.

The array size should be as big as the size of the measured object, otherwise part of the object cannot be measured. The array size will be determined by the lowest analyzed frequency as well. When the object is larger than the microphone array, batch measurement can be used, but only if the object is standstill. Sound varies with speed (such as the engine acceleration sound), or can be a pulse signal (such as a door closing sound), so in a moving vehicle the test should be done multiple times under the same operating conditions.

The distance between the microphones determines the highest analyzed frequency and resolution of the sound source. The closer together the microphones, the higher the analyzed frequency, and the higher the recognition accuracy of the sound source. But they cannot be too close, otherwise the sound fields would interfere with each other.

The NAH method can be used to measure the sound near the surface of complex structures, and the tested results have very high accuracy across the entire frequency band. Therefore, it is widely used in the identification, localization, and contribution analysis of automotive sound sources. The NAH method is mainly used to measure low and middle frequency noise. Figure 3.38 shows a microphone array being used to test the sound radiation of a vehicle body. Figure 3.39 shows the sound source distribution of the body at 192 Hz, and the areas with a high level of noise are the mirror, A-pillar, and rear quarter window.

Although NAH is a powerful method for identifying sound sources, it has some shortcomings. First, it requires a microphone array that is of a similar size to the tested object. Batches of measurements can be processed for a large object, but the multiple measurements will cause many errors, especially under non-standstill working conditions. Second, the microphones must be densely placed for high frequency analysis, but in routine testing work, it is very hard to organize so many microphones and test channels. In addition, over-dense microphone arrangement will affect the sound field.

However, the beamforming method overcomes these drawbacks. In NAH measurement, sound waves are regarded to have a plane wave propagation, whereas in beamforming, the sound wave propagates in a spherical wave.

Figure 3.38 A microphone array is used to test the sound radiation of a vehicle body.

Figure 3.39 Sound source distribution of the body at 192 Hz.

3.4.3.5 Beamforming Method

Beamforming is a technology for identifying sound sources in the far field with a microphone array. In a sound field, the far field is the area where there exists a planar wave relationship between particle velocity and sound pressure. In the far field, the sound wave propagates in the form of a plane sound wave.

A microphone array is placed in the far field. The sound waves generated by the sound source are transmitted to each microphone of the array in the form of a plane wave or spherical wave. The distance between the sound source and each microphone and their relative positions are different, so the time needed to receive the sound waves for each microphone is different. By using this positional difference and the time difference, enhancing the signals in the selected directions and weakening the signals in other directions, the sound signals can be reconstructed in the microphone array and the sound source can be identified.

The construction of an array affects the recognition accuracy. Different arrays have different directivity, and the array structure has a significant impact on the recognition results. Arrays are divided into planar arrays and spatial arrays. Planar arrays include line arrays, cross arrays, rectangular arrays, circular arrays, spiral arrays, and random arrays, as shown in Figure 3.40. Spatial arrays include rectangular arrays, diamond arrays, and spherical arrays. The most commonly used arrays are spherical arrays, as shown in Figure 3.41. The array shape and microphone arrangement can be either regular, as in a rectangular array and circular array, or irregular, as in a random array and Archimedes spiral array. Regular arrays induce spatial aliasing problems that could affect the identification accuracy. The identification accuracy of a well-designed irregular array is higher than that of a regular array, but array design is very difficult because a lot of engineering experience and data are needed.

(a)   (b)   (c)   (d)

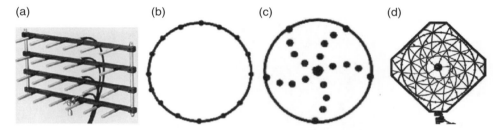

Figure 3.40 Configurations of planar array beamforming: (a) rectangular array, (b) circular array, (c) spiral array, (d) random array.

Figure 3.41 Spherical array beamforming.

In beamforming testing, an array is placed in the far field, so this method is particularly suitable for testing the noise of an object when microphones cannot be placed close to it: for example, when measuring the wind noise outside a vehicle during testing in a wind tunnel. The sound wave propagation is independent of the array size, so beamforming can be used to measure a large objects: for example, a circular array with a diameter of only 0.5 m can be used to measure the outside noise of a whole vehicle. In addition, the beamforming method can be used under both standstill and non-standstill working conditions.

However, the beamforming method also has drawbacks. The theory assumes that the sound source consists of unrelated point sound sources, and the resolution depends on

Figure 3.42 A spherical array placed inside a vehicle to test interior noise.

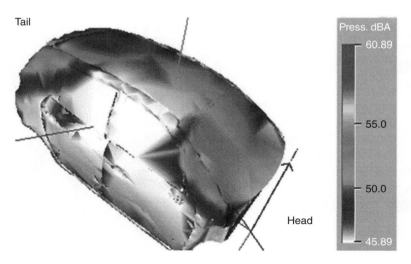

Figure 3.43 Display of the tested interior noise.

the distance between the array and the sound source, the array size, and the sound wavelength; therefore, the accuracy of sound source identification at low frequency is poor. This technique is suitable only for identifying high-frequency noise, mainly above 1000 Hz.

The beamforming method is widely used in the automotive industry, and there are many beamforming products (also known as acoustic cameras) on the market. As in NAH measurement, many signals are recorded simultaneously with microphones, so the measured noise can be quickly displayed in the form of images. The vehicle's exterior noise and interior noise can be tested by a planar array and spherical array, respectively. Figure 3.42 shows a spherical array placed inside a vehicle to test the interior noise. Figure 3.43 shows a test result.

3.5 Damping Control for Structural Vibration and Sound Radiation

3.5.1 Damping Phenomenon and Description

A standstill object will vibrate when subjected to an external excitation. After the excitation is released, the vibration disappears slowly because of the damping effect. Figure 3.44 shows a beam without damping and a beam with damping, and their free vibration motions when they are subjected to an initial input. For the beam without damping, the vibration will last for a very long time, but for the beam with damping, the vibration will gradually decay. From the perspective of NVH, damping is defined as the capability to consume a system's vibration energy. The system's vibration energy or sound energy is transmitted into heat energy or other energies and dissipates, so the

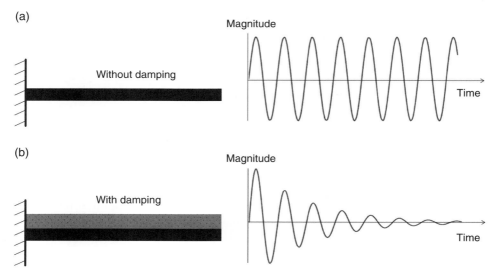

Figure 3.44 (a) A beam without damping and its motion, and (b) a beam with damping and its motion.

system's vibration is suppressed and the radiated noise is reduced. Damping makes the amplitude of the free vibration continually decay until the vibration stops. In forced vibration, the damping consumes a portion of the work done by the excitation and forces the vibration amplitude to decrease, and in particular suppresses resonance of the system.

All objects and structures have intrinsic damping properties. Damping can be divided into internal damping and external damping. Internal damping is generated by internal friction, resulting in energy loss when the internal molecules or metal grains move. External damping is generated by the friction between an object and an outside viscous fluid or friction between the solids, resulting in energy loss.

Every object or structure has internal damping, some more and some less. The internal damping of steel metals is very small, and the loss factor is between 0.0001 and 0.0006. The damping loss factors for metal alloy materials, however, are much higher than for metal materials, reaching to 0.2. But the damping for macromolecule polymers is very high, with a loss factor of up to 0.8.

The scope of research on damping control includes the study of damping mechanisms, the establishment of damping models, and the analysis of the characteristics of damping materials and damping structures, and their application. Body damping research focuses on damping structure design, the selection of damping materials, the optimization of damping control, and so on.

3.5.2 Damping Models

3.5.2.1 Types of Damping Models

Damping exists in many parts of a vehicle body, such as the inherent damping of the body sheet metals, damping inside damping materials, and damping induced by vibration isolation elements. Body damping is divided into three categories. The first category is the internal damping of body structures, such as the body panels and viscoelastic materials. The second category is the damping generated by friction between adjacent components. The third category is damping induced by the airflow when a vehicle moves.

To study structural damping and material damping, scientists and engineers have built many damping models, including viscous damping models, structural damping models, proportional damping models, and so on. In terms of body damping, the models can also be divided into three categories.

The first category is structure damping models, including viscous damping models, hysteretic damping models, and structural damping models. These are used to analyze the body panels' structure and the damping materials. This book focuses on the viscoelastic damping model.

The second category is dry friction damping models, and the most important one is the Coulomb damping model. Coulomb damping is an interface damping between contacted objects with a relative motion or moving tendency. The friction force (F_d) is proportional to the surface normal pressure (N), opposite to the movement direction, and independent of speed amplitude, i.e. $F_d = -\mu N$, where μ is a friction coefficient. Friction forces exist between the body sheet metals, between the metal panels and the rubber parts, and between the connecting joints.

The third category is air damping models, and a typical one is the aerodynamic damping model. A vehicle is subjected to air resistance when it moves. The resistant force (F_d), an external resistance, is proportional to the square of the vehicle's speed: that is, $F_d = \beta u^2$, where β is the air friction damping coefficient, and u is the vehicle speed.

This book focuses on body panel damping that is used to control panel vibration and sound radiation.

A single DOF system, as shown in Figure 3.45, is used to describe the equations and characteristics of a system with damping. The dynamic equation of the system is expressed as follows:

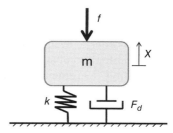

Figure 3.45 A single DOF system with damping.

$$m\ddot{x} + F_d + kx = f, \tag{3.53}$$

where F_d, m, and k are the damping force, mass, and stiffness of the system, respectively; x and f are the system displacement and external force, respectively.

Based on the single DOF model, the first category of damping models (the viscoelastic model, hysteresis model, and structural damping model) are briefly described below.

3.5.2.2 Viscous Damping Model

The viscous damping model is the most commonly used one. Viscoelastic damping force is proportional to the speed of a moving object: that is, $F_d = c\dot{x}$, where c is the damping coefficient. The viscous damping model is expressed as

$$m\ddot{x} + c\dot{x} + kx = f. \tag{3.54}$$

A concept of critical damping (c_{cr}) is introduced, which is expressed as $c_{cr} = 2\omega_n m = 2\sqrt{km}$. $\omega_n = \sqrt{k/m}$ is the natural frequency of the system. The damping ratio (ξ) is defined as

$$\xi = \frac{c}{c_{cr}}. \tag{3.55}$$

For free vibration, Eq. (3.54) can be rewritten as

$$\ddot{x} + 2\xi\omega_n\dot{x} + \omega_n^2 x = 0. \tag{3.56}$$

In engineering, the most damping is underdamped: i.e. $0 < \xi < 1$. Solving Eq. (3.56), the free vibration response of the system is obtained as follows:

$$x = e^{-\xi\omega_n t}\left(X_0 \cos\sqrt{1-\omega_n^2}\,t + \frac{\dot{X}_0 + \xi\omega_n X_0}{\sqrt{1-\omega_n^2}} \sin\sqrt{1-\omega_n^2}\,t\right), \tag{3.57}$$

where X_0 is the initial displacement of the system.

Figure 3.46 shows a damped oscillation of the underdamped free vibration system. Its amplitude gradually decays over time, and the attenuation depends on the damping ratio. In order to evaluate the damping influence on the amplitude attenuation speed, a logarithmic ratio of amplitudes of two adjacent cycles can be used, which is defined as logarithmic decrement, and expressed as

$$\delta = \ln\frac{X_i}{X_{i+1}} = 2\pi\xi\frac{1}{\sqrt{1-\xi^2}}. \tag{3.58}$$

For cases of very small damping, i.e. $\xi \ll 1$, Eq. (3.58) can be simplified as

$$\delta = 2\pi\xi. \tag{3.59}$$

From Eqs. (3.58) or (3.59), the damping ratio determines how quickly the free vibration amplitude will decay.

The viscous damping model is the simplest damping model, and the damping force is linear. It is convenient to use this model to analyze and solve the vibration equation. Other complex damping models can be simplified as a viscoelastic damping model through the principle of equivalent energy loss.

3.5.2.3 Hysteretic Damping Model

In the viscosity model, the consumed energy is proportional to the vibration frequency (described later in this section). This conclusion does not match some structural vibration energy consumption, so the hysteresis damping model is introduced to eliminate the influence of frequency: i.e. the hysteresis damping force is $F_d = \frac{h}{\omega}\dot{x}$, where h is the damping hysteresis coefficient. The hysteretic damping model is

$$m\ddot{x} + \frac{h}{\omega}\dot{x} + kx = f. \tag{3.60}$$

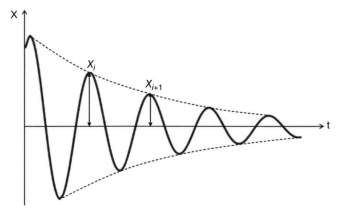

Figure 3.46 Damped oscillation of underdamped free vibration system.

3.5.2.4 Structural Damping Model

Structural damping is where the damping force inside the structure is proportional to the elastic force, and opposite to the speed: i.e. the damping force is $F_d = ivkx$, where v is the structural damping coefficient. The structural damping model is

$$m\ddot{x} + ivkx + kx = f. \tag{3.61}$$

Combine the damping force and the elastic force in Eq. (3.61) to one term, forming a total internal force that is expressed as

$$F_i = (1+iv)kx. \tag{3.62}$$

$(1+iv)k$ is called the complex stiffness of the system.

3.5.2.5 Proportional Damping and Equivalent Damping

For some very complex damping systems, it is almost impossible to obtain their analytical solutions using the models introduced above, so the systems have to be simplified. The simplified models generally include the proportional damping model and equivalent damping model. The damping is regarded to have a proportional relationship with mass and stiffness: i.e. the damping term is broken down into a mass term and a stiffness term. Because the mass term and stiffness term satisfy the orthogonal principle, a multi-DOF system can be discretized into many single DOFs by modal transformation.

3.5.3 Loss Factor

Here, the viscous damping model is used as an example to discuss loss factor. For free vibration, the damping effect can be evaluated by decay of the vibration amplitude. For a forced vibration, the standard for evaluating the damping effect is the energy dissipation. In one cycle, the ratio of energy consumed by damping to the total mechanical energy is used to measure the damping effect.

Assuming that the system's motion is harmonic, the excitation force and displacement are expressed as follows:

$$f = F \sin \omega t \tag{3.63}$$

$$x = X_0 \sin(\omega t + \varphi), \tag{3.64}$$

where F and X_0 are the magnitudes of the excitation force and displacement, respectively, and ω is the circular frequency.

The system velocity is

$$\dot{x} = X_0 \omega \cos(\omega t + \varphi). \tag{3.65}$$

For the viscous damping model, the energy consumed in one vibration cycle is

$$E_d = \int_0^{2\pi} F_d dx = \int_0^{2\pi} c\dot{x}dx = \int_0^{2\pi} c(\dot{x})^2 dt = \int_0^{2\pi} cX_0^2 \cos^2(\omega t + \varphi)dt = \pi cX_0^2 \omega. \tag{3.66}$$

The energy consumed by the damping is proportional to the vibration frequency.

Mechanical vibration energy includes kinetic energy and potential energy. At any time, the kinetic energy is

$$E_k = \frac{1}{2}m\dot{x}^2 = \frac{1}{2}mX_0^2\omega^2\cos^2(\omega t + \varphi). \tag{3.67}$$

The potential energy is

$$E_p = \frac{1}{2}kx^2 = \frac{1}{2}kX_0^2\sin^2(\omega t + \varphi). \tag{3.68}$$

At resonance, $\omega = \omega_n$, $\omega_n^2 = k/m$, and the total mechanical energy in one cycle is

$$E = \int_0^{2\pi}(E_k + E_p)dt = \pi kX_0^2. \tag{3.69}$$

The ratio of the energy consumed by the damping to the total mechanical energy in one cycle is defined as the loss factor, which is expressed as follows:

$$\eta = \frac{E_d}{E} = \frac{c\omega}{k}. \tag{3.70}$$

At resonance,

$$\eta = \frac{E_d}{E} = \frac{c\omega_n}{k} = \frac{c}{k}\sqrt{\frac{k}{m}} = \frac{c}{\sqrt{km}}. \tag{3.71}$$

Substitute the critical damping, $c_{cr} = 2\sqrt{km}$, and the damping ratio, $\xi = \frac{c}{c_{cr}}$, into Eq. (3.71), and the loss factor at resonance is

$$\eta = 2\xi. \tag{3.72}$$

Equation (3.72) gives a numerical relationship between the damping ratio and loss factor, but their physical meanings are totally different. The damping ratio describes decay of vibration amplitude, whereas the loss factor describes system energy dissipation.

The loss factors for the hysteresis damping model and the structural damping model can be obtained in a similar way to the above method for the viscous damping model.

3.5.4 Characteristics of Viscoelastic Damping Materials

As a macromolecule polymer, the damping effect of a viscoelastic material depends on many factors, such as its molecular structure, the friction and motion between the molecules, and additive components.

During motion, an elastic material repeatedly stores energy and releases energy, without consuming energy. The main property representing elastic material is Young's modulus. A viscous material cannot store energy, but it can consume energy. The main property of a viscous material is its loss factor. Viscoelastic materials simultaneously possess the features of both elasticity and viscosity, so their properties are described in terms of Young's modulus and loss factor. The energy loss of steel is independent of temperature and frequency, but the damping characteristics of a viscoelastic material vary with temperature and frequency.

Figure 3.47 shows typical curves of Young's modulus and loss factor varying with temperature for a viscoelastic material. According to the characteristics of the material's Young's modulus and loss factor, the temperature range can be divided into three regions: the low temperature region, middle temperature region, and high temperature region.

1) Low temperature region. The material exhibits a glass state and a large Young's modulus. The intermolecular binding force is large, and the molecules are not active, so the material's loss factor is low. Therefore, the low temperature region is also known as the "glass state region."
2) Middle temperature region. Starting from the border between the low temperature region and the middle temperature region (referred to as the critical temperature, T_g), the intermolecular binding force reduces and the molecules become more active as the temperature increases. Therefore, the Young's modulus decreases rapidly, and the loss factor quickly increases and reaches a peak value at a certain temperature. This region is also known as the "transition state region."
3) High temperature region. In this region, the material exhibits a rubbery state, and even a flow state. Thus, the molecules are still active and the Young's modulus continues to decline, but the declining tendency slows down, and the loss factor reduces as well. The high temperature region is also referred to as the "rubbery state region."

In the transition state region, the materials have the optimal damping effect. The optimal temperature for damping materials is generally between 20° and 60°. The wider the optimal temperature range, the higher the loss factor, and the range can be increased by

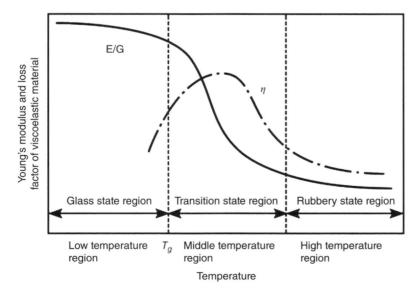

Figure 3.47 A typical curve of Young's modulus and loss factor varying with temperature for a viscoelastic material.

adjusting the material's composition. The optimal temperature range for a free damping structure is lower, close to the glass state region, as shown in Figure 3.48a. The optimal temperature range for constrained damping structure is higher, close to the rubbery state region, as shown in Figure 3.48b.

Figure 3.49 shows curves of the Young's modulus and loss factor for a typical viscoelastic material varying with frequency at a fixed temperature. As the frequency increases, the Young's modulus increases, while the loss factor increases at the beginning and reaches a peak, after which it declines. The optimal frequency range of viscoelastic damping materials is between 200 and 500 Hz.

In the design of a viscoelastic damping structure, the varying characteristics of Young's modulus and loss factor with temperature and frequency must be considered. For example, the working temperatures and excitation frequencies for body structures and engine structures are different, so different materials should be chosen. Materials with a high loss factor across a wide temperature and frequency range are regarded as the optimal damping materials.

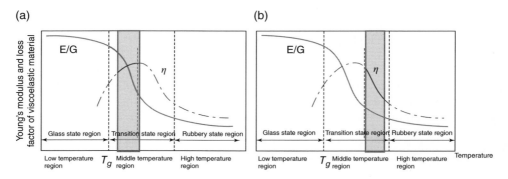

Figure 3.48 Optimal temperature range for (a) a free damping structure and (b) a constrained damping structure.

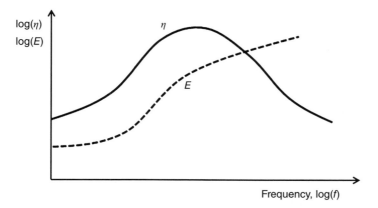

Figure 3.49 Young's modulus and loss factor of a typical viscoelastic material varying with frequency.

3.5.5 Classification of Body Damping Materials and Damping Structures

3.5.5.1 Classification of Body Damping

Body damping includes four categories: viscoelastic damping materials and structures, high damping alloy structures, composite damping materials, and intelligent damping materials and structures.

1) Viscoelastic damping materials. Viscoelastic material is a macromolecule polymer with both viscous liquid and elastic solid characteristics. Viscosity means that this material can consume energy when subjected to an external excitation. Elasticity means that the energy can be stored during the motion. That is, during the motion, part of the energy is converted into thermal energy or other forms of energy and consumed, while the other part of the energy is stored in the form of potential energy. The loss factor of viscoelastic damping material varies with temperature, amplitude, and frequency. A viscoelastic material and a base plate can be combined to form a composite structure in order to achieve both good damping and good elasticity.

2) Damping alloys. Damping alloy is a special metal material. When subjected to an alternating stress and strain excitation, its energy is consumed as a result of a magnetoelastic effect, grain boundary effect, and so on. For most metals, the loss factor is very small, and there is almost no damping effect. But the loss factor of damping alloys is relatively large, so they have good damping properties.

3) Composite damping materials. A composite damping material is a special material made from two or more constituent materials, including a matrix material and additives, to achieve high damping properties. The matrix materials include polymer matrix materials and metal matrix materials. By changing the fibrous structure of the polymer matrix and adding special particles to the metal matrix, the composite materials can attain high strength and high damping simultaneously.

4) Intelligent damping materials and structures. An intelligent damping material or structure varies its characteristics in response to an external input to achieve high damping. If only changes in the external environment (such as vibration) alter the damping of the material or system, the structure is called a semi-active control structure. A typical semi-active damping material is piezoelectric damping material. Conductive materials (such as piezoelectric particles) are added to the macromolecule material; when subjected to an excitation, the piezoelectric particles transfer vibration energy into electrical energy, which is then converted into heat energy that dissipates, thereby suppressing the vibration and noise radiation.

Active control is a method of changing the structure's properties, such as damping, by an external action. Figure 3.50 shows a special constrained damping structure. Unlike a traditional damping structure, the intermediate material is not a viscoelastic material; instead, it is a special liquid called an electro-rheological fluid that contains many tiny polar particles in a dielectric medium. When subjected an external electric field, the material changes from a liquid to a solid, and its loss factor could be over 10, providing tremendous damping. Because of the need for an external energy input, such as an electric field, the generated damping is called active damping.

The intelligent damping materials used on vehicles are mainly electro-rheological fluids (described above) and magneto-rheological fluids. Magneto-rheological fluid contains magnetic particles that change the fluid's properties when subjected to a magnetic field, increasing the loss factor.

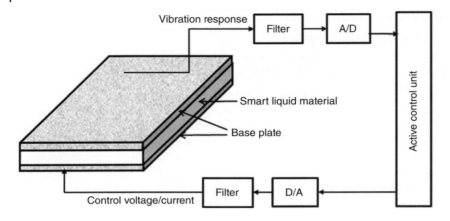

Figure 3.50 An active damping structure.

3.5.5.2 Classification of Viscoelastic Damping Material

Viscoelastic damping materials are divided into three categories: asphalt-based damping materials, rubber-based damping materials, and water-based damping materials. The damping material is composed of three parts: the matrix material, the infilling material, and additives. Each of the three categories of viscoelastic damping material uses a different matrix material: asphalt, rubber, or resin.

Asphalt-based damping material is a composite material in which asphalt is the matrix, into which graphite, sawdust, paraffin, toluene, and other organic and inorganic fillers are mixed. Asphalt itself has considerable damping properties. Graphite can improve the thermal conductive properties of the material, so aging of the material due to overheating and degradation due to sunlight can be avoided, and at the same time, the material's stiffness is increased and the loss factor peak is widened. Sawdust reduces the material's density, thus reducing its weight.

Rubber-based damping materials use rubber as matrix, and mica, carbon black, graphite, etc. are added. The main rubbers used include butyl rubber, chlorinated butyl rubber, and nitrile rubber. Rubber damping has good waterproofing, oil-resistance, and adhesion properties. The rubber used can be natural or synthetic. Natural rubber mainly comes from *Hevea brasiliensis*; after the bark has been peeled, white latex leaks out. Synthetic rubber is a highly elastic synthetic polymer. The damping performance of natural rubbers is better than that of synthetic rubbers, but the cost is higher. The most frequently used damping materials on vehicles are synthetic rubbers.

Water-based damping material uses resin as matrix, and mica, calcium carbonate, and graphite infilling material are added. Before adding the infilling material, water and resin are mixed to form a water-soluble resin liquid. Water acts as a solvent, so the damping material is known as water-based damping material. Mica is the main additive, and its lamellar structure increases the shear deformation of the polymer; thus, mica can improve the damping properties of the material.

Asphalt is obtained mainly from petroleum byproducts, and synthetic rubber uses unsaturated hydrocarbons extracted from petrochemical products as a raw material, which are then subjected to a complex chemical reaction. When heated, the asphalt and rubber volatize some substances that are harmful to humans and the environment. In winter, the materials become harder, and their elasticity and damping effects decrease.

Resin is extracted from natural macromolecule polymers. Water-based damping material is tasteless, non-toxic, and its organic solvents are non-volatile, so it does not pollute the environment and is harmless to humans. Water-based material is a safe and reliable damping material, and the trend in the automobile industry is towards more and more use of environmentally friendly water-based damping material.

3.5.5.3 Configuration of Damping Material

Damping materials are mainly configured as damping sheets and spray damping. Damping sheets are divided into three categories – fusion damping sheets, magnetic damping sheets, and foil constrained sheets – according to the way they are connected to the vehicle body. Figure 3.51 shows the configurations of damping.

1) Fusion damping sheet. At room temperature, this damping material is a solid plate. After the plate has been placed on a metal sheet and baked in a high temperature environment, it melts and then firmly adheres on the metal sheet after cooling down. Because this damping material is baked and then bonded with the metal sheet at high temperature, it is called a fusion damping sheet. Fusion damping sheets are mainly used on the vehicle floor. The damping sheet is placed on the floor and adheres to it after passing through the baking and painting assembly line.

2) Magnetic damping sheet. Magnetic powders are added into the viscoelastic material to give it magnetic properties. The magnetic damping sheet is attached to a metal sheet by magnetism, but because the binding force between them is not strong, the sheets are subsequently placed in a high temperature oven to firmly bond the damping sheet with the metal sheet. Magnetic damping sheets are commonly used on vertical, tapered, or arced panels, such as the door panels, side panels, roof, wheelhouses, and so on.

3) Foil constrained sheet. The damping material is pasted on the surface of a foil sheet. A layer of pressure-sensitive glue (fusion adhesive) is pasted on the surface of the damping material, and then this is covered by a protective sheet. In application, the protective sheet is removed, and the damping material is directly attached to the body panels. This damping sheet can be used on a variety of body panels, and also on non-metal sheets (such as the surface of the air filter box). Because of their easy application, foil constrained sheets are widely used in the vehicle development process.

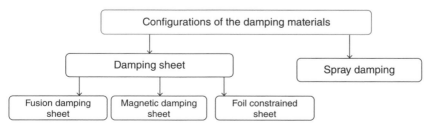

Figure 3.51 Configurations of damping materials.

4) Spray damping. Unlike solid damping sheets, spray damping material is a liquid. The material is sprayed or brushed onto the body panels and bonds with them as it dries. Spray damping is mainly used on the wheelhouses and outside floor, among other locations.

3.5.5.4 Types of Damping Plate Structure

Damping structures are divided into free damping structures and constrained damping structures. Free damping structures are those in which the damping material is freely pasted on a plate, as shown in Figure 3.52. Constrained damping structures are those in which the damping material is sandwiched between two plates, as shown in Figure 3.53.

A free damping structure consists of a base plate and a damping layer. The damping material is pasted or sprayed onto the plate surface and is not subjected to any constraints, so it is called a free damping layer. In Figure 3.52a, only one surface has a damping layer, so this is called a one-sided damping structure, whereas in Figure 3.52b, both surfaces have damping layers, so this is called a two-sided damping structure. When the plate is subjected to a bending vibration, the plate and damping layer generate compressed and tensile deformations, and the damping material is subjected to alternating stress and strain actions; part of this mechanical energy is converted into heat energy, thus suppressing the plate's vibration.

The loss factor of a free damping structure depends on the material's characteristics and the ratio of the damping layer's thickness to the base plate's thickness. Figure 3.54 shows a loss factor curve for a free damping structure varying with the ratio of the damping layer thickness (H_2) and the steel plate thickness (H_1). In the low ratio region, the loss factor is very small; then, with the increase of the ratio, the loss factor increases rapidly. When the ratio is larger than 6, the loss factor continues to increase, but at a slower rate; when the ratio is larger than 9, the loss factor increases very little. Therefore, if the damping layer is too thin, there is no obvious damping effect on the structure; however, if the damping layer is too thick, the damping effect is not noticeably increased,

(a)

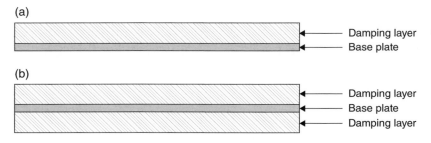

Damping layer
Base plate

(b)

Damping layer
Base plate
Damping layer

Figure 3.52 Free damping structure.

Constrained layer
Damping layer
Base plate

Figure 3.53 Constrained damping structure.

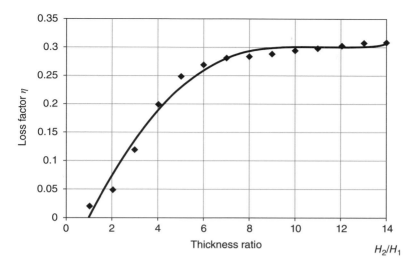

Figure 3.54 A loss factor curve for a free damping structure varying with the ratio of damping layer thickness (H_2) and steel plate thickness (H_1).

but the cost increases. Typically, the ratio of damping layer thickness to steel plate thickness is between 2 and 5.

The constrained-layer damping structure shown in Figure 3.53 consists of a base layer, a damping layer, and a constrained layer. The damping material is pasted on the base plate, and then another plate with high stiffness (usually steel) is pressed onto it to form a "sandwich" structure. When the structure is subjected to bending vibration, the constrained layer and the base layer slide against each other, because the elastic modulus of the base layer and the constrained layer is much higher than that of the damping layer. This results in shear motion of the damping layer, which converts part of the vibration mechanical energy into heat energy. The thickness of the constrained layer is almost the same as the thickness of the base layer; compared with free damping structures, the ratio of the constrained damping layer thickness to the base layer thickness is smaller.

3.5.6 Measurement of Damping Loss Factor

Damping materials and structures are measured to obtain their loss factor, Young's modulus, and other parameters, and the characteristic curves of these parameters vary with temperature and frequency. Methods to test and analyze damping materials and structures include the half-power bandwidth method, peak resonance method, and free attenuation method. In automobile engineering, the most commonly used method is the half-power bandwidth method outlined by the Society of Automotive Engineers (SAE) in document J1637, "Laboratory Measurement of the Composite Vibration Damping Properties of Materials on a Supporting Steel Bar."

One end of a steel bar is clamped on a test fixture, while the other end is free, forming a cantilever beam, as shown in Figure 3.55. The fixture should have large enough stiffness to separate its modal frequencies from those of the tested sample. The damping

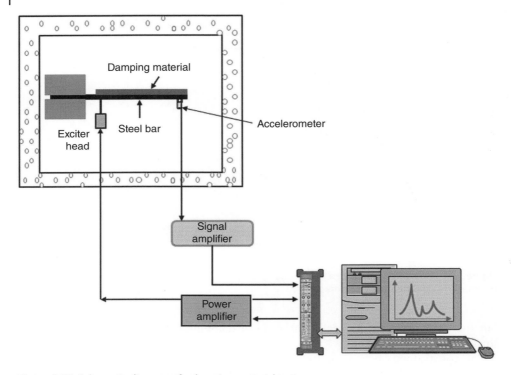

Figure 3.55 Schematic diagram of a damping material test.

material is pasted on the steel bar, with a gap above the test fixture in order to ensure that the fixture does not interfere with the measurement.

The bar is then excited by an exciter; the excitation signal could be a sine sweep or random signals. After the excitation and response signals have been measured, the transfer functions between the responses and the excitation are obtained. Owing to low weight of the tested specimen, the additional weight of sensors may affect the accuracy of the test, so the best method for obtaining the excitation and response signals is to use non-contact sensors. If a contact sensor has to be used, its weight should be less than 0.5 g.

The testing fixture is placed in a temperature-controlled chamber, so the damping performance can be evaluated at different temperatures. The temperatures used should range between −20 and 55 °C and the vibration frequencies should range between 100 and 1000 Hz.

The method for analyzing the loss factor in the SAE J1637 standard is the half-power bandwidth method. On a tested transfer function (as shown in Figure 3.56), the resonant frequency (f) for a given mode is determined, and then the locations, called half-power points, where the values are 3 dB lower than the peak at the resonant frequency are chosen. The frequencies corresponding to the two half-power points are obtained, which are called the lower frequency (f_l) and upper frequency (f_u), respectively.

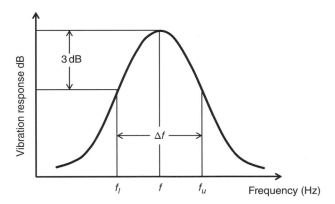

Figure 3.56 Resonant frequency and lower and upper frequencies on a transfer function.

The difference between the upper frequency and the lower frequency is called the frequency bandwidth, and is expressed as $\Delta f = f_u - f_l$. The loss factor at this resonant frequency is

$$\eta_c = \frac{f_u - f_l}{f}. \tag{3.73}$$

There are other measurement methods in addition to the SAE J1637 standard. ASTM E756, the "Standard Test Method for Measuring Vibration-Damping Properties of Materials," is a standard of the American Society for Testing and Materials. SAE J1637 was developed on the basis of ASTM E756. The two methods have many similarities, but also some differences: for example, the tested specimen and installation conditions are different.

3.5.7 Application of Damping Materials and Structures on Vehicle Body

Macromolecule polymers have high damping properties, but with low stiffness, so they cannot be used alone in engineering. But if they are combined with steel sheets, the resulting composite structure will have the high stiffness properties of steel and the high damping properties of the macromolecule polymers.

Damping materials are pasted or sprayed onto the locations with the largest strain energy in order to achieve the best results. The most effective frequency range in which to use damping materials and structures is between 200 and 500 Hz.

3.5.7.1 Application of Damping Sheet

A damping sheet is a flat plate of damping material that is solid at room temperature. It could be a soft, sticky layer attached to foil, which can be stuck to a metal sheet; it could also be a hard damping plate, which is placed on a metal sheet and then melts and bonds with it after being baked in a high-temperature oven. The most commonly used damping sheets include foil constrained layers and expandable damping sheets.

1) Foil constrained layer. The foil constrained layer is a soft structure in which the damping material is pasted onto foil and then covered by a piece of paper. At room temperature, the damping material is soft and sticky. In application, the covering paper is removed, then the damping material is directly attached to the body panels. Figure 3.57 shows the foil layers used on the door and the roof. Because they are lightweight and can be easily cut into any shape and attached, foil constrained layers have found widespread use in the vehicle NVH development process.

2) Expandable damping sheet. At room temperature, this damping sheet is a hard plate. In application, it is placed on a body panel, and then melts and bonds with the panel after being baked in high-temperature environment. Expandable damping sheets have many advantages, such as their effective damping effect, excellent adhesive properties, high stiffness, and ability to prevent metal corrosion. They are widely used on dash panels, wheelhouses, and consoles, as shown in Figure 3.58. The thickness of the expandable damping sheet is generally between 2 and 4 mm.

Asphalt plate is a type of expandable damping sheet. Because it is cheap and has a good damping effect, asphalt is widely used on vehicle floors. But because it is heavy, asphalt plate is difficult to use on vertical, inclined, and arced surfaces at room temperature, so it is rarely used on locations other than the floor.

(a) (b)

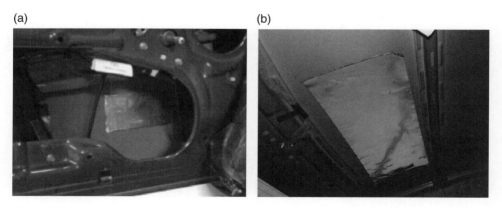

Figure 3.57 Foil layers used on (a) a door and (b) a roof.

(a) (b) (c)

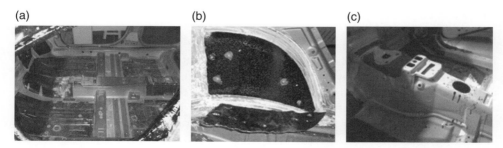

Figure 3.58 Application of expandable damping sheets (a) on the vehicle floor, (b) on a wheelhouse, and (c) on the console.

3.5.7.2 Application of Damping Coating

Damping coating uses damping materials in liquid form, which are sprayed or brushed onto the body surface. Whereas damping sheets appear smooth, damping coating looks uneven, with many grains on its surface.

Damping coating possesses excellent damping properties and is usually applied on the outside of the body, such as on the bottom of the floor and on the wheelhouses, as shown in Figure 3.59. Damping coating can effectively attenuate the impact of small stones thrown up from the road and the percussion of splashing water, reducing body panel vibration and sound radiation.

3.5.7.3 Composite Damping Steel Plate

Composite damping steel plate is a constrained damping structure in which a layer of viscoelastic damping material is sandwiched between two layers of steel. The steel plate has high strength, and the damping material and the shear motion between the constrained steel plates have a high damping effect. Because of these combined properties, composite damping steel plates are widely used in vehicle bodies and engines.

Among the body panels, composite damping plate is most commonly used on the dash panel. In some vehicles, the entire dash panel is made of composite damping plate. In addition to its damping effect, the plate provides better sound insulation than a traditional dash panel structure. Figure 3.60a shows a conventional dash panel, with the sound absorption and insulation layers on both sides. Figure 3.60b shows a composite

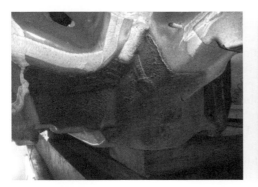

Figure 3.59 Application of damping coating.

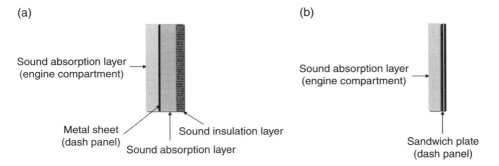

Figure 3.60 (a) A conventional dash panel; (b) a composite damping dash panel.

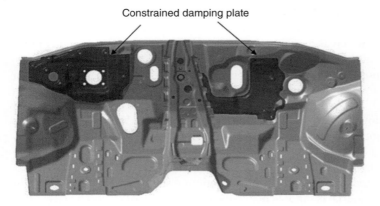

Constrained damping plate

Figure 3.61 Two pieces of composite damping plate are attached on two locations of a dash panel.

damping plate dash panel and a sound absorption layer. The composite dash panel has fewer insulation and sound absorption layers and weighs 1–3 kg less than the conventional dash panel, but it achieves the same sound insulation effect.

Composite damping plate is mostly used on local body structures: that is, the plate is attached to some local panels to suppress local vibration. Figure 3.61 shows two pieces of composite damping plate attached to two locations on a dash panel to increase the local damping. Of course, the local stiffness and sound insulation are increased as well.

Vibration of the dash panel is a major contributor to a vehicle's interior noise. The severe vibration locations and corresponding frequencies can be identified by modal analysis, and then composite damping plates are attached to these locations. Figure 3.62a shows the vibration curve comparison at a point on the dash panel for a steel panel and a composite structure. The dash panel vibration is reduced across the entire speed range when composite damping plates are used; in particular, the peak vibration at around 2400 rpm is significantly reduced. Figure 3.62b is the corresponding interior noise. The noise is reduced with composite plates by around 1–3 dB across the entire speed range, and at 3400 rpm, the noise is lowered by 5 dB (A).

3.6 Stiffness Control for Body Panel Vibration and Sound Radiation

The modal separation principle is fundamental to controlling body panel vibration and sound radiation: i.e. the panel modal frequencies must be separated from the excitation frequencies, and the modal frequencies for adjacent parts must be separated.

The body panels are welded to the frames, so the frames can be regarded as the boundaries of the panels. For example, the floor is divided into several panels by side frames, cross members, and tunnel frames, whereas the dash panel is surrounded by A-pillars, a cross member under the windshield, and a front floor cross member. The frames have sufficient stiffness and elasticity, so the boundaries constituted by the frames represent neither a clamped boundary condition nor a simply

(a)

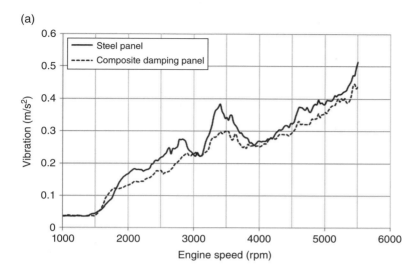

(b)

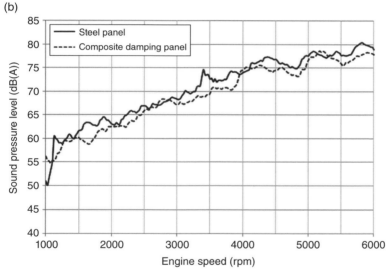

Figure 3.62 Comparison of (a) vibration on a dash panel and (b) interior noise between a steel panel and composite damping panel.

supported boundary condition; instead, the panel's boundary condition is between the two cases.

Panel vibration consists of different modal responses. Methods for suppressing panel vibration and sound radiation include adding damping and/or adding mass on the panel, and increasing its stiffness. The purpose of adding mass and/or increasing stiffness is usually to attenuate a narrow-band, low-frequency vibration response by shifting the panel's modal frequency to separate it from the excitation frequency and/or from the acoustic cavity modal frequency. For a single frequency vibration problem, the panel can be simplified as a single DOF, as shown in Figure 3.45.

3.6.1 Mechanism of Stiffness Control

A panel's sound radiation energy depends on its vibration velocity. By analyzing the transfer functions between velocity responses and an input force, methods for controlling the panel's vibration can be obtained. Here, a single DOF system is used to illustrate how to obtain the transfer functions.

Assume displacement is $x = X_0 e^{j\omega t}$, where X_0 is the displacement magnitude. The excitation force is $f = F e^{j\omega t}$, where F is the force magnitude. Differentiate the displacement, obtaining

$$\dot{x} = j\omega X_0 e^{j\omega t} \tag{3.74a}$$

$$\ddot{x} = -\omega^2 X_0 e^{j\omega t}. \tag{3.74b}$$

Using Eqs. (3.74a) and (3.74b), the relationship among the displacement, velocity, and acceleration can be obtained, expressed as follows:

$$x = \frac{\dot{x}}{j\omega} \tag{3.75a}$$

$$\ddot{x} = j\omega \dot{x} \tag{3.75b}$$

$$\ddot{x} = -\omega^2 x. \tag{3.75c}$$

The transfer function is the ratio of the output to the input, and its concept and analysis are introduced in detail in Chapter 5. The vibration response can be expressed by displacement, velocity, or acceleration, so the corresponding transfer functions are the ratio of the displacement, velocity, or acceleration outputs to the force input. The panel's sound radiation directly relates to its vibration velocity, so the transfer function between the velocity and the force input is used to represent the capacity of sound radiation.

Substitute Eqs. (3.75a), (3.75b), and (3.75c) into Eq. (3.54), obtaining

$$\left[\left(k - m\omega^2 \right) + jc\omega \right] x = f \tag{3.76a}$$

$$\left[\frac{\left(k - m\omega^2 \right) + jc\omega}{j\omega} \right] \dot{x} = f \tag{3.76b}$$

$$\left[\frac{-j\left(k - m\omega^2 \right) + c\omega}{j\omega^2} \right] \ddot{x} = f. \tag{3.76c}$$

According to Eqs. (3.76a), (3.76b), and (3.76c), the transfer functions for displacement, velocity, and acceleration to the excitation force are obtained as follows:

$$H_d(\omega) = \frac{x}{f} = \frac{1}{\left(k - m\omega^2 \right) + jc\omega} \tag{3.77a}$$

$$H_v(\omega) = \frac{\dot{x}}{f} = \frac{j\omega}{(k - m\omega^2) + jc\omega} \tag{3.77b}$$

$$H_a(\omega) = \frac{\ddot{x}}{f} = \frac{-\omega^2}{(k - m\omega^2) + jc\omega}, \tag{3.77c}$$

where H_d, H_v, and H_a represent the transfer functions of displacement, velocity, and acceleration to the excitation force, respectively.

The magnitudes of the above transfer functions are

$$\left| H_d(\omega) \right| = \frac{1}{\sqrt{(k - m\omega^2)^2 + (c\omega)^2}} \tag{3.78a}$$

$$\left| H_v(\omega) \right| = \frac{\omega}{\sqrt{(k - m\omega^2)^2 + (c\omega)^2}} \tag{3.78b}$$

$$\left| H_a(\omega) \right| = \frac{\omega^2}{\sqrt{(k - m\omega^2)^2 + (c\omega)^2}}. \tag{3.78c}$$

Here, the examples in Table 3.5 are used to represent the impact of stiffness on the transfer functions. For mode $(1, 1)$, the simply supported rectangular plate is equivalent to a single DOF system. When the kinetic energy and potential energy of the plate's vibration are equal to the kinetic energy and potential energy of the vibration of a lumped-mass system, respectively, the plate can be simplified as a single DOF system, and the corresponding equivalent mass and equivalent stiffness are:

$$m_{e(1,1)} = 0.25 m_p \tag{3.79}$$

$$k_{e(1,1)} = \frac{1}{4} abD_0 \left[\left(\frac{\pi}{a} \right)^2 + \left(\frac{\pi}{b} \right)^2 \right]^2, \tag{3.80}$$

where $m_{e(1,1)}$ and $k_{e(1,1)}$ are the equivalent mass and stiffness, respectively. m_p is the plate mass. For the example in Table 3.5, the equivalent mass and stiffness are 0.2355 kg and 16 938.4 N m^{-1}, respectively. Assume the damping is 10 N·s m^{-1}.

Figure 3.63 shows the transfer functions between displacement, velocity, and acceleration and the force for different stiffness values. With the increase in stiffness, the system's natural frequency increases, the magnitude of the displacement transfer function decreases, the magnitude of the velocity transfer function doesn't change, and the magnitude of the acceleration transfer function increases. Thus, the major function of increasing stiffness is to increase the frequency.

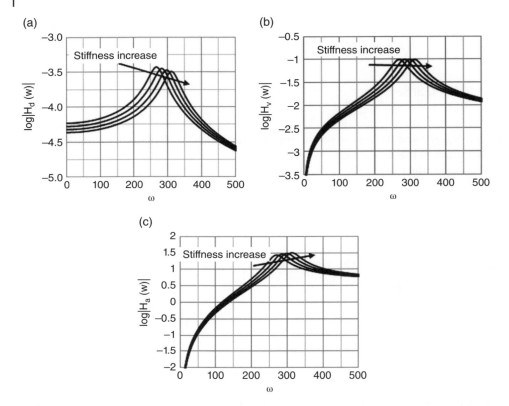

Figure 3.63 Influence of stiffness on transfer functions between (a) displacement, (b) velocity, and (c) acceleration and input force.

3.6.2 Tuning of Plate Stiffness

3.6.2.1 Approaches to Increase Plate Stiffness

For a simply supported square plate with side length L, according to Eq. (3.5), the modal frequency of its mode (1, 1) is

$$\omega_{(1,1)} = \frac{2\pi^2}{L^2}\sqrt{\frac{D_0}{\rho h}}. \tag{3.81}$$

Substitute Eq. (3.2) into Eq. (3.81), obtaining

$$\omega_{(1,1)} = \sqrt{\frac{\pi^4 E h^3}{3L^4\left(1-\mu^2\right)\rho h}}. \tag{3.82}$$

The plate mass is $m = \rho L^2 h$, and the equivalent mass for mode (1, 1) is $m_{e(1,1)} = 0.25\rho L^2 h$, The natural frequency can be expressed by the relationship between mass and stiffness, $\omega = \sqrt{k/m}$. Substitute equivalent mass and stiffness into Eq. (3.82), and the stiffness can be expressed as follows:

$$k_{(1,1)} = \frac{\pi^4 E h^3}{12\left(1-\mu^2\right)L^2}. \tag{3.83}$$

For the simply supported square plate with side length L, i.e. $a = b = L$, the stiffness for mode (1, 1) can be also obtained by Eq. (3.80), and the result is the same as the one from Eq. (3.83).

According to Eq. (3.83), the plate stiffness is proportional to the third power of the side thickness and the Young's modulus of the material, and is inversely proportional to the square of the length. To increase the plate stiffness, the plate length should be reduced and the thickness should be increased, or materials with a high Young's modulus should be used. Obviously, due to limitations of design, space, weight, and cost, it is very hard to change the vehicle panel's length, thickness, and material. Therefore, other methods must be found in order to increase the stiffness. In engineering, the approaches shown in Figure 3.64 can be used to increase the plate stiffness, that is:

1) Stepped or fluted plates.
2) Beaded or curved plates.
3) Plates supported by supporting beams or reinforcement adhesives.

3.6.2.2 Influence of Stiffness Changes on Structure Modes and Responses
To illustrate the impact of stiffness on modes, the simply supported square plate shown in Figure 3.65a is used here as an example. The plate's stiffness can be increased in three

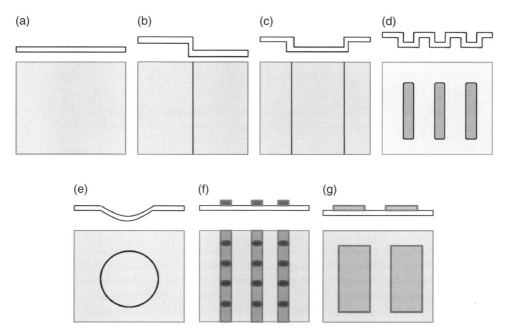

Figure 3.64 Flat plate and reinforced plates. (a) Flat plate. (b) Stepped plate. (c) Fluted plate. (d) Bead plate. (e) Curved plate. (f) Plate with supporting beams. (g) Plate with reinforcement adhesives.

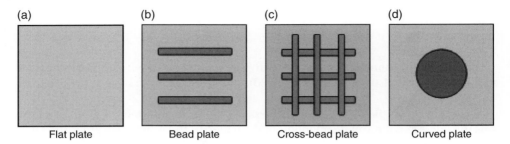

(a) Flat plate (b) Bead plate (c) Cross-bead plate (d) Curved plate

Figure 3.65 Square flat plate and stiffened plates. (a) Flat plate. (b) Bead plate. (c) Cross-bead plate. (d) Curved plate.

Table 3.11 Modal frequencies (Hz) for a flat plate and stiffened plates.

Modal no.	Flat plate	Bead plate	Cross-bead plate	Curved plate
1	54.55	88.03	127.7	98.76
2	136.4	166	184.4	177.7
3	136.4	197	236.1	177.7
4	217.8	293.3	349.2	514.1
5	272.8	305.9	387.6	559.3

ways: with a bead plate, a cross-bead plate, or a curved plate, as shown in Figure 3.65b, Figure 3.65c, and Figure 3.65d, respectively.

Table 3.11 lists modal frequencies for the simply supported square plate (side length 300 mm, thickness 1 mm) and the three stiffened plates. The stiffness of the bead plates and curved plates is much higher than the flat plate, and the cross-bead plate has a higher frequency than the bead plate. The first modal frequencies for the flat plate, bead plate, cross-bead plate, and curved plate are 54.5, 88, 127.7, and 98.7 Hz, respectively. The modal frequency of the first acoustic cavity for most sedans is between 40 and 60 Hz, so the structural mode of the flat plate has great potential to couple with the cavity mode. After the plate has been stiffened by the above methods, its structural mode will be separated from the cavity mode.

Figure 3.66 shows the first five modes for the flat plate, bead plate, cross-bead plate, and curved plate, respectively. Compared with the flat plate, the stiffened plates have the same first modal shape, but their second and higher modal shapes are different. The change of the modal shapes will influence the sound radiation of the plates.

To compare the transfer functions between the velocity responses and the excitation force, any point on the plate can be chosen as the excitation point, and another point is picked up as the response point. Most of the loads that the body is subjected to are applied to the frames, so the excitation point is chosen on the plate's edge. The magnitude of the modal shape for the first mode is at the middle point, so this is picked up as the response point. Figure 3.67 shows the excitation point and the response point.

Figure 3.68 shows the transfer functions between the velocity and the force for the flat plate, bead plate, cross-bead plate, and curved plate, respectively. After the plate is

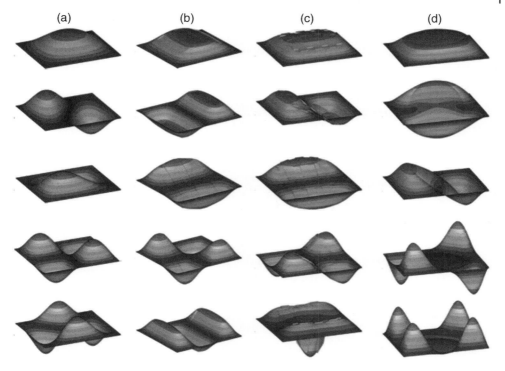

Figure 3.66 First five modes for a (a) flat plate, (b) bead plate, (c) cross-bead plate, and (d) curved plate.

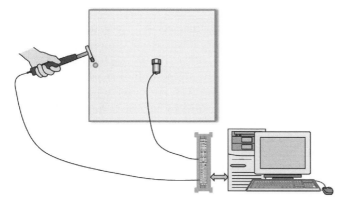

Figure 3.67 Excitation point and response point on a plate.

stiffened, the first modal frequencies increase, but the magnitudes of the transfer functions could increase or decrease.

The objective is to increase the plate's frequency to separate it from the excitation frequency and/or the cavity modal frequency. The first modal frequency of a body panel is usually low, so it could potentially couple with the excitation frequencies from the engine, road, etc. After the panel has been stiffened, its frequency is higher than the excitation frequencies, so it will be difficult to excite. The modal frequencies for the

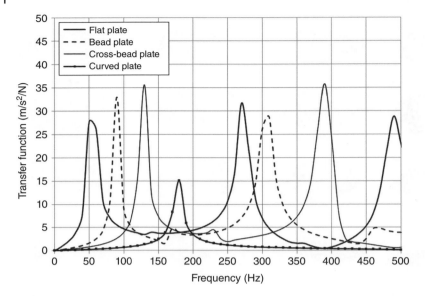

Figure 3.68 Transfer functions between velocity and force for a flat plate, bead plate, cross-bead plate, and curved plate.

second and higher modes are usually higher than the excitation frequencies, so the possibility of these modes being excited is very low.

3.6.3 Influence of Plate Stiffness Tuning on Sound Radiation

Tuning a plate's stiffness will change its sound radiation capacity. Here, to compare the radiation capacity of different plates, sound pressure levels at 1 m above the plates are compared.

The example in Figure 3.67 is used to calculate the sound radiation when the plates are excited, and Figure 3.69 shows the radiated sound pressure levels for the four plates. For the first mode, the sound radiation increases with the increase in plate stiffness. According to the sound radiation theory, the higher the plate stiffness, the stronger the radiation capacity. The modal shape for the first mode (1, 1) is simple, and the tendency of sound radiated energy is consistent with that of the plate's velocity. So, the objective of increasing the plate stiffness is not only to separate its frequency from the excitation frequencies, but also to separate them from those of the cavity mode. The sound radiation of the second and higher-order modes is a superposition of the sound radiation of each structural mode, and the phase has an important impact on the result: i.e. superposition for the same phases, and cancellation for opposite phases.

3.6.4 Case Study of Body Stiffness Tuning

Increasing stiffness is a common method of attenuating body panel vibration and sound radiation. Two ways can be used to achieve this. One way is to redesign the panel structure: for example, the panel can be designed as a curved surface, a stepped shape, a

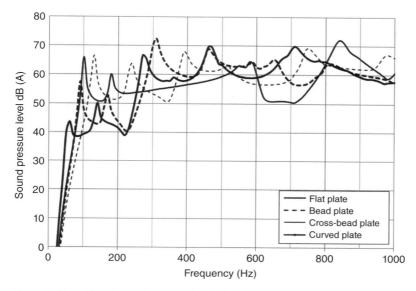

Figure 3.69 Radiated sound pressure levels for a flat plate, bead plate, cross-bead plate, and curved plate.

(a)   (b)

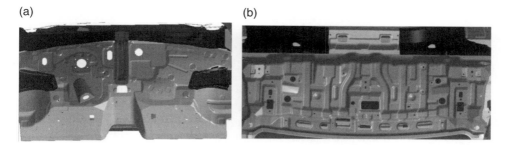

Figure 3.70 (a) A few beads on a panel surface; (b) many beads punched on a panel surface.

beaded shape, and so on. The other way is to add supporting structures to the panel, such as supporting beams, reinforcement adhesives, and so on.

3.6.4.1 Increasing Body Panel Stiffness by Structure Design

Meticulous redesign of a dash panel is the most typical way to increase the structure's stiffness. The dash panel is a big-surface metal sheet, and the length and height of a mid-sized sedan are about 1200 and 500 mm, respectively, so its first modal frequency is very low. In order to increase its modal frequency, the panel must be designed in two ways. The first way is to design the dash panel with many different stepped planes or surfaces, and the second way is to punch beads or ribs on the surface. Figure 3.70 shows two dash panels. The modal frequency for the panel in Figure 3.70a is low because it has few beads on its surface, whereas the modal frequency for the panel in Figure 3.70b is much higher because many beads are punched on the surface.

(a)

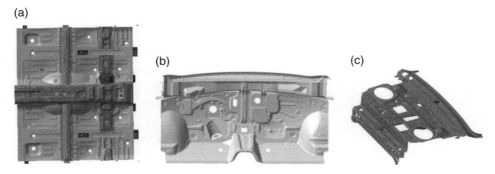

(b)   (c)

Figure 3.71 Beads punched on body panels. (a) Beads on floor. (b) Beads on dash panel. (c) Beads on rear package tray.

(a)   (b)

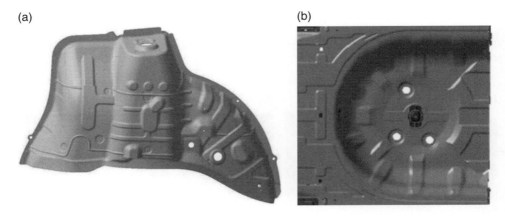

Figure 3.72 Arc-shaped body panels. (a) Wheelhouse. (b) Luggage trunk.

In a similar way to the dash panel, the stiffness of many other body panels is increased by changing the structure's design. Figure 3.71 shows punched ribs on the floor, dash panel, rear package tray, and so on. Some panels are deformed to be arc shapes, such as side panels, wheelhouses, the luggage trunk, and other structures, as shown in Figure 3.72. The arc shapes meet the body styling requirements and increase the panels' stiffness as well.

3.6.4.2 Increasing Body Panel Stiffness by Adding Supporting Structures

Due to the limitations of body styling, ribs cannot be punched on some outer body panels, or the panels cannot be designed as several surfaces, such as the door outer panels, fenders, roof, engine hood, and trunk lid. If the large panels are not reinforced, their modal frequencies will be very low. Only the internal surfaces of the exposed panels can be reinforced, and three ways are used to achieve this: adding supporting plates, adding supporting beams, and using reinforcement adhesive.

1) Additional supporting plate. Additional plates are attached on the internal surfaces of the outer body panels to increase their stiffness. Figure 3.73 shows supporting plates added onto the internal surfaces of an engine hood (a), trunk lid (b), and rear side panel (c).
2) Supporting beam. One or more beams are attached on the internal surfaces of the outer body panels to increase their stiffness. Figure 3.74a shows a side-impact beam firmly glued onto the internal surface of an outer door panel. This beam not only enhances the collision resistance capacity of the door, but also increases its stiffness. Figure 3.74b shows several supporting beams on a roof.
3) Reinforcement adhesive. In some locations where it is difficult to place supporting beams or plates, reinforcement adhesive can be pasted onto the internal surfaces of the outer body panels to increase their stiffness. Reinforcement adhesive is an epoxy resin composite material, comprising an adhesive layer, a foam layer, and a separation layer, as shown in Figure 3.75. Reinforcement adhesive can be easily attached to the surface of a steel sheet, and it becomes very stiff, like a metal sheet, after it is baked at high temperature (160 °C), so the panel's local stiffness is effectively

(a)   (b)   (c)

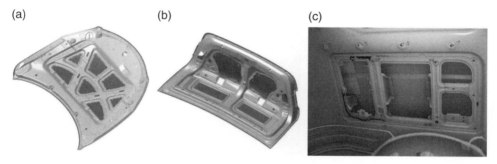

Figure 3.73 Supporting plates added on the internal surfaces of (a) an engine hood, (b) a trunk lid, and (c) a rear side panel.

(a)   (b)

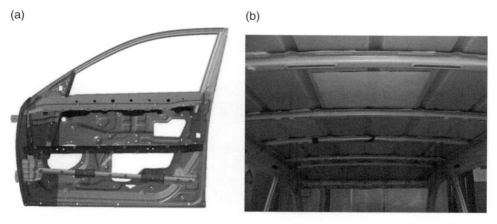

Figure 3.74 Supporting beams on the internal side of outer body panels: (a) side-impact beam on the internal surface of an outer door panel; (b) supporting beams on a roof.

increased. Reinforcement adhesive has many applications on the vehicle body, such as the door panels, wheelhouses, roof, and so on. Figure 3.76 shows reinforcement adhesive used on the internal surface of a door outer panel.

Below is an example to show the application of reinforcement adhesive. A huge interior booming is perceived for a sedan during acceleration, and the corresponding engine speed range is from 3200 to 3500 rpm. The booming is induced by vibration of the roof and the rear wheelhouses. Reinforcement adhesives are pasted onto the roof and the wheelhouses, as shown in Figure 3.77. Figure 3.78 shows a comparison of the interior sound pressures with and without the reinforcement adhesive. The interior booming is significantly reduced by 3–4 dB in the above speed range.

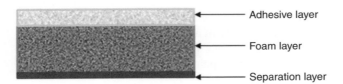

Adhesive layer

Foam layer

Separation layer

Figure 3.75 Structure of reinforcement adhesive.

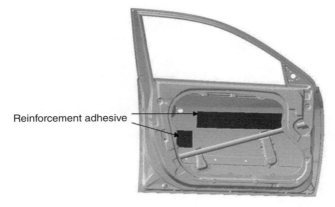

Reinforcement adhesive

Figure 3.76 Reinforcement adhesive used on the internal surface of an outer door panel.

Figure 3.77 Reinforcement adhesives pasted on a roof and wheelhouse.

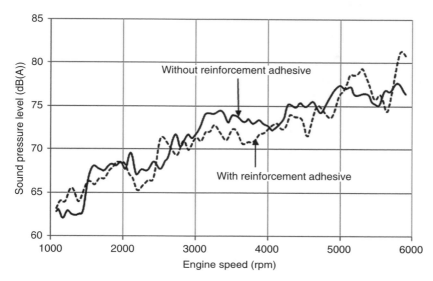

Figure 3.78 Comparison of interior sound pressures with and without reinforcement adhesives.

3.7 Mass Control for Body Panel Vibration and Sound Radiation

3.7.1 Mechanism of Mass Control

The mechanism of mass control is similar to that for stiffness control. By analyzing the transfer function between a response and input, the influence of the mass on the transfer function can be obtained.

Equation (3.77) provides the transfer functions between the displacement, velocity, and acceleration and the force for a single DOF system. The data used in Section 3.6 is used again to plot the transfer functions for different mass, as shown in Figure 3.79. From the plots, it is easy to see that with the increase in frequency, the magnitude of the displacement transfer function increases, the magnitude of the velocity transfer function keeps constant, and the magnitude of the acceleration transfer function decreases.

In some cases where the structures cannot be modified, mass can be added to attenuate panel vibration and sound radiation. The weight of the mass and the optimal location to place it can be determined by analyzing or testing the panel modal shapes and sound radiation characteristics. After the mass is added, the panel modal frequency decreases and the magnitude of the transfer function changes, so the panel vibration and sound radiation are suppressed. Sometimes, the added mass is called a "mass damper".

3.7.2 Application of Mass Control

Mass dampers are widely used in vehicle bodies. In addition to using measured or analyzed modal characteristics to choose a mass damper, the "trial and error" method is widely used in engineering to quickly identify and solve problems such as interior

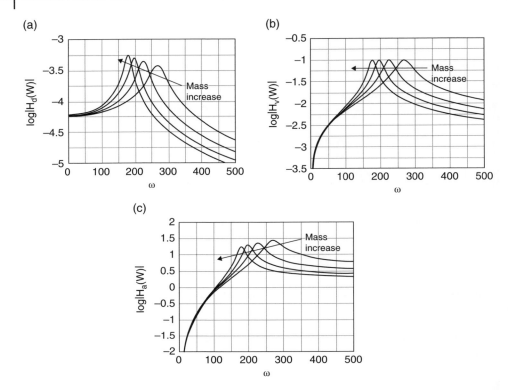

Figure 3.79 Transfer functions between (a) displacement, (b) velocity, and (c) acceleration and input force for different masses.

booming noise and/or structural resonance. The mass dampers used in trial and error processing include sandbags, magnets, and steel boxes. The following two examples illustrate the application of a mass damper.

3.7.2.1 Mass Damper Used on Body Panel

A passenger car is driven on a coast road at a speed of $60\,km\,h^{-1}$, and an unacceptable interior booming is perceived. The test results show that a huge peak of sound pressure at 46 Hz is the cause of the booming. The road excitation frequency corresponding to the speed of $60\,km\,h^{-1}$ is 46 Hz, and the frequency of the first acoustic cavity mode is 46.3 Hz. Therefore, it can be concluded that the booming is generated by coupling between a panel's structural mode and the cavity mode, and the panel is excited by road input.

The acoustic cavity modes must be analyzed. The sound pressure variation of the first cavity mode is along the longitudinal direction of the vehicle body and has almost no change in the transverse and vertical directions. The panels that are perpendicular to the longitudinal direction, including the dash panel, trunk lid, and front and rear windshields, must be analyzed. The modal shapes and frequencies of the panels can be obtained by FE analysis or testing. Sometimes, in order to quickly find the panel generating the booming, trial and error is used: i.e. a mass is placed on one panel at a time, and the change in booming is evaluated.

After two masses of 1.5 kg each are added to the trunk lid, as shown in Figure 3.80, the booming disappears. Figure 3.81 shows a comparison of the interior sound pressure level with and without the masses. After the masses are added, the peak at 46 Hz reduces 9 dB, so the trunk lid is identified as the most important contributor to the booming.

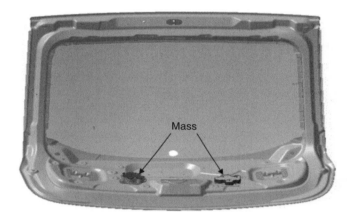

Figure 3.80 Two masses of 1.5 kg each are added to a trunk lid.

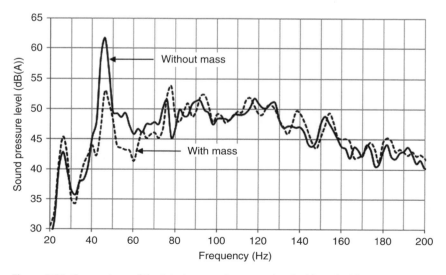

Figure 3.81 Comparison of the interior sound pressure level with and without masses.

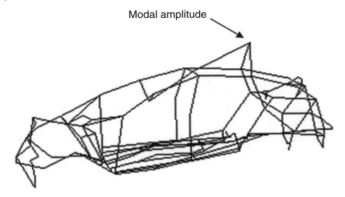

Modal amplitude

Figure 3.82 Tested body modal shape at 53 Hz.

Mass

Figure 3.83 A 2 kg mass mounted on a rear roof bow.

3.7.2.2 Mass Damper Used on Body Frame

A passenger sedan is driven down the road at a speed of $50\,km\,h^{-1}$, and an obvious booming is perceived inside the vehicle. The test results shows that the booming peak is at 52 Hz. The cause of low-frequency booming is usually coupling between the first cavity mode and a panel's structural mode, so the analysis focuses on the panels perpendicular to the longitudinal direction.

This example is very similar to the previous one. The first cavity modal frequency is 53 Hz. The windshields need to be tested for vibration, but it is impossible to add a mass directly onto the windshield. FE analysis and modal testing of the body show that the amplitude of the structural modal shape in the longitudinal direction is at the interface between the rear windshield and rear roof bow, as shown in Figure 3.82. Thus, a 2 kg mass is mounted on the rear roof bow, as shown in Figure 3.83, and the booming peak is greatly reduced from 60 dB (A) at 53 Hz to 51 dB (A), as shown in Figure 3.84, and the booming is eliminated.

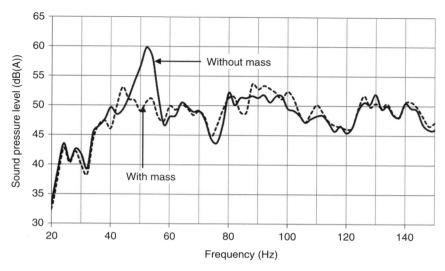

Figure 3.84 Comparison of interior sound pressure levels with and without mass.

3.8 Damper Control for Body Vibration and Sound Radiation

3.8.1 Mechanism of Dynamic Damper

A dynamic damper consists of a spring-damping-mass primary system and an additional spring-damping-mass system, as shown in Figure 3.85. The additional system added to the primary system generates a vibration with a 180° phase difference from the main system, offsetting vibration of the primary system at a certain frequency. A single DOF system (also called the primary system) is used as an example to illustrate the damper's application. m_1, k_1, and c_1 are the mass, stiffness, and damping of the primary system, respectively. m_2, k_2, and c_2 are the mass, stiffness, and damping of the additional system (the damper), respectively.

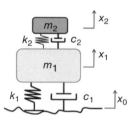

Figure 3.85 A dynamic damper and a primary system.

The primary system and the additional system constitute a two DOF system, and the dynamic equations can be written as follows:

$$m_1\ddot{x}_1 = -k_1(x_1 - x_0) - c_1(\dot{x}_1 - \dot{x}_0) - k_2(x_1 - x_2) - c_2(\dot{x}_1 - \dot{x}_2) \tag{3.84a}$$

$$m_2\ddot{x}_2 = -k_2(x_2 - x_1) - c_2(\dot{x}_2 - \dot{x}_1), \tag{3.84b}$$

where x_1 and x_2 are the displacements of the primary system and the additional system, respectively. x_0 is the base displacement.

Assume the systems move harmonically; the above two equations are transformed into the frequency domain, obtaining

$$\left[(k_1 + k_2) + j\omega(c_1 + c_2) - m_1\omega^2\right]X_1 - (k_2 + j\omega c_2)X_2 = (k_1 + j\omega c_1)X_0 \tag{3.85a}$$

$$-\left(k_2 + j\omega c_2\right)X_1 + \left(k_2 + j\omega c_2 - m_2\omega^2\right)X_2 = 0, \tag{3.85b}$$

where X_1, X_2, and X_0 are the displacement magnitudes of the primary system, additional system, and the base, respectively.

The Eq. (3.85b) is changed to

$$X_2 = \frac{k_2 + j\omega c_2}{k_2 + j\omega c_2 - m_2\omega^2}X_1. \tag{3.86}$$

Substitute Eq. (3.86) into Eq. (3.85a), the displacement magnitude ratio of the primary system to the base is

$$\frac{X_1}{X_0} = \frac{\left(k_1 + j\omega c_1\right)\left(k_2 + j\omega c_2 - m_2\omega^2\right)}{\left[\left(k_1 + k_2\right) + j\omega\left(c_1 + c_2\right) - m_1\omega^2\right]\left(k_2 + j\omega c_2 - m_2\omega^2\right) - \left(k_2 + j\omega c_2\right)^2}. \tag{3.87}$$

Figure 3.86 shows the transmissibility between the primary mass response and the base input with and without the dynamic damper. There is a huge peak for the single primary DOF system, but the peak is significantly reduced after the dynamic damper is added. There are two peaks for the dual-mass system, and they are on two sides of the huge peak of the single primary DOF system, but their magnitudes are much lower. The frequency of the dynamic damper can be adjusted by tuning the mass weight and spring stiffness of the additional system.

There are many modes for a body panel, but interior booming is basically caused by the first structural mode. If only a panel's vibration at its fundamental frequency is of concern, it can be regarded as a single DOF system. A damper is added to the panel to constitute a dual-mass system, so the vibration at this frequency and the sound radiation can be suppressed.

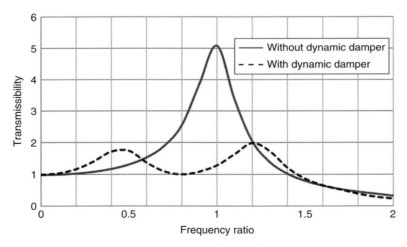

Figure 3.86 Transmissibility of primary mass response and base input with and without a dynamic damper.

3.8.2 Application of Dynamic Damper to Attenuate Interior Booming

Dynamic dampers are widely used in automobiles: for example, they are placed on the engine mount bracket, exhaust pipe, chassis brackets, and so on. Of course, dynamic dampers are also used to suppress a panel's vibration at a certain frequency and to reduce the coupling between the body panel and an acoustic cavity mode. Below, an example is used to illustrate the application of dampers to the body.

In the first example in Section 3.7, where mass dampers are used to reduce interior booming, two masses each weighing 1.5 kg are added onto the trunk lid (Figure 3.80). The interior booming at 46 Hz is eliminated, and the corresponding peak is reduced by 9 dB (as shown in Figure 3.81). However, it is a challenge to add 3 kg of weight onto the body.

Based on the modal characteristics of the trunk lid at 46 Hz, a dynamic damper can be added to attenuate the booming. The 3 kg mass is removed, and instead, two 46 Hz dynamic dampers weighing 0.5 kg each are used. The peak at 46 Hz is reduced by 9 dB and, subjectively, the interior booming cannot be perceived. Figure 3.87 shows a comparison of the interior sound pressure in the three cases: the original case, with the dynamic dampers, and with the masses.

The interior booming at 46 Hz can be eliminated by adding masses or dynamic dampers. The two methods achieve the same results, but the dynamic damper is lighter than the mass, which is important for reducing the body weight. Of course, the dynamic damper includes a mass block and a damping rubber or metal spring, so its structure is complex and the cost is higher than a simple mass. Therefore, the cost, weight, manufacturing, and so on must be comprehensively considered when choosing a mass damper or dynamic damper.

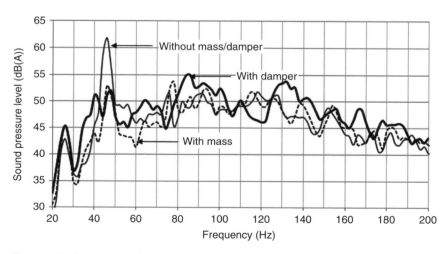

Figure 3.87 Comparison of interior sound pressure for three cases: the original case, with dynamic dampers, and with masses.

3.9 Noise and Vibration for Body Accessory Components

The frames, panels, windshields, and sound package are the main structures of a vehicle body, whereas the brackets, steering system, mirrors, seats, etc. are the accessory structures. These accessories are welded or riveted to the body, and are subjected to excitations from the engine and road. When they are resonant with the excitations, a lot of NVH problems will be generated: for example, the bracket's resonance will be transmitted to the body and then to the steering wheel, so the driver could perceive the vibration, and the mirrors could shake.

Proper accessory design is very important. The accessories described in this section are divided into three categories: the brackets, steering system, and seats.

3.9.1 Bracket Mode and Control

3.9.1.1 Classification of Brackets

Many brackets are used to connect the body and other systems. For example, an engine is connected with the body frames or sub-frame by mount brackets; an exhaust system is connected with the body by hangers; a battery is mounted on the body by a tray; a mirror is connected with the body or door by a bracket; a motor is connected with the body by a bracket; an antilock brake system (ABS) is connected with the body by a bracket; and so on.

Brackets are divided into two categories in this book: strongly excited brackets and weakly excited brackets.

Strongly excited brackets are those that are strongly related to power train excitation and frequently generate noise and vibration problems; they include engine mount brackets and exhaust hangers. These brackets are excited directly by the engine, so serious noise and vibration problems are commonly generated. The bracket must be carefully designed in order to meet NVH requirements.

Weakly excited brackets are those that are not directly or not strongly related to the power train excitation and do not frequently generate noise and vibration problems. These brackets can be further divided into two types. The first type are brackets related to power train excitation; they include air filter holders, resonator holders, and fuel pipe fasteners. Although these brackets are excited by the power train, the excitation they are subjected is much weaker than that experienced by the engine brackets. For example, the air filter box is subjected to internal airflow pulsation, and it transmits this vibration to its holder, while the fuel pipe is subjected to internal fuel pulsation, and it transmits this vibration to the fasteners. However, the excitations are relatively weak. The second type are brackets that are not directly related to power train excitation; they include battery holders, mirror brackets, air-conditioning pipe brackets, and wiring supporters. These brackets are mainly subjected to road excitation or indirect excitation from the power train.

Weakly excited brackets are only subjected to low level excitations, but sometimes the generated problems cannot be ignored. For example, the fuel pipe could induce the body floor to vibrate, and then the floor will radiate noise to the interior. During engine idling, the interior noise could increase by 1–3 dB due to the flow excitation inside the pipe. Another example is a mirror bracket with low stiffness; when subjected to an external excitation, the image on the mirror will be blurred.

3.9.1.2 Noise and Vibration Problems of Brackets

A bracket is a "bridge" to connect the body and another system; therefore, it is a transmission path for a noise and vibration source. Usually, brackets can bring three types of noise and vibration problems.

The first problem is that a bracket's structural mode is excited, making it a vibration transmission path. For example, an engine mount bracket with a first modal frequency of 300 Hz could be resonant with the engine excitation at 4500 rpm because the fourth order excitation frequency for a four-cylinder engine is 300 Hz. The engine vibration passes through the bracket, which transmits it to the body, generating the interior noise.

The second problem is that the bracket radiates noise. In a similar way to a metal sheet, when the bracket is resonant with an external excitation, it radiates sound. For example, a resonant engine mount bracket could directly generate sound.

The third problem is that the bracket and its connected component constitute a spring-mass system in which the bracket and the component are similar to a spring and mass, respectively. Resonance could occur when it is subjected to an external excitation. For example, a battery and its supporter constitute a spring-mass system. When the frequency of an external excitation (such as road excitation or engine excitation) is consistent with the spring-mass system's frequency, resonance occurs and the vibration is transmitted to the vehicle body.

3.9.1.3 Principles of Bracket Design

In order to design a good bracket and meet NVH requirements, the following four principles must be followed.

The first principle is that a bracket's frequency should be separated from the main excitation frequencies. For example, the main excitation orders for a four-cylinder engine are the second and fourth orders. At 6000 rpm, the excitation frequency for the fourth order is 400 Hz. To achieve the frequency separation principle, the engine bracket frequency should be 20% higher than the excitation frequency, so the bracket frequency is set to 480 Hz.

Figure 3.88a shows an engine mount bracket, and its first modal frequency is 300 Hz. At 4500 rpm, the engine's fourth order excitation frequency is consistent with the bracket's frequency, so the bracket resonates and induces an interior booming. After the

(a)   (b)

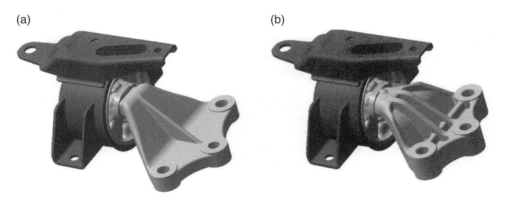

Figure 3.88 Engine mount bracket: (a) low modal frequency; (b) high modal frequency.

(a)   (b)

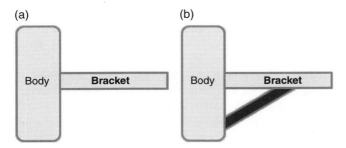

Figure 3.89 (a) Cantilever beam; (b) cantilever beam with supporter.

(a)   (b)

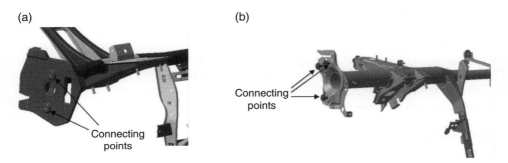

Figure 3.90 Two connection methods between steering CCBs and A-pillars: (a) with two connecting points, and (b) with three connecting points.

bracket is modified, as shown in Figure 3.88b, its frequency increases to 520 Hz, so that separates it from the engine excitation frequency, and the resonance and interior booming are eliminated.

The second principle is that the bracket should be designed to be as short as possible. A bracket is similar to a cantilever beam, as shown in Figure 3.89a. There are two ways to increase the cantilever frequency. One is to add a bracket to support its free end, as shown in Figure 3.89b. The other is to punch ribs on the bracket.

The third principle is that the connecting points between the bracket and the body should be reasonably distributed. An inadequate number of connections will reduce the system's overall stiffness. Figure 3.90 shows two connections between the steering CCBs and A-pillars. There are only two connecting points in Figure 3.90a, but three connecting points in Figure 3.90b. It is apparent that the connection stiffness in Figure 3.90a is lower.

If a bracket has three connecting points, they must be distributed as a triangle; a straight line distribution must be avoided. Figure 3.91a shows an engine bracket with three mounting points, but the points are almost distributed in a straight line, which leads to insufficient bracket stiffness. Figure 3.91b shows another engine bracket where the three mounting points are triangularly distributed; therefore, its stiffness is much higher than that in Figure 3.91a.

The fourth principle is that strong structure locations on the body should be chosen as the connecting points: i.e. the locations with high driving point dynamic stiffness,

(a)　　　　　　　　　　　　　　　　　　　(b)

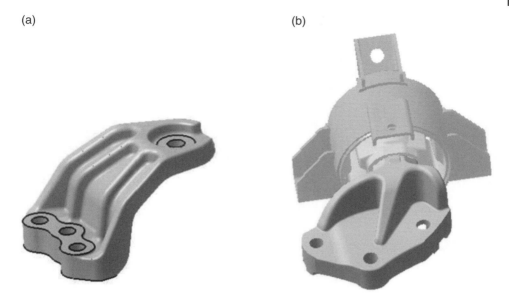

Figure 3.91 Mounting points of engine brackets: (a) straight line distribution, and (b) triangular distribution.

such as frames. The concept of driving point dynamic stiffness is described in Chapter 5. The brackets cannot be connected to weak structure locations: for example, the exhaust hangers should not be placed on the floor panel.

3.9.2　Control of Steering System Vibration

3.9.2.1　Structure of Steering System and Vibration Problems

The steering system consists of a steering shaft system and CCB system (also known as an IP reinforcement tube system), as shown in Figure 3.92. Figure 3.93a shows a steering shaft system that is connected with a body by two brackets. The steering wheel is mounted on the upper end of the steering shaft, and the airbag is installed in the middle of the steering wheel. Figure 3.93b is a CCB system. The two ends of the beam are connected to the A-pillars, a big bracket is connected to the middle of the beam and the floor, and several brackets on the beam are connected to the dash panel.

The steering system is connected with the body, so the vibrations transmitted to the body are likely to be transmitted to the steering wheel. For example, the engine vibration passes through the mounting system to the subframe, and then transmits to the steering shaft, finally reaching the steering wheel; the driveline vibration is transmitted to the steering knuckle arm, then the relay rod, rack and pinion system, and steering shaft, and finally to the steering wheel; and vibration of the front cooling module is transmitted to the body, and then to the steering wheel.

Drivers can directly perceive steering wheel vibration, which is one of customers' most complained about NVH problems. Steering wheel vibration includes three categories: idle vibration, shaking during acceleration, and nibble at high speed cruising or during acceleration.

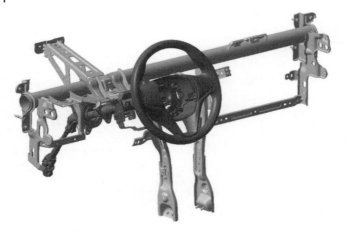

Figure 3.92 A steering system.

(a)   (b)

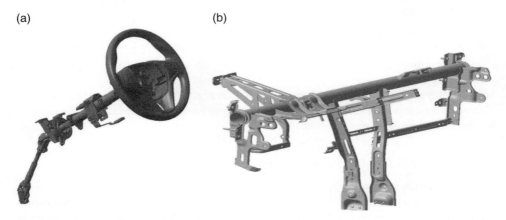

Figure 3.93 (a) A steering shaft system. (b) A CCB system.

The idle vibration frequency is relatively low, generally from 20 to 35 Hz. The reason for this vibration is that the natural frequency of the steering wheel is coupled with the excitation frequencies of the engine, cooling fan, and so on. Idle vibration is a very common NVH problem, especially in the AC-on condition because the engine speed increases compared with the AC-off condition, which could easily cause the engine firing frequency to be coupled with the steering wheel modal frequency. For example, a vehicle with a four-cylinder engine has a 650 rpm idle speed, so the corresponding engine firing order frequency is 21.7 Hz. The natural frequency of the steering wheel is 30 Hz, so it is separated from the firing frequency. However, in the AC-on condition, the engine speed increases to 900 rpm, and the corresponding firing frequency is 30 Hz, so it is coupled with the natural frequency of the steering wheel, resulting in the steering wheel vibration.

Nibble is steering wheel vibration in its circumferential direction during cruising or acceleration, and the frequency range is usually from 10 to 15 Hz. The nibble source is the unbalanced force of the tires, especially when the phases of the left and right tires are opposite. The vibration is transmitted to the steering wheel through the suspension system, so the driver perceives the nibble.

Nibble relates to the tires, wheels, suspension, and steering wheel, so it is necessary to control each component in order to eliminate or reduce the nibble. When nibble occurs, we need to check the dynamic unbalance of the tires, the subframe stiffness, the stiffness and damping of the suspension bushings, the suspension's moving trajectory, and the friction and damping of the steering wheel system.

3.9.2.2 Frequency of Steering System

The steering system frequency depends on its geometrical structure, inertia, and stiffness. The frequency is determined by the frequencies of the steering shaft system and the CCB system. The steering system can be assumed to be a series connection of the two subsystems, which can be expressed as follows:

$$\frac{1}{f} = \frac{1}{f_1} + \frac{1}{f_2}, \tag{3.88}$$

where f_1 and f_2 are the frequencies of the steering shaft system and the CCB system, respectively.

According to Eq. (3.88), only if frequencies of the two subsystems are high enough can the frequency of the steering system achieve the desired value. The frequencies of the two subsystems are analyzed in the following sections.

The frequency of the steering system must avoid the engine and road excitation frequencies, and must also be separated from the body modal frequencies by at least 5 Hz.

3.9.2.3 Structural Control of Steering Shaft System

Figure 3.94 shows the structure of a steering shaft system, including the steering shaft, steering wheel, and two brackets. It is similar to a cantilever beam, and its frequency is determined by the mounting locations of the brackets, the shaft stiffness, the wheel mass, and the airbag mass.

Typically, the steering shaft is connected to the body by two brackets, an upper bracket and lower bracket. Each bracket has two mounting points, so there is a total of four mounting points, as shown in Figure 3.94. The mounting point locations have significant influence on the shaft stiffness. The influence of factors such as the distance

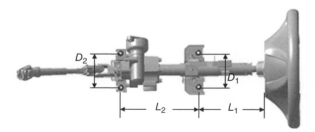

Figure 3.94 Structure of a steering shaft system.

between the mounting locations, the shaft diameter, and the wheel mass on the frequency are briefly described below.

1) Distance between the upper bracket and the steering wheel. The section between the upper bracket and the steering wheel is similar to a cantilever beam. L_1 represents the beam's length. The shorter the length, the higher the frequency. To increase the frequency, the upper bracket must be mounted as close to the steering wheel as possible.
2) Distance between the upper bracket and lower bracket. The section between the brackets forms a torsional spring, and the longer the distance, the higher the torsional stiffness. Therefore, the lower bracket must be installed as far down as possible to increase the distance.
3) Distances between the mounting points (D_1 and D_2) for each bracket. The shorter the distance, the higher the frequency of the steering shaft, because a shorter distance can force the brackets to hold the shaft more tightly. Therefore, the mounting points should be as close to the shaft as possible.
4) The shaft diameter. The shaft bending stiffness is proportional to the fourth power of its diameter, so increasing the shaft diameter will greatly enhance the frequency.
5) The wheel mass. Being a cantilever beam, the smaller the wheel mass, the higher the frequency of the steering system. To reduce the wheel's weight, lightweight materials such as aluminum or composite material can be used to make the wheel. The airbag should be as light as possible, and it should be installed on the bottom of the steering wheel in order to maintain a short distance between the airbag and the upper mounting point. In addition, the airbag must be solidly connected to the wheel.
6) Openings on the steering shaft. The steering system frequency will be reduced if there are openings on the shaft; therefore, such openings must be avoided. In cases where openings cannot be avoided, they must be as small as possible.

3.9.2.4 Control of CCB System

Figure 3.95 shows a CCB system, including a main beam and supporting brackets. The beam is connected with the A-pillars, and the brackets are connected with the floor, dash panels, and so on.

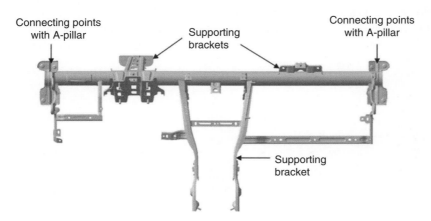

Connecting points with A-pillar

Supporting brackets

Connecting points with A-pillar

Supporting bracket

Figure 3.95 A CCB system.

The CCB frequency is determined by its stiffness, the boundary conditions of its connections with the A-pillars, and the stiffness of the supporting brackets.

1) The CCB diameter. The beam is connected with the body at two ends. If the connection is very tight, it can be regarded as a fixed-boundary beam, so its stiffness is determined by its length and diameter. The length is determined by the vehicle layout, so the way to increase its stiffness is to increase its diameter. However, increasing the beam diameter is constrained by spatial arrangement, weight, and cost.
2) Connection with the A-pillar. This connection determines the CCB's boundary conditions. A clamped boundary condition guarantees the highest stiffness. Therefore, more connecting points should be used between the beam and the A-pillars, and the connecting points should be as close to the beam as possible.
3) Supporting brackets. Adding supporting brackets at the middle of the CCB is equivalent to dividing the beam into several sections, so that each section length becomes shorter. The bracket connecting the CCB to the floor is the most important one, and the larger the distance between the two supporting points, the higher the stiffness of the CCB system.

3.9.2.5 Dynamic Damper of Steering Wheel

In the later phases of vehicle product development, if steering wheel vibration is still a problem, especially in the AC-on condition, a dynamic damper should be considered. For example, the modal frequency of a steering system is 31 Hz. The idling speeds for a four-cylinder engine are 650 and 900 rpm for the AC-off and AC-on conditions, respectively, and the corresponding firing order frequencies are 21.7 and 30 Hz, respectively. Because the excitation frequency is too close to that of the steering system, resonance of the steering wheel is easily produced. If the problem is identified in a late phase of vehicle development, it is impossible to modify other systems, so the best way to eliminate the resonance is to install a dynamic damper. The damper is designed as a flat shape so that it can fit the space of the middle part of the steering wheel, as shown in Figure 3.96.

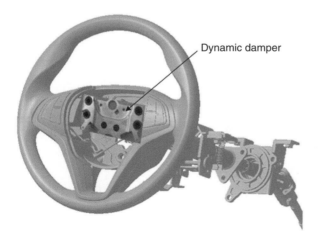

Figure 3.96 A dynamic damper installed in a steering wheel.

3.9.3 Control of Seat Vibration

The seat is an important accessory. Human bodies are in direct contact with the seats, which transmit vehicle vibration to them, so careful design of the seat's dynamic performance is very important. The outside of the seat consists of the headrest, seat pad and cushion, seatback cover and cushion, side shields, and so on. The seat interior is composed of the seat cushion frame, seatback frame, zigzag springs, tilt recliners and tracks, and so on, as shown in Figure 3.97.

Seat vibration analysis is divided into two parts: seatback vibration analysis and seat comfort analysis.

3.9.3.1 Control of Seatback Vibration

The seatback can be regarded as a cantilever beam, and the clamped point of the beam is the joint of the seatback frame and seat cushion frame. The stiffness of the seatback is determined by the frames' stiffness and the stiffness at the clamped joint.

The seatback frequency is usually close to the excitation frequencies of the engine, road, and cooling fan, so the seatback could resonate in response to these excitations. The modal frequency of the seatback frame is mainly determined by the structures of the frame and the zigzag springs. To test the seatback modes, small pieces of seatback covers and foams are peeled off, and the corresponding frames are exposed; then accelerometers are placed on the frames, as shown in Figure 3.98a. Alternatively, sensors are directly placed on the whole exposed metal frame, as shown in Figure 3.98b.

The two major modes that cause the seatback to vibrate are the first longitudinal bending mode and the first lateral bending mode, as shown in Figure 3.99. The modal frequencies can be adjusted by modifying the seatback stiffness. The seatback is similar to a cantilever beam, therefore its upper end will amply the vibration of the lower end, which means low level vibration at the seat track will be amplified on the seatback. Therefore, sufficient stiffness of the seatback is very important for controlling its vibration.

Backrest stiffness control includes frame stiffness control and joint stiffness control. Figure 3.100 shows two seatback frame structures. In the skeleton in Figure 3.100a, the zigzag springs are sparse, so its stiffness is low, whereas the skeleton in Figure 3.100b

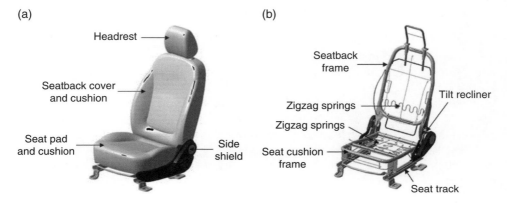

Figure 3.97 Seat structure: (a) outside structure, (b) interior structure.

(a)　　　　　　　　　　　　　　　　(b)

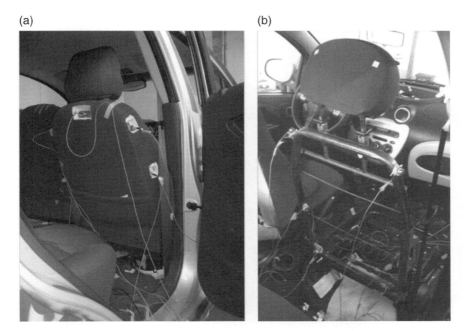

Figure 3.98 Accelerometer layout for modal testing of the seatback. (a) Frame covered by foams; (b) exposed metal frame.

(a)　　　　　　　　　　　　　　　　(b)

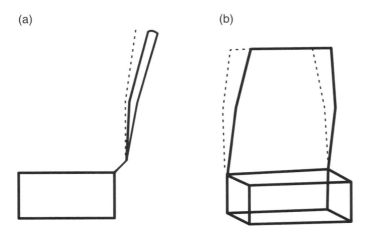

Figure 3.99 Modal shapes of a seatback: (a) first longitudinal bending mode; (b) first lateral bending mode.

has strong supporting brackets on both sides and dense net springs, so its stiffness is high. In seatback design, both the stiffness of the skeleton frame and the distribution of the springs must be sufficient to avoid vibrations.

The stiffness of the joint structure between the seatback frame and the seat cushion frame determines the cantilever beam boundary. Only when the joint is strong enough

(a) (b)

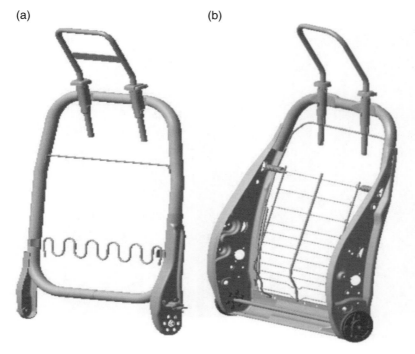

Figure 3.100 Two seatback frame structures: (a) weak structure, (b) strong structure.

(a) (b)

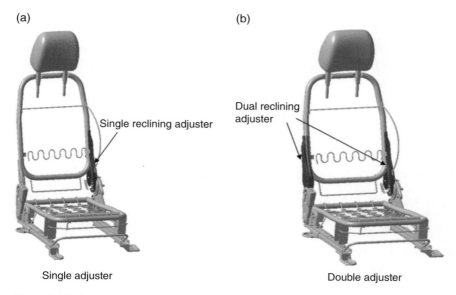

Single adjuster Double adjuster

Figure 3.101 Joint structure between backrest frame and seat cushion frame. (a) Single adjuster; (b) double adjusters.

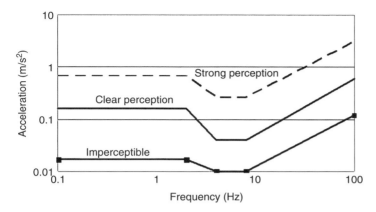

Figure 3.102 Acceleration sensitivity of the human body in a vertical direction.

can the seatback be regarded as a cantilever beam; if the joint can freely rotate, the seatback is not a cantilever beam. The main component in determining joint stiffness is the reclining adjuster. The boundary will only have high enough stiffness if the adjuster firmly holds the two frames. Figure 3.101 shows two reclining adjusters. The joint using double adjusters in Figure 3.101b has higher stiffness than the joint using a single adjuster in Figure 3.101a.

3.9.3.2 Seat Comfort Analysis

Seat comfort refers to the perception of riding comfort in the absence of resonance. Riding comfort is mainly determined by the vertical vibration on the seat cushion and horizontal vibration on the seatback. When sitting, an occupant is mainly subjected to vertical loads. The most sensitive frequency range of the human body in the vertical direction is from 4 to 8 Hz, as shown in Figure 3.102. The structures that determine riding comfort are mainly the seat cushion, seatback cushion, and zigzag springs.

The most fundamental index for analyzing and evaluating seat comfort is the vibration transfer function between the cushion and the seat track. The responses on the cushion and on the track represent the human body response and the input to the seat, respectively. The transfer function can be tested on a vehicle or on a test bench. Figure 3.103 shows a vibration platform and an installed seat. The white noise excitations, where the vibration range is from 0.05 to 0.45 g, are inputted to the platform, and the vibration signals on the platform and on the seat cushion are tested. Figure 3.104 shows the transfer functions between the seat cushion and the platform for different excitations for a luxury vehicle seat. Two interesting characteristics were found from the tested transfer functions: (i) the peak frequencies decrease with the increase in excitation; (ii) the peak amplitudes decrease as the excitation increases. For a linear structure, the transfer function is independent of input and output, and superposition law is applicable. The test results reveal that the seat cushion structure is a nonlinear system, including both nonlinear stiffness and nonlinear damping.

For seat dynamic design, generally the seat cushions should attenuate the vibration transmitted from the platform or the vehicle body: that is, the magnitudes of the transfer functions within 4–8 Hz should be less than 1 in order to achieve good ride quality.

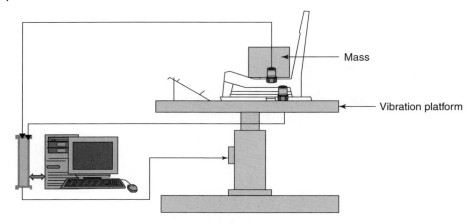

Figure 3.103 A vibration platform and an installed seat.

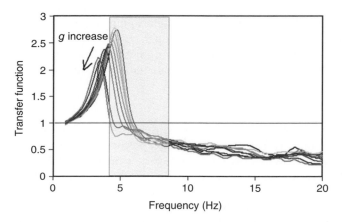

Figure 3.104 Transfer functions between a seat cushion and platform for a luxury vehicle seat.

For low level excitations (0.05–0.1 g), which vehicles experience most of the time, the transfer functions within the sensitive frequency range are less than 1, as shown in Figure 3.104. However, for sports vehicles, which can experience high level excitations when driven at high speed, the cushions not only fail to attenuate the vibration, but actually amplify it. Figure 3.105 depicts the transfer functions of a typical sports vehicle seat. Within the frequency range of 4–8 Hz, the magnitudes of the transfer functions are larger than 1: that is, within the human-body-sensitive frequency range, the vibration is amplified, so the drivers experience excitement and driving fun.

In addition to vertical vibration of the seat cushion, lateral vibration of the seatback is very important. The most sensitive frequency range of the human body in the lateral direction is between 8 and 12 Hz. The analysis method for seatback vibration is the same for seat cushion analysis, and the structural frequency can be obtained by testing or modal analysis. By tuning the cushion parameters, a comfortable seatback structure and cushion can be designed.

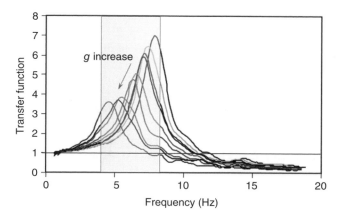

Figure 3.105 Transfer functions of a typical sports vehicle seat.

Bibliography

Abreu R.D.A., Moura F. Operational Modal Analysis techniques used for Global Modes identification of vehicle body excited from a vehicle in idle engine. SAE Paper 2012-36-0639; 2012.

Ahmadian M., Jeric K.M. The Application of Piezoceramics for Reducing Noise and Vibrations in Vehicle Structures. SAE Paper 1999-01-2868; 1999.

Aubert A., Howle A. Design Issues in the Use of Elastomers in Automotive Tuned Mass Dampers. SAE Paper 2007-01-2198; 2007.

Balasubramanian, M. and Shaik, A. (2015). Optimizing Body Panels for NVH Performance. *SAE International Journal of Passenger Cars – Mechanical Systems* 8 (3): doi: 10.4271/2015-01-22652015.

Bennur M. Vehicle Acoustic Sensitivity Performance Using Virtual Engineering. SAE Paper 2011-01-1072; 2011.

Bertolini C., Gaudino C., Caprioli D. Improved NVH Performance Via Genetic Optimization of Damping and Shape of Vehicle Panels. SAE Paper 2005-01-2329; 2005.

Beranek, L.L. (1971). *Noise and Vibration Control*. New York: McGraw Hill.

Bayod J.J. FEM Evaluation of Elastic Wedge Method for Damping of Structural Vibrations at Low Frequencies. SAE Paper 2011-01-1689; 2011.

Beranek, L.L. (1996). *Acoustics*. Acoustical Society of America.

Black M.D., Rao M.D. Material Damping Properties: A Comparison of Laboratory Test Methods and the Relationship to In-Vehicle Performance. SAE Paper 2001-01-1466; 2001.

Comesaña D.F., Wind J., de Bree H.E. A Scanning Method for Source Visualization and Transfer Path Analysis Using a Single Probe. SAE Paper 2011-01-1664; 2011.

Comesana D.F., Grosso A., Bree H.E., Wind J., Holland K. Further Development of Velocity-based Airborne TPA: Scan & Paint TPA as a Fast Tool for Sound Source Ranking. SAE Paper 2012-01-1544; 2012.

Courjal A., Balachandran K. Advanced Modeling Approaches for the Evaluation of Interior Vehicle Acoustics over the Full Range of Frequencies. SAE Paper 2012-01-1546; 2012

Crocker, M.J. (2007). *Handbook of Noise and Vibration Control*. John Wiley & Sons.

Cunefare, K.A. and Currey, M.N. (1994). On the exterior acoustic radiation modes of structures. *Journal of Acoustical Society of America* 96 (4): 2302–2312.

Currey, M.N. and Tunefare, K.A. (1995). The radiation modes of baffled finite plates. *Journal of Acoustical Society of America* 98 (3): 1570–1580.

Danti M., Vigè D., Nierop G.V. A Tool for the Simulation and Optimization of the Damping Material Treatment of a Car Body. SAE Paper 2005-01-2392.

Davidsson P. Structure-acoustic Analysis; Finite Element Modelling and Reduction Methods. Doctoral Thesis. Lund University; 2004

Doyle, J.F. (2007). *Wave propagation in structures*. New York: Springer-Verlag.

Ebbitt G., Hansen M. Mass Law - Calculations and Measurements. SAE Paper 2007-01-2201; 2007.

Elliott, S.J. and Johnson, M.E. (1993). Radiation modes and the active control of sound power. *Journal of Acoustical Society of America* 94 (4): 2194–2204.

Everest, F.A. (2001). *The Master Handbook of Acoustics*. McGraw-Hill.

Ewins, D.J. (1984). *Modal Testing: Theory and Practice*. Research Press.

Fahy, F.J. (2003). Some Applications of the Reciprocity Principle in Experimental Vibroacoustics. *Acoustical Physics* 49 (2): 217–229.

Fahy, F. and Gardonio, P. (2007). *Sound and Structural Vibration: Radiation, Transmission and Response*. Elsevier.

Fard M., Liu Z. Automotive Body Concept Modeling Method for the NVH Performance Optimization. SAE Paper 2015-01-0012; 2015.

Flint J. A Review of Theories on Constrained Layer Damping and Some Verification Measurements on Shim Material. SAE Paper 2003-01-3321; 2003.

Frank E., Moon C., Rae J., Popovich M. Optimization of Test Parameters and Analysis Methods for Fuel Tank Slosh Noise. SAE Paper 2013-01-1961; 2013.

Grosso A., Comesana D.F., Bree H.E.D. Further Development of the PNCA: New Panel Noise Contribution Reference-Related (PNCAR). SAE Paper 2012-01-1539; 2012.

Guidati S., Sottek R. Advanced Source Localization Techniques Using Microphone Arrays. SAE Paper 2011-01-1657; 2011.

Gwaltney G.D., Blough J.R. Evaluation of Off-Highway Vehicle Cab Noise and Vibration Using Inverse Matrix Techniques. SAE Paper 1999-01-2815; 1999.

Giovanni V. Passive Control and Local Structural Modification of a Mechanical System Through the Application of High Stiffness Double Layer Dampers. SAE Paper 2007-01-2351; 2007.

He H., Zhang Q., Fridrich R.J. Vehicle Panel Vibro-Acoustic Behavior and Damping. SAE Paper 2003-01-1406; 2003.

Hou H., Zhao W., Hou J. Internal Pressure Characteristics when Evaluating Dynamic Door Blow Out Deflection. SAE Paper 2015-01-2327; 2015.

ISO2631-1. Mechanical vibration and shock – Evaluation of human exposure to whole-body vibration. SAE Paper 1975-07-15.

Jain C.P., Balachandran P. Reduction of Seat Back Vibrations in a Passenger Car- An Integrated CAE - Experimental Approach. SAE Paper 2011-01-0497; 2011.

Jun Y.D., Park B.H., SeoK K.S., Kim T.H., Chae M.J. Objective Evaluation of Hold Feeling for Passenger Car Seats. SAE Paper 2015-01-2271; 2015.

Kato T., Hoshi K., Umemura E. Application of Soap Film Geometry for Low Noise Floor Panels. SAE Paper 1999-01-1799; 1999.

Kavarana F., Schroeder A. A Practical CAE Approach to Determine Acoustic Cavity Modes for Vehicle NVH Development. SAE Paper 2012-01-1184; 2012.

Khan H., Sergiyenko S., Reis C. Sheet Dampers vs. Spray-On Dampers: Current Status and Prospective Applications. SAE Paper 2005-01-2280; 2005.

Kim H.S., Yoon S.H. A Design Process using Body Panel Beads for Structure-Borne Noise. SAE Paper 2007-01-1540; 2007.

Knechten T., Coster C., Linden P.V. Improved High Frequency Isolation and Sound Transfer Measurements on Vehicle Bodies. SAE Paper 2014-01-2077; 2014.

Koners G. Panel Noise Contribution Analysis: An Experimental Method for Determining the Noise Contributions of Panels to an Interior Noise. SAE Paper 2003-01-1410; 2003.

Kulkarni V., Tiwari A. CAE based Study of Vehicle Floor Beading Patterns for Low Frequency Noise and Vibration Reduction. SAE Paper 2011-26-0021; 2011.

Kung S.W., Singh R. Determination of Viscoelastic Core Material Properties Using Sandwich Beam Theory and Modal Experiments. SAE Paper 1999-01-1677; 1999.

Lee S., Park K., Sung S.H., Nefske D.J. Boundary Condition Effect on the Correlation of an Acoustic Finite Element Passenger Compartment Model. SAE Paper 2011-01-0506; 2011.

Leppington, F.G., Broadbent, E.G., and Heron, K.H. (1982). The Acoustic Radiation Efficiency of Rectangular Panels. *Proceedings of the Royal Society of London* 382 (1783): 245–271.

Lewis T., Jackson P., Nwankwo O. Design and Implementation of a Damping Material Measurement/Design System. SAE Paper 1999-01-1675; 1999

Li F., Sibal S.D., McGann I.F., Hallez R. Radiated Fuel Tank Slosh Noise Simulation. SAE Paper 2011-01-0495; 2011.

Li W.L. Modeling the Vibrations of and Energy Distributions in Car Body Structures. SAE Paper 2011-01-1573; 2011.

Lilley K.M., Fasse M.J., Weber P.E. A Comparison of NVH Treatments for Vehicle Floorpan Applications. SAE 2001-01-1464; 2001.

Lin J.Z., Lanka S., Ruden T. Physical and Virtual Prototyping of Magnesium Instrument Panel Structures. SAE Paper 2005-01-0726, 2005.

Liu J., Zhou L., Herrin D.W. Demonstration of Vibro-Acoustic Reciprocity including Scale Modeling. SAE Paper 2011-01-1721; 2011.

LMS. Airborne Sources Quantification. LMS International. 2007.

Man P.D., Schaftingen J.J.V. Prediction of Vehicle Fuel Tank Slosh Noise from Component-Level Test Data. SAE Paper 2012-01-0215; 2012.

Mazzarella L., Godano P., Horak J. Reciprocal Powertrain Structure-borne Transfer Functions Synthesis for Vehicle Benchmarking. SAE Paper 2007-01-2354; 2007.

Mead, D.J. and Yaman, Y. (1991a). The Response of Infinite Periodic Beams of Point Harmonic Forces: A Flexural Wave Analysis. *Journal of Sound and Vibration* 144 (3): 507–530.

Mead, D.J. and Yaman, Y. (1991b). The Harmonic Response of Rectangular Sandwich Plates with Multiple Stiffening: A Flexural Wave Analysis. *Journal of Sound and Vibration* 145 (3): 409–428.

Min K.J., Kang C.W., Seo K.H., Youn H.J. The Experimental Study on the Body Panel Shape to Minimize the Weight of the Damping Material. SAE Paper 2003-01-1715; 2003.

Oblizajek K.L., Sopoci J.D. Small Amplitude Torsional Steering Column Dynamics on Smooth Roads: In-Vehicle Effects and Internal Sources. SAE Paper 2011-01-0560; 2011.

Onsay T., Akanda A., Goetchius G. Vibro-Acoustic Behavior of Bead-Stiffened Flat Panels: FEA, SEA, and Experimental Analysis. SAE Paper 1999-01-1698; 1999.

Pang, J., Mo, C., Dukkipati, R., and Sheng, G. (2003). Automotive seat cushion nonlinear phenomenon: experimental and theoretical evaluation. *International Journal of Vehicle Autonomous Systems* 1 (3/4): 421–435.

Pang, J., Qatu, M., Dukkipati, R. et al. (2004). Model identification for nonlinear automotive seat cushion structure. *International Journal of Vehicle Noise and Vibration* 1 (1/2): 142–157.

Pang, J., Qatu, M., Dukkipati, R., and Sheng, G. (2005). Nonlinear seat cushion and human body model. *International Journal of Vehicle Noise and Vibration* 1 (3/4): 194–206.

Park H.C., Yoon S.H. Contribution Analysis of Vehicle Interior Noise Using Air-borne Noise Transfer Function. SAE Paper 2007-01-2359; 2007.

Pelinescu I., Christie A. Measuring Damping Loss Factors of High Performance LASD Coatings. SAE Paper 2011-01-1632; 2011.

Rocha T., Calçada M., Ribeiro Y. The Use of Piezoelectric Resonators to Enhance Sound Insulation in a Vehicle Panel. SAE Paper 2012-36-0613; 2012.

Pol A.D., Naganoor P. Torsion Mode Achievement on BIW of Next Generation Land Rover – Freelander. SAE Paper 2014-01-0005; 2014.

Prasanth B., Wagh S., Raghuvanshi J. Body Induced Boom Noise Control by Hybrid Integrated Approach for a Passenger Car. SAE Paper 2013-01-1920; 2013.

Rejlek J., Priebsch H.H. On the Use of the Wave Based Technique for a Three-Dimensional Noise Radiation Analysis of Coupled Vibro-Acoustic Problems. SAE Paper 2011-01-1713; 2011.

Ross D., Kerwin E., Ungar, E. *Damping of Plate Flexural Vibration by Means of Viscoelastic Laminae, in Structural Damping*. American Society of Mechanical Engineers; 1959.

SAE International Surface Vehicle Recommended Practice. Laboratory Measurement of the Composite Vibration Damping Properties of Materials on a Supporting Steel Bar. SAE standard J1637, Rev. 2008.

Saha P., Deshpande S.P., Fisk J., Owen D.E. Damping Performance Using a Panel Structure. SAE Paper 2013-01-1938; 2013.

Saha P., Hussaini A.S. A Graduated Assessment of a Sprayable Waterborne Damping Material as a Viable Acoustical Treatment. SAE Paper 2003-01-1588; 2003.

Sakthivel A., Sriraman S., Verma R.B. Study of Vibration from Steering Wheel of an Agricultural Tractor. SAE Paper 2012-01-1908; 2012

Sanderson M.A., Onsay R.T. CAE Interior Cavity Model Validation using Acoustic Modal Analysis. SAE Paper 2007-01-2167; 2007.

Shin S.Y., Lee S.D., Go B.C. Study of Vehicle Seat Vibration Characteristics through Sensitivity Analysis. SAE Paper 2014-01-0032; 2014.

Shorter P.J., Gardner B.K., Bremner P.G. A Review of Mid-Frequency Methods for Automotive Structure-Borne Noise. SAE Paper 2003-01-1442; 2003.

Silva, C.W. (2007). *Vibration Damping, Control and Design*: CRC Press.

Silva G.A.L., Nicoletti R. Sheet Metal Bending Pattern Optimization For Desired Natural Frequencies. SAE Paper 2012-36-0630; 2012.

Sophiea D., Xiao H. A New Light Weight, High Performance, Spray Applied Automotive Damping Material. SAE Paper 1999-01-1674; 1999.

Sottek R, Sellerbeck P., Klemenz M. An Artificial Head Which Speaks from Its Ears: Investigations on Reciprocal Transfer Path Analysis in Vehicles Using a Binaural Sound Source. SAE Paper 2003-01-1635; 2003.

Stotera D, Connelly T., Gardner B., Seifferlein E., Alvarez R.A. Testing and Simulation of Anti-Flutter Foam and High Damping Foam in a Vehicle Roof Structure. SAE Paper 2013-01-1944; 2013.

Sung S., Chao S., Lingala H., Mundy L. Structural-Acoustic Analysis of Vehicle Body Panel Participation to Interior Acoustic Boom Noise. SAE Paper 2011-01-0496; 2011.

Tathavadekar P., Onsay T., Liu W. Damping Performance Measurement of Non-uniform Damping Treatments. SAE Paper 2007-01-2199; 2007.

Terashi S., Asal M., Nalto J. Damping Aanlysis of Body Panels for Vehicle Interior Noise Reduction. SAE Paper 891135; 1989.

Thomas A.J., Gosain A., Balachandran P. Vehicular Cabin Noise Source Identification and Optimization Using Beamforming and Acoustical Holography. SAE Paper 2014-01-0004; 2004.

Thomas, W.T. and Dahleh, M.D. (1993). *Theory of Vibration with Application. Prentice Hall.*

Tsuji H., Enomoto T., Maruyama S., Yoshimura T. A Study of Experimental Acoustic Modal Analysis of Automotive Interior Acoustic Field Coupled with the Body Structure. SAE Paper 2012-01-1187; 2012.

Tsuji H, Maruyama S, Yoshimura T., Takahashi E. Experimental Method Extracting Dominant Acoustic Mode Shapes for Automotive Interior Acoustic Field Coupled with the Body Structure. SAE Paper 2013-01-1905; 2013.

Tudor J. Determination of Dynamic Properties and Modeling of Extensional Damping Materials. SAE Paper 2003-01-1433; 2003.

Ver, I.L. and Beranek, L.L. (2006). *Noise and Vibration Control Engineering: Principles and Applications.* John Wiley & Sons.

Vlahopoulos N., Wang A. Modeling of Stiffened Panels Using the Energy Finite Element Analysis. SAE Paper 2011-01-1696; 2011.

Watanabe T., Yoshimura T. Identification of Vibro-Acoustic Coupled Modes for Vehicle. SAE Technical Paper 2014-01-0031; 2014.

Wolff O. Fast panel noise contribution analysis using large PU sensor arrays. Inter-Noise 2007. 28–31 August, Istanbul, Turkey; 2007.

Wolff O., Sottek R., Panel Contribution Analysis – An Alternative Window Method. SAE Paper 2005-01-2274; 2005.

Yabe K., Inagaki T., Kondo T. Adoption of Floating Seat in a Vehicle to Reduce Seat Vibration. SAE Paper 2015-01-1122; 2015.

Yan L., Jiang W., Huang Z. A Research on the Sound Quality Contribution of Vehicle Body Panel. SAE Paper 2014-01-0896; 2014.

Yang B., Nunez S.W., Welch T.E., Schwaegler J.R. Laminate Dash Ford Taurus Noise and Vibration Performance. SAE Paper 2001-01-1535; 2001.

4

Sound Package

4.1 Introduction

4.1.1 Transfer of Airborne-Noise to Passenger Compartment

Structural-borne noise refers to the sound inside the passenger compartment radiated by body panels that are excited by external vibration sources. Structural-borne noise can be controlled by adjusting the panels' stiffness, by adding mass, and by damping, which are described in detail in Chapter 3.

Airborne noise refers to the sound transferred into the passenger compartment directly from outside sound sources. There are two paths for airborne noise transmission. Figure 4.1 shows the power plant noise transferring into the vehicle interior. One path involves the noise passing through holes and openings on the body into the interior. In the other path, the noise penetrates the panels and passes into the interior.

Figure 4.2 shows the frequency distribution of a typical interior noise. Structural-borne noise dominates the low-frequency range, whereas airborne noise dominates the middle- and high-frequency ranges. Human ears are very sensitive to middle- and high-frequency sound, and can easily perceive even low levels of sound in these ranges.

Outside noise can penetrate through small holes or openings on the body and be heard by the passengers inside. For example, when a vehicle cruises at high speeds, passengers may be able to perceive the outside wind noise. They may complain that the vehicle has a "wind leaking" problem, and may even doubt the vehicle's reliability. Therefore, good sealing is essential for high-quality sound performance, and is also the foundation of noise, vibration, and harshness (NVH) control.

In some cases in which the body is perfectly sealed, the passengers can still hear noise transmitting from the outside, which indicates that the sound insulation of the body is not effective. The direct result of poor sound insulation is high levels of interior noise. Reduction of interior noise is the first step of vehicle NVH control, and also a base for improving the vehicle's sound quality. Therefore, sound insulation is very important in body design.

The control of the airborne noise should follow two principles. The first is to prevent outside noise from being transmitted directly into the interior: i.e. holes and leakages should be avoided on the body. The second principle is to reduce the level of outside noise that penetrates through the body: i.e. the body should have good sound insulation, which can be realized with metal panels, sound absorptive structures, and sound insulation structures.

Noise and Vibration Control in Automotive Bodies, First Edition. Jian Pang.
© 2019 China Machine Press. All rights reserved. Published 2019 by John Wiley & Sons Ltd.

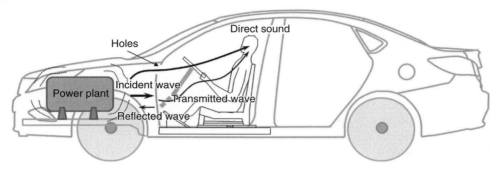

Figure 4.1 Two transfer paths of airborne noise into vehicle interior.

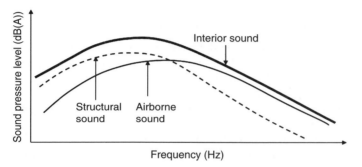

Figure 4.2 Frequency distribution of a typical interior noise.

A thick steel panel is a good sound insulation structure. However, body panels are generally quite thin, usually between 0.8 and 1.2 mm. Therefore, additional sound insulation structures must be added onto either the inside or outside of the body.

Sound absorption is of equal importance to sound insulation. Sound absorption is a common method for reducing interior noise, and it works by transferring sound energy into heat energy, which is dissipated.

4.1.2 Scopes of Sound Package Research

Traditionally, the technologies and non-metal materials and structures that are used to control body vibration and sound are called the sound package. The materials and structures in effect "wrap up" the body, hence the use of the word package. The sound package can be divided into the special sound package and general sound package. Usually, the scope of the special sound package includes body sealing, sound absorptive materials and structures, and sound insulation materials and structures. The general sound package refers to the special sound package in addition to the damping materials and structures, and the reinforcement materials and structures.

Body sealing is the foundation of the sound package: that is, the sealing of holes on the body in order to prevent air leakage. If the body has serious air leakage, there is little point in implementing sound insulation and absorption treatments. Body sealing is divided into static sealing and dynamic sealing, and the scope of sealing research includes sealing implementation, sealing measurement, sealing design, and cavity baffling.

The principle of sound absorption is that no reflective wave is produced when a sound strikes a material or a structure. The sound sources can be reduced if sound absorption materials are installed on the outside of the body, while the interior noise can be reduced and the sound quality can be improved if they are installed inside the vehicle. Sound absorption research includes the definition and measurement of sound absorption coefficients, factors influencing the coefficients, and sound absorption structures and their application.

The principle of sound insulation is that when a sound wave strikes a structure, a fraction of the sound energy is reflected and the reminder penetrates the structure: i.e. part of the sound wave is insulated by the structure. Sound insulation research includes the definition and measurement of sound insulation coefficients, factors influencing the coefficients, and sound insulation structures and their application.

Damping materials are usually non-metal materials that are attached to the surfaces of metal panels or plates in order to suppress their vibration and reduce sound radiation. Damping materials and structures are usually used to attenuate structural-borne noise at low- and middle-frequency ranges.

A reinforcement sheet or material is made of non-metal materials and is attached to the surface of a metal panel in order to increase its stiffness. Reinforcement sheets are usually used to increase the body's local stiffness.

The mechanisms and applications of damping materials and reinforcement materials are introduced in detail in Chapter 3. Generally speaking, they come under the scope of the general sound package. However, this chapter only covers the special sound package.

4.2 Body Sealing

4.2.1 Importance of Sealing

4.2.1.1 Holes and Apertures on the Body

There are many holes and apertures on a vehicle body, as shown in Figure 4.3. They are classified into three categories: function holes, manufacturing process holes, and error-state holes.

Function holes are those that are deliberately opened in body panels in order to realize some designed functions. For example, the steering column, wires, and air-conditioning pipe must pass through the dash panel, so corresponding holes on the panel must be opened, as shown in Figure 1.26.

Manufacturing process holes are opened for certain manufacturing processes and then sealed after the processes have been finished. For example, in the electrophoresis process, a body-in-white (BIW) is immersed in an electrophoresis fluid tank. After the process is finished, the liquid is drained out of the body through holes, which are later sealed. There are many holes belonging to this category, such as the holes on the vehicle floor shown in Figure 4.4.

(a)   (b)

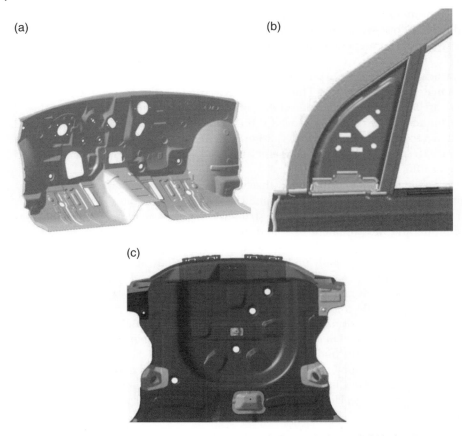

(c)

Figure 4.3 Holes and apertures on a vehicle body: (a) holes on dash panel; (b) hole on mirror supporter; (c) holes on spare tire floor.

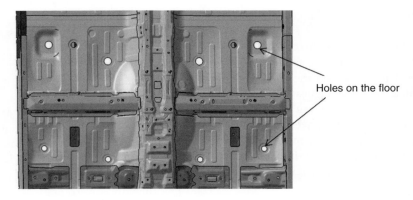

Holes on the floor

Figure 4.4 Manufacturing process holes on floor.

Figure 4.5 An error-state hole at a junction area of three panels.

Aperture

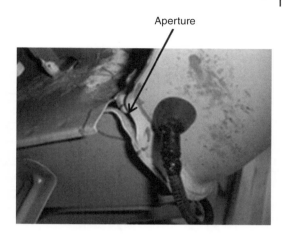

Error-state holes and apertures are the result of design error and/or manufacturing error. Figure 4.5 shows a junction area of three panels. The three panels cannot form a closed surface and leave an aperture. This sort of un-designed hole is sometimes called a "mouse hole," which means the hole is so big that a mouse could pass through it. There could be many error-state holes on a vehicle body if it is poorly designed, such as an aperture between a door handle and the door body due to poor attachment.

4.2.1.2 The Problems Generated by Holes

The holes on the body have a significant influence on the sound insulation. The bigger the holes, the worse the body sound insulation. Opening rate is defined as the ratio of the opening hole area to total body area, expressed as follows:

$$\gamma = \frac{A_{hole}}{A_{body}}, \tag{4.1}$$

where A_{body} is the total body area and A_{hole} is the opening hole area.

The bigger the opening rate, the easier it is for outside noise to be transmitted into the interior. When the opening rate is high enough, the outside and the interior are directly connected with each other.

The combined effectiveness of sound insulation and sound absorption can be expressed by sound transmission loss (STL), which is described later in this chapter. The higher the STL, the less sound is transmitted from the outside to the interior, and the better the effectiveness the sound insulation. If there are holes on the body, the STL will be reduced. The opening rate is a very important factor influencing the STL, and the corresponding STL can be expressed as follows:

$$TL_{hole} = -10\log\left[\gamma + 10^{-TL/10}\left(1 - \gamma\right)\right], \tag{4.2}$$

where TL is the STL without holes, and TL_{hole} is the STL with opening rate γ.

According to Eq. (4.2), the relationship between the STLs with different opening rates and without holes can be directly plotted, as shown in Figure 4.6. First, the influence of the opening rates on STL can be illustrated on the plot. For example, fix a point on the horizontal axis, such as 30 dB, then plot a vertical line that will be

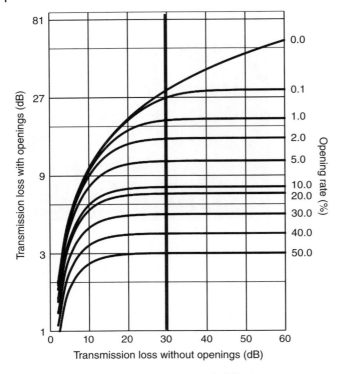

Figure 4.6 Relationship between STLs with different opening rates and without holes.

intersected with STL curves of different opening rates. The STLs at the opening rates 0, 0.1, 1, 2, 5, 10, and 20% are 30, 27, 20, 15, 11, 7.7, and 7 dB, respectively. The STLs drop rapidly with the increase in opening rates. Second, the relationship between STLs and the sealing can be analyzed. For example, when the opening rate is 1%, the STL drops from 30 dB when there are no holes to 20 dB. If the opening rate is only 0.1%, the corresponding STL is 27 dB. The plot shows that even if a vehicle body has very good sound insulation, the overall sound insulation effectiveness will be signifi-cantly reduced if there is air leakage.

When holes exist on the body, the STL is reduced more at high frequency than at low frequency, as shown in Figure 4.7. The straight line represents the STL without holes, which linearly increases with frequency increase. The dotted line represents the STL with holes, which indicates that the STL at high frequencies is significantly reduced compared with the STL without holes.

4.2.1.3 Necessity of Sealing

Sealing is the foundation of automotive noise control. If the openings on a body are large enough, the interior sound will be out of control, even if the outside noise sources are very well attenuated. For example, the sound pressure level in an engine compart-ment is only 55 dB (A) in the idling condition (a relatively quiet noise level), but due to poor sealing on the dash panel, the interior noise is 48 dB (A), which represents a noisy

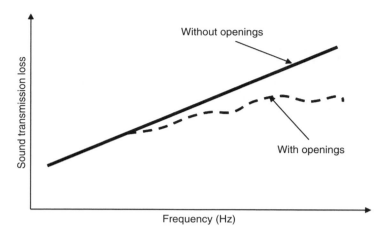

Figure 4.7 Influence of opening holes on STL.

vehicle. After the holes on the dash panel have been well sealed, the interior noise is reduced to 36 dB (A), which represents a very quiet vehicle.

Holes in the body not only influence the STL, but also reduce the effectiveness of air-conditioning and let outside dust and pollution into the interior. Only if the opening rate is well controlled and the body has good sealing can the sound package and vehicle noise control be implemented.

4.2.2 Static Sealing and Dynamic Sealing

Body sealing is divided into two categories: static sealing and dynamic sealing. Static sealing refers to the body sealing when the vehicle is at a standstill, whereas dynamic sealing refers to sealing when vehicle is running. When the vehicle is moving, the body, doors, and other components are excited by the engine and road, and their relative deformations are different, so apertures could appear at some locations where there are no holes in the standstill condition. The outside noise, especially wind noise, will pass through the apertures and radiate into the interior. Dynamic sealing is introduced in Chapter 6, and only static sealing is described in this chapter.

The tasks of static sealing are to seal the above-mentioned three categories of holes and to reduce the opening rate in the standstill condition. The first task is to seal the function holes. For example, the steering column hole on a dash panel must be well sealed. The second task is to seal the manufacturing process holes. For example, the holes on the floor used for the electrophoresis process must be sealed by plugs. The third task is to seal error-state apertures resulting from design and manufacture errors with sealing adhesives.

4.2.3 Measurement of Static Sealing

The purpose of static sealing measurement is to identify the locations of the holes and to describe the size of the leakage area. This can be done with the smog method, ultra-sound method, or air leakage method.

4.2.3.1 Smog Method

A smog generator is placed inside a vehicle and releases smog. Some of the smog will flow to the outside if there are holes in the body. People standing outside the vehicle observe the smog flowing from the interior to identify leakage locations.

Usually, the smog flow rate is slow due to the small pressure difference between the inside and outside of the body. To speed up the smog flow rate, a blower can be used together with the smog generator. Sometimes, colored smog is used so it can be clearly observed.

The smog method is a traditional one, and it is simple and easy to perform. However, the method cannot provide the leakage quantity of the holes, and it is harmful to the environment as well, so it is seldom used today.

4.2.3.2 Ultrasound Method

Ultrasound is a sound wave with a frequency higher than the upper limit of the human hearing range. One characteristic of ultrasound is its directivity, so it is easily prevented or masked.

An ultrasonic leak detector system includes a generator and a receiver. The generator and receiver are placed inside and outside of a vehicle body, respectively. A detector installed on the receiver is used to identify leakage locations, as shown in Figure 4.8. The generator and the receiver can be also placed in reverse: i.e. the generator is placed outside and the receiver is located inside. As a result of the characteristics of the directivity and masking of ultrasound, the leakage locations and quantities can be easily read on the receiver screen. In some complicated testing environments, a plastic sound-focusing detector can be used to narrow the identification area and eliminate noise disturbance.

The leakage quantity can be digitally displayed on the receiver screen, so the ultrasound method can be used to quantitatively determine the leaking area.

The leakage quantity can not only be read on the receiver screen, but also heard. People cannot hear ultrasound, but the ultrasound can be transferred into a sound-frequency signal by a differential method, bringing it into the human hearing range.

Because it is simple and independent of disturbance, the ultrasound method is widely used in engineering.

4.2.3.3 Air Leakage Method

In the air leakage method, the principle of flow pressure difference is used to measure the body leakage.

A blower and the tested body are connected by a flexible pipe, as shown in Figure 4.9, and the blower blows air into

Figure 4.8 Receiver of an ultrasonic leak detector.

Figure 4.9 A blower and a tested body are connected by a flexible pipe.

the body. When it stops blowing, air will flow out of the body if there are holes and crevices in it, and the pressure inside the body will drop. In this case, in order to keep the interior pressure constant, the blower has to continuously blow air into the body to keep the inflowing and outflowing air balanced. After the pressures of the throttle and the interior pressure have been measured, the relationship between the airflow amount and the leak area can be calculated, and expressed as follows:

$$Q = a_D A \sqrt{\frac{2(P_i - P_0)}{\rho_0}},$$ (4.3)

where Q is the volume flow amount (m^3/h); a_D is the flow coefficient; A is leak area (cm^2); P_i, P_0 are the pressures inside and outside the body, respectively (Pa); and ρ_0 is the air density (kg/m^3).

The flow rate and pressure are easily measured. According to Eq. (4.3), the leakage area that represents the body leakage size can be calculated. The larger the area, the more the air leakage.

Figure 4.10 shows a leakage test device that consists of a blower, a flexible pipe, a speed meter, a throttle, a pressure recorder, and so on. The flow rate, pressures, and so on can be displayed on the device.

This testing method can accurately provide the amount of body leakage. If the leaking areas are sealed individually, the amount of each leak can be tested as well. Table 4.1 lists a body overall leakage area, and the leakage area of each leaking portion.

The table shows that the body overall leakage area is $133\,cm^2$, and eight leaking portions are identified, such as a chassis draining hole, door draining hole, and so on.

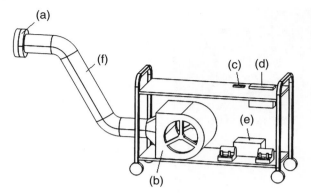

(a)

(f)

(c) (d)

(e)

(b)

Figure 4.10 A leakage test device: (a) throttle, (b) blower, (c) speed meter, (d) controller, (e) pressure recorder, (f) flexible pipe.

Table 4.1 Overall body leakage and leakage area of each leaking portion.

Leakage area (cm^2)	Leakage contribution ($\triangle$/cm^2)	Note
133	—	Overall body
131	2	Chassis draining hole is sealed
113	18	Door check and door frame are sealed
109	4	Door handles are sealed
97	12	Door draining holes are sealed
92	5	Glass runs and belt seals are sealed
85.5	6.5	Door seals are sealed
62.5	23	Passing-through hole for air conditioning pipe is sealed
60.5	2	Passing-through hole for brake pedal bracket is sealed

The holes for the air-conditioning pipe and door check are the greatest contributors to the overall leakage, where the leakage areas are 23 and 18 cm^2, respectively.

This method can quantitatively provide the leakage areas, and is simple and accurate, so it is widely used in engineering.

4.2.4   Control of Static Sealing

Static sealing control follows three principles: the design principle, digital mock up (DMU) checking, and test verification. The design principle means determining the sealing locations, sizes, and sealing methods for the function holes and manufacturing process holes. DMU checking involves examining whether the sealing design on a 3D digital vehicle satisfies the design requirements. Test verification involves testing a vehicle or prototype to determine the leakage locations and quantities.

4.2.4.1 Sealing for Three Categories of Holes

Before implementing body sound insulation and sound absorption, the holes and apertures must be perfectly sealed. The principles and methods for sealing the three categories of holes and apertures – function holes, manufacturing process holes, and error-state holes – are introduced below.

Function holes are divided into two categories. One category is holes where the components passing through, such as wires, do not induce the body to vibrate. The other category is holes where the components passing through cause the body to vibrate: for example, the shifter cable, which transmits the transmission's vibration to the body.

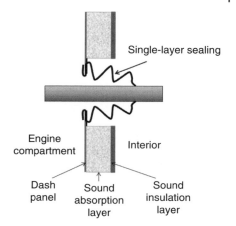

Figure 4.11 A pipe passing through a dash panel and single-layer sealing.

To seal the function holes, three steps should be followed. The first step is air sealing: i.e. preventing air from flowing through the holes. The second step is sound insulation: i.e. the transmitted sound should be as low as possible. The third step is to control vibration transmitted to the body by the components that pass through the hole: i.e. the components must be well isolated from the body.

Figure 4.11 shows a pipe passing through a dash panel. A rubber sealing ring is inserted between the pipe and the panel, so air cannot flow down then side of the pipe. However, sound can easily pass through the rubber ring due to its poor sound insulation effectiveness.

Two methods can be used to improve the sound insulation effectiveness. One is to increase the ring's thickness, as shown in Figure 4.12a, and the other is to use two rings, as shown in Figure 4.12b.

Many components pass through a vehicle body. Taking a dash panel as an example, components that pass through it include the selection cable, control shifter cable, air-conditioning pipe, cooling pipe, brake fluid tube, steering column, and so on. These components must be well isolated, otherwise the body will be excited. For example, poor isolation between the control shifter and the dash panel will induce the cable to transfer vibration from the transmission to the body, which will radiate sound to the interior.

Figure 4.13 shows a vibration comparison for two dash panels where the hardness of the isolation rubbers between the compressor pipe and the panel is different. An interior booming at 192 Hz is identified. A 192 Hz peak on the panel vibration data is found for the original rubber with a hardness of 65 HA and a small volume. The vibration on the compressor has a peak at 192 Hz, which is transmitted to the panel through the pipe. After the rubber has been replaced by a new one with a hardness of 55 HA and a bigger volume, the vibration peak on the panel is significantly reduced and the interior booming disappears.

Manufacturing process holes have unique functions during manufacture and must be sealed after their function is complete. These holes should be designed to be as small as

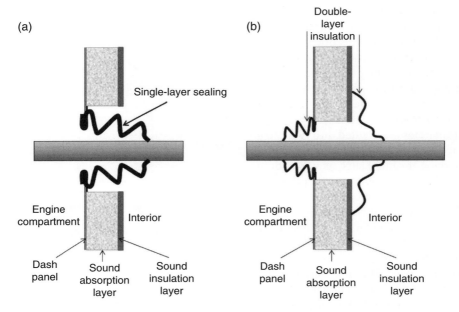

Figure 4.12 Improved sound insulation: (a) single ring with increased thickness; (b) two rings form double-layer insulation.

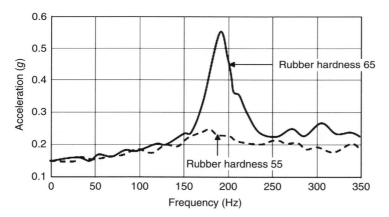

Figure 4.13 Vibration comparison on a dash panel for two cases in which the hardness of the isolation rubbers between the compressor pipe and panel is different.

possible, and only big enough to meet the manufacturing requirements. There are three ways to seal these holes: with patches, plugs, and thermoforming material, shown in Figure 4.14.

Generally, the manufacturing process holes should be smaller than 10 mm. For small holes, sealing with patches usually meets the sound insulation requirements. For holes

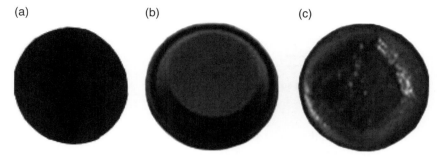

Figure 4.14 Sealing methods for manufacturing process holes: (a) patch, (b) plug, (c) thermoforming material.

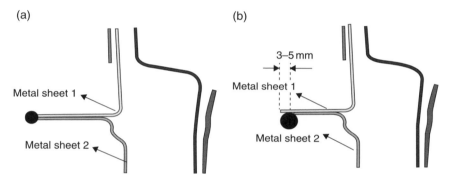

Figure 4.15 Sealing between two metal sheets: (a) sheets of the same length; (b) sheets of different lengths.

larger than 10 mm, the patches are too thin to meet the sound insulation requirements, so a plug or thermoforming material must be used. Among the three sealing methods, thermoforming material has the best sound insulation performance due to its high density, but plugs have higher sound insulation effectiveness than patches.

Error-state holes are usually caused by design error or manufacturing error at the intersection of several metal sheets. When two metal sheets overlap, gaps can emerge between the sheets, which must be sealed. Sealant is usually placed on the interface of the two sheets in order to seal the gaps.

The end of the overlap between the two metal sheets should have enough space for holding the sealant. Figure 4.15a shows that sheet 1 and sheet 2 are placed together and their lengths are the same, so there is no space for placing the sealant. One sheet is shortened to make a difference between the two sheet ends, as shown in Figure 4.15b, so a space is created for placing the sealant. Usually, the difference in length is between 3 and 5 mm in order to provide enough space for the sealant.

In some overlap areas where it is very hard to place the sealant, a special groove can be designed, as shown in Figure 4.16. Before the sheets are put together, the sealant is

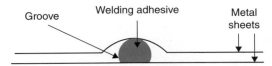

Figure 4.16 A groove for holding welding adhesive.

placed inside the groove. When the sheets are welded, they will simultaneously be sealed by the sealant. The sealant used in the groove is a special welding adhesive. Welding adhesive is divided two categories: expansive adhesive and non-expansive adhesive. Expansive adhesive will expand after being heated in a high-temperature oven, which will achieve an excellent sealing result. If there are big gaps between the sheets, expansive adhesive will achieve a better sealing result than non-expansive adhesive.

4.2.4.2 Sealing Design and Digital Mock Up Checking

In the design phase and prototype phase of vehicle development, hole sealing is one of most important aspects of body NVH control. During the design phase, the sealing requirements must be provided to the design team as an important body design guideline. The requirements include air sealing, sound insulation, and vibration isolation. If these requirements are considered in the body design, the body will achieve good NVH performance, the development costs will be reduced, and the development time will be shortened.

DMU checking follows the initial design. The purpose of DMU checking is to find potential NVH problems in a digital vehicle caused by poor sealing. DMU checking is processed according to the above-mentioned three categories of sealing.

1) Function holes. The checked items include the holes' size, location, and isolation between passing-through components and a vehicle body. Taking the dash panel as an example, the checked holes include those where components such as the steering column, wires, selection cable, shifter cable, water pipe, and so on will pass through the panel. Figure 4.17 shows a digital structure of the shifter cable and selection cable passing through the dash panel. In order to thoroughly check the digital design of the cables passing through the dash panel, three steps should be followed. First, the hole size should be checked to make sure that the size is reasonable. Second, the isolation between the cables and the panel should be checked. If there is an isolation, the volume and hardness of the isolator should be checked. Third, the angles between the cables and the panel should be checked. It is best for the cables to perpendicularly pass through the panel.

Figure 4.18 shows a DMU diagram of the draining holes on a door. The holes' function is to let water inside the door smoothly flow out. These holes directly link the inside and outside of the vehicle. If the holes are big, the water can flow out smoothly, but simultaneously the outside noise can easily be transmitted into the interior, and vice versa if the holes are small. Thus, the holes should be carefully designed to satisfy the requirements for both draining and sound insulation.

Figure 4.17 A digital structure of shifter cable and selection cable passing through a dash panel.

Figure 4.18 A DMU diagram of draining holes on a door.

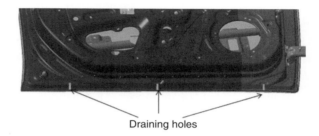

Draining holes

2) Manufacturing process holes. There are two purposes of DMU checking of these holes. The first is to confirm their process functions, and the second is to confirm that the holes are sealed after the required manufacturing processes have been completed.

3) Error-state holes. DMU checking should ascertain whether error-state holes and apertures exist, and if so, ways to seal them should be implemented. In the design phase, such holes and apertures could appear in two situations: when there is an aperture in the overlap area between two metal sheets, and when there is a large mouse hole at the overlap area of several metal sheets. Figure 4.19 shows a DMU diagram of an intersection of three sheets on the wheelhouse. A mouse hole has been found, so the design must be modified or the holes' sizes must be reduced so that the hole can be sealed with sealant.

4.2.4.3 Experimental Control of Sealing

Based on the design requirements and DMU checking, a prototype vehicle is designed and manufactured. The prototype must be tested to check its sealing status. First, the BIW is tested, then the trimmed body or the prototype vehicle is tested.

The purpose of the BIW sealing test is to identify leakage areas due to manufacture error, such as apertures between metal sheets, mouse holes, missing sealant, and so on.

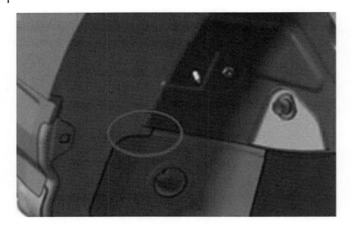

Figure 4.19 A DMU diagram of an intersection of three sheets on a wheelhouse.

First, the windows, function holes, and manufacturing process holes are sealed with plastic covers and adhesive tape. Then, a leakage testing machine is used to test the body, and the leakage areas and amounts are identified. Usually, the total leakage amount of a BIW should be less than $15\,\mathrm{cm}^2$.

The purpose of prototype vehicle leakage checking is to examine the leakage of the body seals and door seals, function holes, manufacturing process holes, and so on. For luxury vehicles, the amount should be less than $30\,\mathrm{cm}^2$. For economy vehicles, the amount should be less than $50\,\mathrm{cm}^2$.

4.3 Sound Absorptive Materials

4.3.1 Sound Absorption Mechanism and Sound Absorption Coefficient

Sound absorption refers to the process of sound energy attenuation when sound waves propagate in a medium or transmit into another medium.

The mechanism of sound absorption is that part of the sound energy is changed into heat energy. During sound transmission processing, the particle velocity at each location is different, which forms a velocity gradient. The velocity gradient induces internal friction forces and adhesive forces among the particles, which prevent the particles' movement and cause the sound energy to continuously transform into heat energy. However, the density and temperature for each particle are different. The density gradient and temperature gradient force the particles to exchange heat, which causes part of the sound energy to transform into heat energy.

Sound absorptive material absorbs part of the sound energy because the sound waves are dampened by friction among the vibrating fibers of the material and by viscous effects due to air vibrating in the pores of the material. Figure 4.20 shows an incident wave with energy E_i travelling in the air and striking a material; part of the energy, E_a, is absorbed by the material, and the rest of the energy, E_n is reflected back to the air. The sound absorption capacity of a material or structure can be expressed by its sound absorption coefficient, which is defined as the ratio of the absorbed energy to the incident energy, and expressed as follows:

$$\alpha = \frac{E_a}{E_i} = 1 - \frac{E_r}{E_i}. \qquad (4.4)$$

When $E_r = E_i$, i.e. $\alpha = 0$, all the incident waves are reflected. When $E_a = E_i$, i.e. $\alpha = 1$, all the incident waves are absorbed. The sound absorption capability for most sound absorptive materials is between the two cases, i.e. the sound absorption coefficient is between 0 and 1. The higher the coefficient, the stronger the material's sound absorption capability. Usually, a material with a coefficient higher than 0.2 is called a sound absorptive material.

The sound absorption coefficient depends not only on the object (a material or a structure), but also on the angle between the incident wave's direction and the object's surface. Sound absorptive materials and structures are divided into two categories: porous sound absorptive materials and resonant sound absorptive structures.

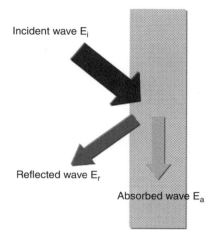

Figure 4.20 Incident wave, absorbed wave, and reflected wave.

4.3.2 Porous Sound Absorptive Material

4.3.2.1 Characteristics of Porous Sound Absorptive Material
A porous sound absorptive material is a material containing many interconnecting pores, as shown in Figure 4.21. The internal structure should have enough pores to effectively absorb sound energy. Porous sound absorptive materials are categorized as fiber materials, foam materials, and particle materials according to the different shapes of their pores.

Figure 4.22 shows a typical curve of sound absorption coefficient for porous materials that varies with frequency. At a low frequency range, the coefficient is low. Usually, the materials have a sound absorption function above 250 Hz. The coefficient increases with the increase in frequency. The coefficient reaches a maximum value at a certain frequency, then fluctuates with frequency increase.

4.3.2.2 Factors Influencing Sound Absorption Coefficient
Factors influencing the sound absorption coefficient include a material's structural parameters, geometrical parameters, and environmental factors. The structural parameters

Figure 4.21 Structure of porous sound absorptive material.

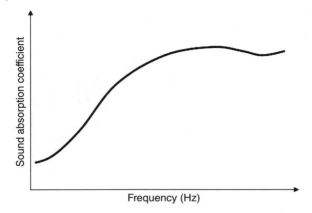

Figure 4.22 Typical curve of sound absorption coefficient for porous materials.

include flow resistance, porosity, and structural factors. The geometrical parameters include the thickness and density of a structure. The environmental factors are temperature and humidity.

4.3.2.2.1 Influence of Structural Parameters
Flow resistance refers to the resistant force that air particles are subjected to when they move through the space inside a material. When air with volume velocity U flows through a material with surface area A, there will be a pressure difference (Δp) between the front end and back end of the material. The resistance phenomenon can be described by flow resistivity (R_f), and is expressed as follows:

$$R_f = \frac{\Delta p A}{U} \tag{4.5}$$

Or, it can be written as another way,

$$R_f = \frac{\Delta p}{u}, \tag{4.6}$$

where u is a linear velocity perpendicular to the material surface, i.e. $u = \dfrac{U}{A}$.

The flow resistance represents the material's permeability. The higher the flow resistance, the worse the material's permeability. Figure 4.23 shows sound absorption coefficients of three materials with high, middle, and low flow resistance, respectively. For a material with low flow resistance, the consumed energy is low due to the material's low internal friction and adhesive forces. Therefore, the corresponding sound absorption coefficient is low at a low frequency range; however, it increases rapidly after a certain frequency. For a material with high flow resistance, the air flow is prevented from penetrating the material, which results in poor sound absorption. Therefore, its sound absorption coefficient is low across the entire frequency range. To achieve good sound absorption performance, the material's flow resistance should fall within a

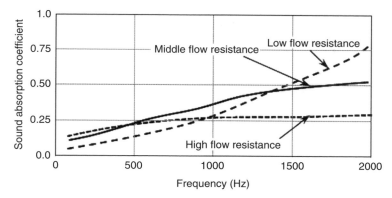

Figure 4.23 Relationship between sound absorption coefficient and flow resistance.

reasonable range. The flow resistivity is usually between 100 and $1000\,\mathrm{Pa \cdot s\,m^{-1}}$. The flow resistivity can be adjusted by tuning the material's density.

Porosity (B) is defined as a ratio of the air volume inside a structure to the total volume of the structure, and is expressed as follows:

$$B = \frac{V_a}{V_m} = 1 - \frac{m_s}{V_s \rho_f}, \tag{4.7}$$

where V_a and V_m are the air volume and total structure volume, respectively; and m_s, V_s, and ρ_f are the mass, volume, and density of the material, respectively.

The porosity of a porous material is usually above 70%. The porosity of slag wool and glass wool are 80 and 95%, respectively. Generally, the higher the porosity, the finer the pore space and the better the sound absorption performance, and vice versa.

The internal micro-structures inside a porous material that influence its sound absorption coefficient are called structural factors. The micro-gaps and arrangement of a real material are very complicated. In order to make theoretical analysis and testing results consistent, a corrected coefficient called a structural factor is introduced. The structural factor for most porous materials is between 2 and 10, and some can be as high as 25. The structural factor has no impact on low frequency sound absorption, but for low flow resistance materials, increasing the structural factor causes the sound absorption coefficients to vary in the middle- and high-frequency ranges.

4.3.2.2.2 Influence of Physical Parameters

Usually, researchers and engineers who work on material research and development are concerned with the influence of structural parameters on sound absorption coefficients. However, NVH engineers are more interested in the influence of the physical or geometrical parameters on sound absorption performance.

Thickness has a significant impact on sound absorption coefficients. Figure 4.24 shows sound absorption coefficients for materials with different thicknesses. The sound absorption coefficients increase with the increase in thickness, especially in the low- and middle-frequency ranges. However, the increase in the coefficients begins to reduce

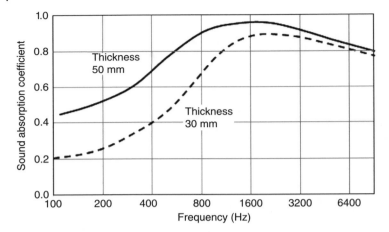

Figure 4.24 Sound absorption coefficients for the same material with different thicknesses.

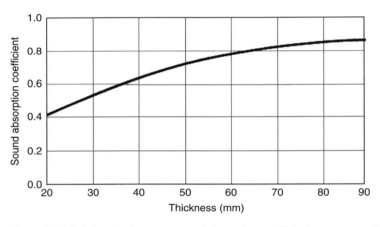

Figure 4.25 Relationship between sound absorption coefficient and increase in thickness.

when the thickness reaches a certain value. Figure 4.25 shows a relationship between thickness and the sound absorption coefficient. When the thickness increases from 20 to 40 mm, i.e. the thickness doubled, the coefficient increases from 0.41 to 0.62, i.e. an increase of 0.21. When the thickness continues to increase by another 20 mm to reach 60 mm, the coefficient increases from 0.62 to 0.78, i.e. an increase of 0.16. The thickness continues to increase to 80 mm, i.e. another 20 mm is added, and the coefficient increases to 0.83, i.e. the increase is only 0.05.

Usually, the thickness of sound package materials used in a vehicle body is less than 30 mm. Within this range, increasing thickness benefits the sound absorption coefficient. In body interior structure design, space should be provided for installing sound package materials.

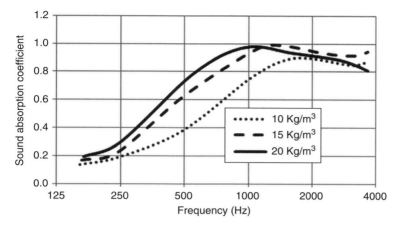

Figure 4.26 Relationship between sound absorption coefficient and the density of glass wool with the same thickness.

The volume density of a structure is related to the material's fibers, the magnitudes of the particles, and so on. The influence of density on the sound absorption coefficient is complicated. It can be described from two perspectives: in one, the same material has different densities, and in the other, different materials have the same density.

Increasing the volume density reduces the porosity inside a material and increases the flow resistance, which results in an increase of the sound absorption coefficient at low frequencies, but a reduction at high frequencies. Figure 4.26 shows the relationship between sound absorption coefficient and density for three samples of glass wool with the same thickness but different densities. As the density increases from 10 to 20 kg m^{-3}, the sound absorption coefficient increases in the low- and middle-frequency ranges, but reduces in the high-frequency range, and the resonant frequencies shift lower.

Sound absorption coefficients for different materials with the same density are different. One material can achieve an optimal sound absorption performance, but another one with the same density cannot, which means each material has its own optimal density. Figure 4.27 shows the optimal density (within 140–160 kg m^{-3}) of a sound absorptive material where the corresponding sound absorption coefficient reaches its maximum value.

4.3.2.2.3 Influence of Environment

At room temperature, there is little impact from temperature on a material's sound absorption coefficients. But temperature changes force the sound speed and wavelength to change, which results in the coefficient shifting with frequency, as shown in Figure 4.28. When the temperature reduces, the coefficient curves shift to a lower frequency. When the temperature increases, the curves shift to a higher frequency.

Increasing humidity will reduce the material's porosity, resulting in a reduction of the sound absorption coefficient. In addition, the humidity will deteriorate the material's properties.

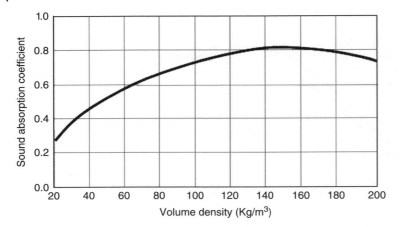

Figure 4.27 Relationship between volume density and sound absorption coefficient.

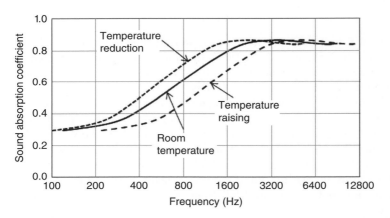

Figure 4.28 Relationship between temperature and sound absorption coefficient.

4.3.3 Resonant Sound Absorption Structure

A resonant sound absorption structure consists of a perforated plate with sound absorption materials behind it, as shown in Figure 4.29. The perforated plate and the air behind can be regarded as many small Helmholtz resonators.

A Helmholtz resonator, as shown in Figure 4.30, is similar to a mass-spring system. The air inside the connecting pipe and the resonator chamber is regarded as a mass and a spring, respectively. After a corrected coefficient of the connecting pipe is included, the frequency of the Helmholtz resonator is calculated as follows:

$$f = \frac{c}{2\pi}\sqrt{\frac{A}{Vt_e}} = \frac{c}{2\pi}\sqrt{\frac{A}{V(t+0.8d)}}, \tag{4.8}$$

where c is the sound speed; A, t_e, and t are section area, effective length, and depth of the connecting pipe, respectively; V is the resonator volume; and d is the pipe diameter.

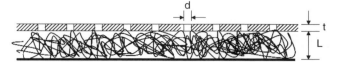

Figure 4.29 Resonant sound absorption structure.

When the frequency of an incident wave is consistent with that of the resonator, the air inside the perforated plate vibrates, and the friction among the molecules is violent. This transfers the sound energy into heat energy, resulting in effective sound absorption. The sound absorption frequency range of the reso-nator depends on the plate thickness, the hole diameters, and the depth of the air behind the plate. The sound absorptive material placed behind the plate influences the absorption effectiveness across a broad frequency range. According to Eq. (4.8), the resonant frequency of the perforated plate is:

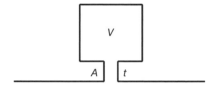

Figure 4.30 Helmholtz resonator.

$$f_0 = \frac{c}{2\pi} \sqrt{\frac{B}{(t+0.8d)L}}, \tag{4.9}$$

where B is porosity, a ratio of the perforated area to the total plate area; L is the depth of the air behind the plate; and d is the diameter of the hole.

It can be seen from Eq. (4.9) that the thicker the plate and the deeper the air behind the plate, the lower the resonant frequency. When the porosity is over 20%, the sound absorption effectiveness will be significantly reduced, and the resonator will lose its function as a Helmholtz resonator.

Perforated resonators are commonly used in vehicles, such as in the engine heat shield shown in Figure 4.31. A perforated aluminum foil is placed on the heat shield cover, and glass fibers are inserted between the sheet and the cover. Figure 4.32 shows a sound absorption coefficient comparison for heat shields with and without the perforated aluminum foil.

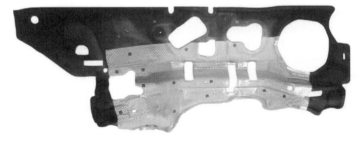

Figure 4.31 Engine heat shield with perforated aluminum foil.

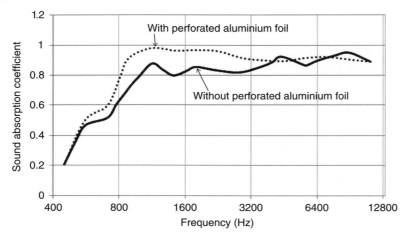

Figure 4.32 Sound absorption coefficient comparison for heat shields with and without perforated aluminum foil.

4.3.4 Measurement of Sound Absorption Coefficient

Measurement of the sound absorption coefficient of a material or structure can be implemented in a reverberation room or inside an impedance tube. A large piece of material is placed in the reverberation room, and the reverberation time is measured; then the sound absorption coefficient can be calculated. Because the directivities of the incident waves are considered, the sound absorption performance can be perfectly tested by the reverberation room method.

When using an impedance tube to measure the sound absorption coefficient, a specimen of the tested material is placed inside the tube. The test results are slightly different from a real case scenario because only perpendicular waves are used in testing, whereas in the vehicle, the incident waves projecting on the material will come from different directions. Impedance tube testing includes single impedance tube testing and dual impedance tube testing. The following sections describe the above two methods in detail.

4.3.4.1 Reverberation Room Method

A reverberation room is a room deigned to create a diffuse or random incident sound field. In a diffuse field, sound energy density is uniformly distributed in space, and the phase of each sound wave is dispersed randomly. The reverberation room has very hard exposed surfaces. A sound wave emitting from a source is reflected when it strikes a wall surface. Part of the energy is dissipated and another part is reflected. The reflected wave travels across the room and then strikes another wall surface. The process repeats until the sound energy is totally dissipated. The sound energy's decaying speed is described by its reverberation time. The reverberation time is defined as the time required for the sound energy to decay by 60 dB after the sound source is cut-off, as shown in Figure 4.33, and indicated by T_{60}.

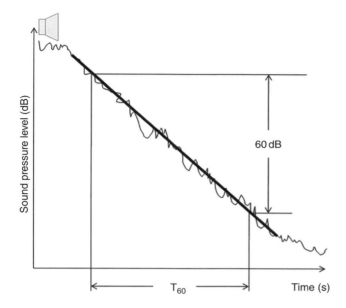

Figure 4.33 Diagram of the definition of reverberation time.

The American acoustician Wallace Clement Sabine discovered that the reverberation time depends on the room volume (V), surface area (S), and average sound absorption coefficient (α), which can be expressed as follows:

$$T_{60} = \frac{0.161V}{S\alpha}. \tag{4.10}$$

Because the wall surfaces of the reverberation room are specially treated, the reverberation effectiveness is excellent and the reverberation time is very long, which results in low sound absorption of the room. In this special environment, the sound absorption coefficient of a material or component can be effectively measured. First, the room's reverberation time is measured. Then, the reverberation time is measured again after the material or component is placed inside the room. The sound absorption coefficient can be calculated based on the two reverberation times, and expressed as follows:

$$\alpha_2 = \frac{0.163V}{S_2} \left(\frac{1}{T_2} - \frac{1}{T_1} \right) + \alpha, \tag{4.11}$$

where T_1 and T_2 are the reverberation times before and after the material or component is placed inside the room, respectively. S_2 is the surface area of the material or component. α is the average sound absorption coefficient of the reverberation room calculated according to Eq. (4.10).

The sound pressures in different locations of the room are different, so 4 to 6 locations should be chosen to test the sound pressures, and then the average value is obtained. The volume of the reverberation room has a significant influence on the accuracy of sound absorption coefficient measurement. Generally, the room's volume should be bigger than $200\,m^3$.

4.3.4.2 Impedance Tube Method

Figure 4.34 shows a diagram of sound absorption coefficient measurement using an impedance tube. A measured specimen is placed on one end of the tube, and on the other end is a speaker that emits a single frequency signal. A microphone is placed inside the tube, and the sound pressure is measured.

Within the analyzed frequency range, the sound wavelengths are much longer than the tube's diameter, so the sound can be regarded as plane wave propagation. When the sound waves travel inside the tube and arrive at one end, some waves are absorbed and some are reflected. The sound pressures for the incident waves, $p_i(x,t)$, and the reflected waves, $p_r(x,t)$, are expressed as follows:

$$p_i\left(x,t\right) = P_i e^{j(\omega t - kx)} \tag{4.12}$$

$$p_r\left(x,t\right) = P_r e^{j(\omega t + kx)}, \tag{4.13}$$

where P_i and P_r are the magnitudes of the incident wave and reflected wave, respectively, k is the wave number, and x is the distance between the microphone and the measured specimen.

The sound wave at any point inside the tube consists of the incident wave and the reflected wave, and the combined sound pressure is expressed as follows:

$$p\left(x,t\right) = P_i e^{j(\omega t - kx)} + P_r e^{j(\omega t + kx)}. \tag{4.14}$$

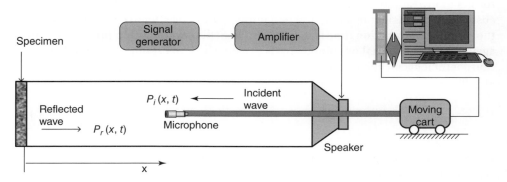

Figure 4.34 A diagram of sound absorption coefficient measurement using an impedance tube.

The reflection coefficient, R, is defined as a ratio of the magnitude of the reflected wave's sound pressure to the magnitude of the incident wave's sound pressure, and is expressed as follows:

$$R = \frac{P_r}{P_i}.$$ (4.15)

Substitute Eq. (4.15) into Eq. (4.14), obtaining

$$p(x,t) = P_i e^{j\omega t} \left(e^{-jkx} + R e^{jkx} \right).$$ (4.16)

The standing wave ratio is defined as the ratio between the maximum and minimum magnitudes of the sound pressures, i.e.

$$n = \frac{p_{max}}{p_{min}},$$ (4.17)

where p_{max} and p_{min} are the maximum and minimum sound pressures, respectively.

The maximum and minimum sound pressures can be obtained according to Eq. (4.16), then substituted into Eq. (4.17), obtaining

$$R = \frac{n-1}{n+1}$$ (4.18)

The sound absorption coefficient is obtained,

$$\alpha = 1 - R^2 = \frac{4n}{(n+1)^2}.$$ (4.19)

In the impedance tube testing method, only a single frequency signal is emitted, and the corresponding sound pressures are obtained. The testing is repeated for each frequency in order to obtain all of the sound pressures in the analyzed frequency range. This method is an accurate way to obtain the sound absorption coefficient, but it is time consuming and somewhat inefficient.

4.3.4.3 Two Microphone Impedance Tube Method

The two microphone impedance tube method overcomes the slow speed of the single microphone impedance tube method, as the sound absorption coefficients for different frequencies can be measured simultaneously. Figure 4.35 shows a diagram of the two microphone impedance tube measurement method. The mechanism is the same as for the single microphone method; the only difference is that two microphones are used.

The sound pressures measured by microphone 1 and microphone 2 are expressed as follows:

$$p_1(x,t) = P_i e^{j(\omega t - kx_1)} + P_r e^{j(\omega t + kx_1)}$$ (4.20)

$$p_2(x,t) = P_i e^{j(\omega t - kx_2)} + P_r e^{j(\omega t + kx_2)},$$ (4.21)

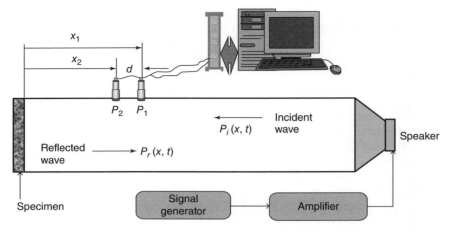

Figure 4.35 A diagram of the two microphone impedance tube measurement method.

where x_1 and x_2 are the distances between corresponding microphones and the measured specimen.

The sound pressure transfer function from microphone 2 to microphone 1 is defined as follows:

$$H_{12} = \frac{p_2(x,t)}{p_1(x,t)} = \frac{P_i e^{j(\omega t - kx_2)} + P_r e^{j(\omega t + kx_2)}}{P_i e^{j(\omega t - kx_1)} + P_r e^{j(\omega t + kx_1)}} = \frac{e^{-jkx_2} + Re^{jkx_2}}{e^{-jkx_1} + Re^{jkx_1}}. \tag{4.22}$$

Assume the distance between the microphones is $d = x_1 - x_2$. The sound pressure transfer function for the incident waves from microphone 2 to microphone 1, H_i, is defined as follows:

$$H_i = \frac{p_i(x_2,t)}{p_i(x_1,t)} = e^{-jk(x_2 - x_1)} = e^{jkd}. \tag{4.23}$$

Similarly, the sound pressure transfer function for the reflected waves from microphone 2 to microphone 1, H_r, is defined as follows:

$$H_r = \frac{p_r(x_2,t)}{p_r(x_1,t)} = e^{jk(x_2 - x_1)} = e^{-jkd}. \tag{4.24}$$

Substitute Eqs. (4.23) and (4.24) into Eq. (4.22), and the reflection coefficient can be obtained, and is expressed as follows:

$$R = \frac{H_i - H_{12}}{H_{12} - H_r} e^{-j2kx_1}. \tag{4.25}$$

After the reflection coefficient has been obtained, the sound absorption coefficient can be calculated by Eq. (4.19). H_i and H_r in Eq. (4.25) include the wave number, i.e. multiple frequencies are included, thus the sound absorption coefficients at multiple frequencies can be measured from one test.

4.4 Sound Insulation Materials and Structures

4.4.1 Mechanism of Sound Insulation and Sound Transmission Loss

Sound insulation means that the sound source is insulated from the receiving environment by special materials and structures.

When the sound wave propagates in the air and strikes an object's surface, part of the sound energy is reflected, and part of the energy penetrates the object and continues to travel in the air, as shown in Figure 4.36. The energy reflected back is called reflected energy, denoted by E_r, and the energy penetrating the object is called transmitted energy, denoted by E_t. Because the transmitted energy is less than the incident energy (E_i), part of the energy is prevented from propagating, and sound insulation is achieved.

The capacity of the sound transmission is represented by the sound transmission coefficient, τ, which is defined as a ratio of the transmitted energy to the incident energy, and is expressed as follows:

$$\tau = \frac{E_t}{E_i}.$$

(4.26)

There are two special cases. The first one is $E_r = E_i$, i.e. $\tau = 0$, which means all of the incident energy is reflected. The second case is $E_t = E_i$, i.e. $\tau = 1$, which means all of the incident energy is transmitted. The transmission performance for all materials is between above two cases, and their transmission coefficients are between 0 and 1. The bigger the transmission coefficient, the more energy is transmitted, and the worse the material's sound insulation.

Sound transmission and sound insulation are opposite concepts, thus the sound insulation can also be expressed by the reciprocal of the sound transmission. Because

Figure 4.36 Incident wave, reflected wave, and transmitted wave.

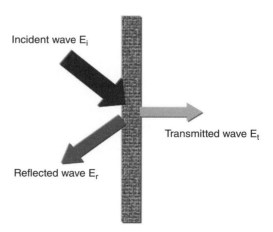

Incident wave E_i

Transmitted wave E_t

Reflected wave E_r

the reciprocal is a huge number, logarithm expression of the number is usually adopted. The sound insulation capacity can be given as the STL, and is expressed as follows:

$$STL = 10\log\left(\frac{1}{\tau}\right) = 10\log\left(\frac{E_i}{E_t}\right) = 10\log E_i - 10\log E_t. \tag{4.27}$$

The larger the STL, the better the sound insulation capacity.

4.4.2 Sound Insulation of Single Plate

Assume the density of a single plate is uniform. The sound insulation performance of the single plate depends on its surface density, stiffness, and damping. Figure 4.37 shows a relationship between the STL and frequency. The STL increases with the increase in frequency. The insulation characteristics can be divided into three regions: the stiffness resonance controlled region, the mass controlled region, and the wave coincidence and stiffness controlled region, which are explained below.

First, the stiffness resonance controlled region. At a very low frequency range that is lower than the plate's resonant frequency, the plate is similar to a rigid body, and its sound insulation quantity decreases with frequency increase. At its resonant frequency, the structure is resonant with the excitation, resulting in rapid reduction of the sound insulation. Above the resonant frequency, the sound insulation increases rapidly due to the mass effectiveness. A plate's resonant frequency is related to its structure and boundary condition.

Second, the mass controlled region. The heavier the mass, the better the sound insulation. The higher the frequency, the better the sound insulation. There are two "6 dB laws" in this region. The first law is that the STL increases by 6 dB when the frequency is doubled: i.e. the STL linearly increases with frequency increase, with a slope of 6 dB/octave, as shown in Figure 4.38. The second law is that the STL increases by 6 dB when the plate's mass is doubled: i.e. the STL line is shifted 6 dB higher. In the mass controlled region, increasing the plate mass is one of the most important ways to increase the sound insulation.

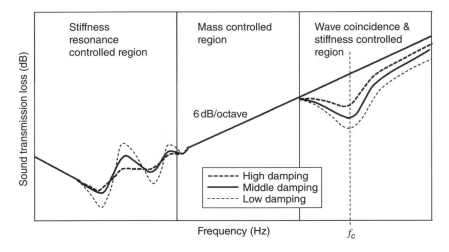

Figure 4.37 Relationship between STL and frequency.

Figure 4.38 Relationship between mass and STL for a single plate.

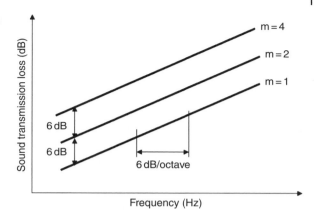

Third, the wave coincidence and stiffness controlled region. When the plate is subjected to sound wave excitation, a forced bending vibration will be generated because it is an elastic structure. The motion of the sound wave and the motion of the plate are coupled, so part of the energy is transmitted through the plate, which results in reduction of its sound insulation. At the early stage of this region, the sound insulation reduces and a valley appears. Above the valley's frequency, the sound insulation increases with a slope of 10 dB/octave, then the increase slows down to 6 dB/octave. Because the increase in sound insulation in the late stage is similar to that in mass controlled region, this stage is also called the extended mass controlled region.

At a certain frequency, the wavelength of the bending wave of the plate is equal to that of the incident wave, so a resonance happens between the two waves. This results in huge energy transmitting into the plate, so its sound insulation significantly reduces. The resonant frequency corresponding to the valley is called the coincident frequency or critical frequency, and is expressed as follows:

$$f_c = \frac{c^2}{2\pi}\sqrt{\frac{M}{K}} = \frac{c^2}{2\pi t}\sqrt{\frac{12\rho}{E}}, \tag{4.28}$$

where c is the sound speed; M, ρ, and t are the plate surface density (kg/m^2), density (kg/m^3), and thickness (m), respectively; E is Young's modulus (N/m^2); and $K = Et^3/12$ is the plate stiffness.

According to Eq. (4.28), the coincident frequency only depends on the plate's thickness if the material is determined. Commonly used materials in the vehicle body include steel and glass. The coincident frequencies for a steel sheet and glass sheet, respectively, are expressed as follows:

$$f_{steel} = \frac{12650}{t} \tag{4.29}$$

$$f_{glass} = \frac{12240}{t}, \tag{4.30}$$

where t is the steel or glass sheet thickness, and the unit is millimeters.

The thickness of the body steel panels is around 1 mm, and the corresponding coincident frequency is above 10 000 Hz. Because few excitation sources have frequencies higher than 10 000 Hz, the coincidence possibility for the steel panels can be ignored. However, the thickness of the glass in the body is usually between 3.5 and 5 mm, and the corresponding coincident frequency is between 2500 and 3500 Hz. Excitation sources within this frequency range are very common, so coincidences are easily generated for the windshields and side glass windows. An important part of sound package work is implementing the sound insulation of body glass.

In order to reduce or avoid influence of the coincidence on the plate's sound insulation, two methods are usually adopted. One is to place the damping materials on the plates. Figure 4.37 shows that the bigger the damping, the smaller the sound insulation reduction, which indicates that adding damping on the plates is a good way to increase the sound insulation. The other method is to avoid the coincident frequency: for example, a plate is designed to have a coincident frequency much higher than the sound wave excitation frequencies. According to Eq. (4.28), the plate coincident frequency can be increased by decreasing its thickness and/or increasing its density.

Assume a plate is uniform, solid, and infinite. When a sound wave projects onto the plate, its STL is

$$STL_0 = 10\log_{10}\left[1+\left(\frac{\pi fM}{\rho_0 c}\right)^2\right], \tag{4.31}$$

where ρ_0 is the air density (kg/m^3).

Because $\dfrac{\pi fM}{\rho_0 c} \gg 1$, Eq. (4.31) can be simplified as

$$STL_0 = 20\log_{10}\left(fM\right) - 20\log_{10}\left(\frac{\rho_0 c}{\pi}\right). \tag{4.32}$$

Substitute the air density $\rho_0 = 1.18 kg/m^3$ and the sound speed $c = 344 m/s$ into Eq. (4.32), obtaining

$$STL_0 = 20\log_{10}\left(fM\right) - 42.5. \tag{4.33}$$

However, it rare for sound waves to perpendicularly project onto a plate. Usually, the sound waves project onto a plate in different directions, and the incident angles are between 0 and 180°. In these cases, the STL is regarded as a constant difference (5 dB) from that of the perpendicular projecting case (STL_0), and is expressed as follows:

$$STL = STL_0 - 5. \tag{4.34a}$$

Or it can be expressed as

$$STL = 20\log_{10}\left(fM\right) - 47.5. \tag{4.34b}$$

Equations (4.33) and (4.34a) imply that the sound insulation has a relation only with surface density and frequency. The equations also reveal the two "6 dB laws" in the mass controlled region.

The equations are derived on the basis of the assumption of an infinite plate, which is an ideal case. Actually, the size of a plate has limitations, and its boundary condition, stiffness, and damping influence the sound insulation effectiveness. Lots of test data show that the sound insulation increases by only 5 dB when a plate's surface density is doubled, and it increases by only 4 dB when the frequency is doubled. Therefore, the theoretical equation must be modified, and an empirical equation for a real plate is obtained as follows:

$$STL = 16\log_{10} M + 14\log_{10} f - 29. \tag{4.35}$$

4.4.3 Sound Insulation of Double Plate

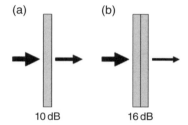

The sound insulation for an ideal single plate follows the mass law: that is, the theoretical sound insulation increases by 6 dB when its thickness is doubled. Figure 4.39a shows that the STL of a plate is 10 dB. Two plates with the same thickness are put together, forming a composite plate, as shown in Figure 4.39b. The STL for the composite plate should be 16 dB. If the two plates are separated, as shown in Figure 4.40. Each plate has 10 dB sound insulation, so ideally, the total STL for the two plates should be 20 dB. Actually, the STL cannot reach 20 dB, and the reason for this is explained later on.

Figure 4.39 (a) A single plate; (b) a composite plate.

Although the STL of the two separated plates is less than 20 dB, it is certainly higher than the 16 dB of the composite plate. A special structure in which the two plates are separated by an air gap is called a double-plate sound insulation structure, as shown in Figure 4.41.

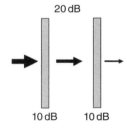

Figure 4.40 Two separated plates.

Figure 4.41 can be divided into three regions. In region I, when a sound wave strikes plate A, part of the sound wave is reflected back, becoming a reflected wave, and the other part passes through the plate, forming a transmitted wave. The transmitted wave propagates in region II. When it strikes plate B, part of the sound wave is reflected back to form reflected wave 2, and the other part of the wave passes through plate B and enters region III, forming transmitted wave 2. In region II, the sound waves propagate repeatedly between plate A and plate B, forming the reflected and transmitted waves.

The sound wave propagates from region I to region III through the two plates. The acoustic energy is greatly attenuated due to the multiple wave reflections and transmissions. Plate A and plate B are not directly connected, and the air between them has elasticity. The vibration of plate A is transmitted to the air, and attenuated by the air, and then transmitted to the plate B. Therefore, the double plate can achieve good sound insulation performance.

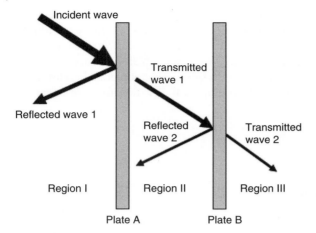

Figure 4.41 A double-plate sound insulation structure.

Air possessing elastic characteristics can be regarded as a spring, so the double-plate sound insulation system can be regarded as a mass-spring-mass system, and its natural frequency (f_0) is:

$$f_0 = \frac{1}{2\pi} \sqrt{\left(\frac{1}{M_1} + \frac{1}{M_2} \right) \frac{\rho_0 c^2}{d}}, \tag{4.36}$$

where M_1 and M_2 are the surface density of the two plates, respectively (kg/m^2); and d is the thickness of the air gap between the plates (m).

Figure 4.42 shows the STL of the double-plate sound insulation system. The curve can be divided into three zones: The low frequency zone, resonance zone, and high frequency sound insulation zone.

1) Low frequency zone. When a sound frequency is lower than the system's natural frequency, the two plates have almost no relative motion, so they move in phase and can be regarded as one plate. The sound insulation follows mass law, and is expressed as:

$$STL = 20 \log_{10} f (M_1 + M_2) - 47.5. \tag{4.37}$$

2) Resonance zone. When a sound frequency is consistent with the system's natural frequency, the double-plate system is resonant and the sound wave easily passes through the plates. The STL is significantly reduced and a valley is formed, and its sound insulation is even lower than that of the single plate. When the surface densities of the two plates are the same, i.e. $M_1 = M_2$, there is only one resonant frequency, and the valley of the sound insulation is very deep. When the surface densities of the two plates are different, i.e. $M_1 \neq M_2$, there are two resonant frequencies and the valleys of the sound insulation are not as deep as when the plates have the same surface densities. The lower the natural frequency, f_0, the better the sound insulation across the whole frequency range, so the system's natural frequency should be made as low as possible.

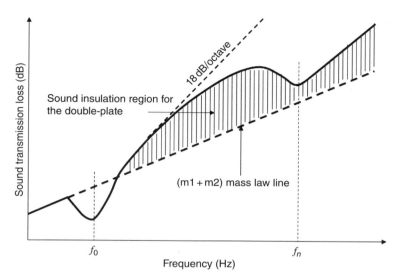

Figure 4.42 STL of a double-plate sound insulation system.

3) High frequency sound insulation zone. When the frequency of a sound wave is higher than the resonant frequency, the STL increases from the valley bottom. As the frequency increases to a certain value, the sound insulation is more than that of the two plates put together. When the frequency further increases, a standing wave effect emerges between the two plates, which reduces the STL. Above the standing wave frequency, the increasing tendency of the sound insulation becomes slower. For a perpendicularly projecting sound wave, the standing wave frequencies are

$$f_n = \frac{nc}{2d},$$ (4.38)

where n is the order of the standing wave (n = 1, 2, 3, …).

When the frequency of a sound wave is higher than $\sqrt{2}\,f_0$, but lower than the standing wave frequency and the coincidence frequency of a single plate, the sound insulation increases at a slope of 18 dB/octave, as shown in Figure 4.42, and the STL is expressed as

$$STL = 20\log_{10} f\left(M_1 + M_2\right) - 47.5 + 40\log_{10}\left(f/f_0\right).$$ (4.39)

Many double-plate structures can be found on the vehicle body. Many body structures consist of a steel panel, sound absorption layers, and sound insulation layers. Taking the dash panel as an example to illustrate the double-plate structure, the steel panel and the insulator can be regarded as two plates, and the middle absorber can be regarded as a spring. When the absorption layer is very soft, its stiffness is lower than the air stiffness, and the system's natural frequency is lower than that calculated by Eq. (4.36). When the stiffness of the absorption layer increases to be higher than the air stiffness, the system's natural frequency is higher than that calculated by Eq. (4.36). Here, we take a dash panel as an example to explain the natural frequency calculation. Assume the thicknesses of the steel panel, the EVA (ethylene vinyl acetate) insulator, and the middle absorber are 1, 1.5–2.5, and 10–20 mm, respectively. The calculated natural frequency of the dash panel is lower than 200 Hz, while the first order standing wave frequency is higher than 8000 Hz. In the middle frequency range, the sound insulation of the dash panel can be estimated by Eq. (4.39).

4.4.4 Measurement of Sound Insulation Materials

Methods of measuring a material's sound insulation performance include the laboratory method and impedance tube method. The laboratory method is divided into the reverberation chamber–reverberation chamber method and the reverberation chamber–anechoic chamber method.

4.4.4.1 Reverberation Chamber–Reverberation Chamber Method

The reverberation chamber–reverberation chamber method uses two connected reverberation chambers to measure the material's STL. In Figure 4.43, two reverberation chambers are connected together, and there is a window between them. The tested sample is placed in the window, and its surrounding is sealed. A ball sound source or a speaker is placed in one reverberation room, which is used as the sound source room. The sound passes through the sample into the other reverberation room, which is used as a receiving room. Microphones are placed in both chambers to measure the sound pressures.

The sound pressures in the two chambers are tested, then the noise reduction (NR) between the two rooms can be obtained, and expressed as follows:

$$NR = SPL_s - SPL_r,\tag{4.40}$$

where SPL_s and SPL_r are the average sound pressures in the sound source room and receiving room, respectively.

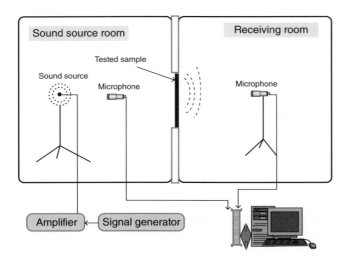

Figure 4.43 Testing diagram of the reverberation chamber–reverberation chamber method.

The sound absorption capacity of the receiving room should be included for calculation of the STL of the sample, which can be expressed as follows:

$$STL = NR + 10\log_{10}(S/\alpha S_r),\tag{4.41}$$

where S and S_r are the surface areas of the sample and the receiving room, respectively. α is the sound absorption coefficient of the receiving room.

The reverberation chamber–reverberation chamber method is relatively simple. Because the STL is calculated by average sound pressures in the two chambers, it is impossible to determine the STL for each component of the sample.

4.4.4.2 Reverberation Chamber–Anechoic Chamber Method

The reverberation chamber–anechoic chamber method uses the connected reverberation chamber and anechoic chamber to measure the STL. A tested sample is installed on the window between the two chambers, as shown in Figure 4.44. A sound source is placed inside the reverberation room, which is used as the sound source room. The anechoic chamber is used as the receiving room.

Inside the reverberation chamber, the sound source randomly strikes on the tested sample. The sound power transmitted to the sample can be calculated by average sound pressure in the reverberation sound field, and expressed as follows:

$$W = \frac{\bar{p}^2}{4\rho c}S,\tag{4.42}$$

where $\bar{p}^2$ is the average squared value of the sound pressures of several tested points and S is the area of the tested sample.

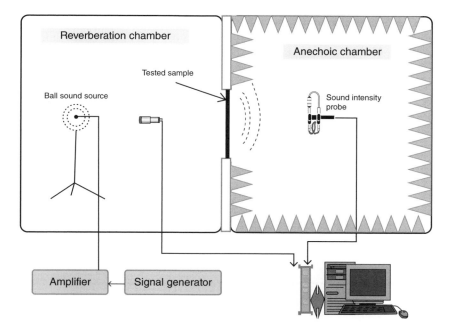

Figure 4.44 Testing diagram of the reverberation chamber–anechoic chamber method.

In the anechoic chamber, the sound power is calculated by the measured sound intensity. There are two methods to measure the sound intensity: one is the scanning method, and the other is a fixed point measurement method. The scanning method involves uniformly scanning the sample to obtain its sound intensity, and the sound power is obtained by multiplying the sound intensity by the sample area, expressed as follows:

$$W_{in1} = I \cdot S, \tag{4.43}$$

where I is the sound intensity of the tested sample.

The fixed point measurement method involves measuring the sound intensity of a small section of the sample each time, and then obtaining its sound power. The overall sound power of the sample is a superposition of that of each section, i.e.

$$W_{in2} = \sum_{i=1}^{n} I_i S_i, \tag{4.44}$$

where S_i and I_i are the area and sound intensity of the ith section, respectively.

Substitute the sound powers of each side of the sample, the air density, and the sound speed into the equation for STL, and the STL of the tested sample will be obtained, expressed as follows:

$$STL = 10\log_{10} \frac{\bar{p}^2 S / (4\rho c)}{IS} = \bar{L}_p - \bar{L}_I - 6, \tag{4.45}$$

where $\bar{L}_p$ and $\bar{L}_I$ are the average sound pressure level in the reverberation room and the average sound intensity level in the anechoic room, respectively.

4.4.4.3 Impedance Tube Method

The impedance tube method involves measuring the sound insulation coefficient by placing a tested sample inside an impedance tube, as shown in Figure 4.45. The tested

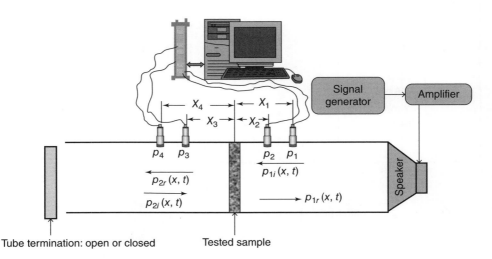

Figure 4.45 Impedance tube method for measuring the sound insulation coefficient.

sample is placed in the middle of the tube, and two microphones are placed on each side of the sample: i.e. two in the sound source tube and two in the receiving tube. The STL can be calculated by measuring the sound pressures of the four microphones.

When an incident wave (p_{1i}) inside the source tube strikes the sample, a partial wave is reflected back and becomes a reflected wave (p_{1r}), and the other partial wave passes through the sample and enters the receiving tube. In the receiving tube, the transmitted wave (p_{2i}) becomes a new incident wave for the sample, and the wave reflected is the new reflected wave (p_{2r}).

For a linear system, there is a linear relationship between the sound waves inside the source tube and the receiving tube, expressed as follows:

$$\begin{Bmatrix} p_{1i}(x,t) \\ p_{1r}(x,t) \end{Bmatrix} = \begin{pmatrix} e_1(\omega) & e_2(\omega) \\ e_3(\omega) & e_4(\omega) \end{pmatrix} \begin{bmatrix} p_{2i}(x,t) \\ p_{2r}(x,t) \end{bmatrix}, \tag{4.46}$$

where e_1, e_2, e_3, and e_4 are the coefficients related to the sound pressures inside the sound source tube and the receiving tube. According to the definition of STL, the STL of the sample is expressed as follows:

$$STL(\omega) = 20\log_{10}|e_1(\omega)|. \tag{4.47}$$

After these coefficients have been obtained, the STL of the sample will be obtained.

According to the previous description, the sound pressure at any point inside a pipe is a superposition of the incident wave pressure and the reflected wave pressure. The sound pressures for the four microphones can be expressed as follows:

$$p_1(x,t) = P_{1i}e^{j(\omega t - kx_1)} + P_{1r}e^{j(\omega t + kx_1)} \tag{4.48a}$$

$$p_2(x,t) = P_{1i}e^{j(\omega t - kx_2)} + P_{1r}e^{j(\omega t + kx_2)} \tag{4.48b}$$

$$p_3(x,t) = P_{2i}e^{j(\omega t + kx_3)} + P_{2r}e^{j(\omega t - kx_3)} \tag{4.48c}$$

$$p_4(x,t) = P_{2i}e^{j(\omega t + kx_4)} + P_{2r}e^{j(\omega t - kx_4)}. \tag{4.48d}$$

$p_1(x,t)$, $p_2(x,t)$, $p_3(x,t)$, and $p_4(x,t)$ are obtained from the testing results of the four microphones. According to Eq. (4.48a–4.48d), the sound pressures (P_{1i}, P_{1r}, P_{2i}, P_{2r}) of the incident waves and the reflected waves of both sides of the tested sample can be obtained by solving the equations.

In Eq. (4.46), there are two equations, but four unknowns (e_1, e_2, e_3, and e_4), so it is impossible to get the solutions for the four unknowns. In order to solve the problem, the termination of the tube is specially designed as two cases: an open termination and closed termination. So Eq. (4.46) is expanded into four equations.

For the open termination, the pressure equations are

$$\begin{Bmatrix} p_{1io}(\omega) \\ p_{1ro}(\omega) \end{Bmatrix} = \begin{pmatrix} e_1(\omega) & e_2(\omega) \\ e_3(\omega) & e_4(\omega) \end{pmatrix} \begin{bmatrix} p_{2io}(\omega) \\ p_{2ro}(\omega) \end{bmatrix}, \tag{4.49}$$

where p_{1io} and p_{1ro} are the sound pressures for the incident wave and the reflected wave in the source tube for the open termination, respectively; and p_{2io} and p_{2ro} are the sound pressures for the incident wave and the reflected wave inside the receiving tube for the open termination, respectively.

For the closed termination, the pressure equations are

$$\begin{Bmatrix} p_{1ic}(\omega) \\ p_{1rc}(\omega) \end{Bmatrix} = \begin{pmatrix} e_1(\omega) & e_2(\omega) \\ e_3(\omega) & e_4(\omega) \end{pmatrix} \begin{Bmatrix} p_{2ic}(\omega) \\ p_{2rc}(\omega) \end{Bmatrix}, \tag{4.50}$$

where p_{1ic} and p_{1rc} are the sound pressures for the incident wave and the reflected wave in the source tube for the closed termination, respectively; and p_{2ic} and p_{2rc} are the sound pressures for the incident wave and the reflected wave inside the receiving tube for the closed termination, respectively.

According to Eq. (4.48), the sound pressures for the four microphones at open and closed terminations are measured, and then the corresponding sound pressures for the incident wave and the reflected wave are calculated. Eqs. (4.49–4.50) include four equations, so the four coefficients e_1, e_2, e_3, and e_4 can be calculated. According to the Eq. (4.47), the sample's STL or noise reduction can be calculated.

4.5 Application of Sound Package

Sound package materials and structures are widely used in a vehicle body, as shown in Figure 4.46. In fact, they are used almost everywhere, such as on the dash panel, floor, roof, seats, trunk, pillars, central console, wheelhouses, and so on. Moreover, with increasing customer demands for better sound quality in recent years, sound package

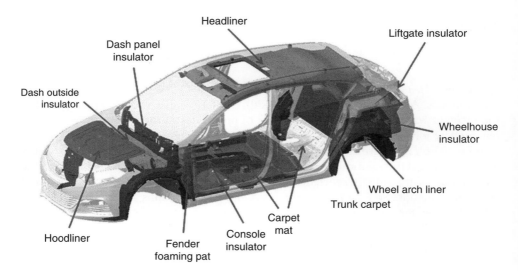

Figure 4.46 Sound package materials and structures in a vehicle body.

materials have been applied more and more often. Sound package applications can be summarized into three categories, which are summarized below.

First, the application of sound absorption material. This is when only sound absorptive materials are used, including those used outside and inside of the body. For example, an engine hoodliner is a sound absorptive material that absorbs middle- and high-frequency noise radiated by the engine. In another example, sound absorptive materials are placed inside a door trim in order to absorb noise transferring from the outside.

Second, the combined application of sound insulation structures and absorption material. This means that sound insulation and sound absorptive materials are combined together to reduce interior noise and improve sound quality. For example, the dash panel is a combined structure including absorbers and insulators. Except for on windshields and side windows, sound insulation structures are rarely used alone. The majority of the sound package is the combination of sound absorbers and insulators.

Third, the application of baffling material. This means that baffling material is installed inside a body beam cavity in order to prevent noise passing through the barrier and subsequently being transmitted into the vehicle interior.

The application of the three categories of sound package materials and structures is introduced in detail in this section.

4.5.1 Application of Sound Absorptive Materials and Structures

4.5.1.1 Application Locations and Types of Sound Absorptive Materials

Figure 4.47 shows the main application areas of absorptive materials on the body. The locations in which sound absorptive materials are used independently include the instrument panel, doors, pillars, roof, wheelhouses, trunk, central console, engine hood, and seats.

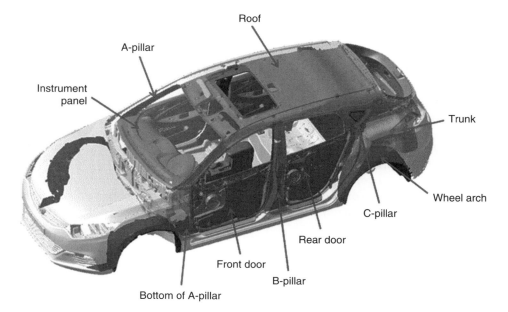

Figure 4.47 Application areas of absorptive materials on a vehicle body.

The application of sound absorptive materials can be divided into three categories. The first category is sound absorption structures with fixed shapes, such as the engine hoodliner and dash absorber. The second category is freely placed materials or structures that do not have strictly fixed shapes: for example, the sound absorptive materials in the wheelhouses, inside the door trim, inside the instrument panel, in the headliner, and inside the pillar trims. The third category is the seat. The seat is a special sound absorption structure, which is extremely important for interior sound absorption.

There are two types of sound absorptive materials used in automobiles: foam absorptive material and fiber absorptive material.

The sound absorption coefficient of foam absorptive material is very high. In particular, in the range between 2000 and 4000 Hz, the coefficient can reach up to 0.7. Polyurethane (PU) foam material is a porous absorbent material, and is widely used in the body, such as on the dash panel, carpet, and so on. However, its cost is relatively high, so it is mainly used in mid-range and luxury vehicles.

Fiber absorptive materials include glass fiber, needle fiber felt, cotton fiber felt, thermoplastic fiber felt, resin fiber, and so on. The sound absorption coefficients of fiber materials increase with the increase in frequency, but the coefficient can reach 0.7 only at very high frequencies. For example, cotton fiber felt is a widely used fiber material on dash panels, carpets, and other places. Due to their low cost, fiber absorptive materials are widely used in economy vehicles.

4.5.1.2 Sound Absorption Structures with Fixed Shape

The typical examples of sound absorption structures with fixed shapes are the engine hoodliner and the dash absorber. They have certain shapes and are installed on the metal sheets with bolts or clips. Figure 4.48 shows an engine hoodliner, including three layers: a layer of sound absorptive material in the middle and two layers of heat insulation fabric. Its function is not only to insulate the heat transfer from the engine, but also to absorb the engine noise.

The absorptive materials in the middle layer include glass fiber, PU foam, and thermoplastic fiber felt. Glass fiber has good performance in thermal insulation and sound absorption, and the advantage of low cost, but it is harmful to human health, so it is mainly used in economy vehicles. PU foams are mainly used in mid-range and luxury vehicles. The cover material is usually non-woven fabric. In application, usually two layers of fabric, just like films, are placed on both sides of an absorptive material, and its function is to increase sound insulation at low and middle frequencies, but it sacrifices a little sound absorption at high frequencies.

Figure 4.49 shows a comparison of noise above an engine hood with and without a hoodliner in the idling condition for a passenger car. With the hoodliner, the noise reduces across a wide frequency range; in particular, there is a 5–10 dB noise reduction in the middle- and high-frequency ranges, which indicates that the hoodliner has good effectiveness of sound insulation.

Figure 4.50 shows an interior noise comparison with and without the hoodliner in the wide open throttle (WOT) condition for a passenger car. After the hoodliner is installed, the interior sound is reduced across the whole frequency range, especially between 2000 and 4000 Hz, where the noise is reduced by 2 dB.

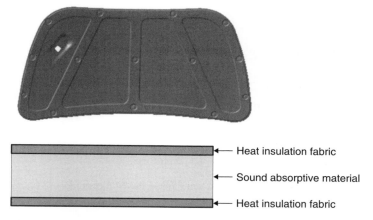

Figure 4.48 An engine hoodliner.

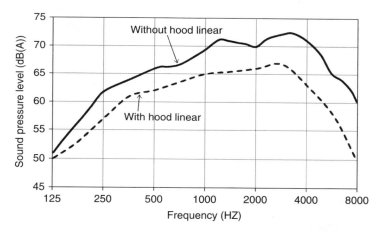

Figure 4.49 Noise comparison above an engine hood with and without a hoodliner during idling.

4.5.1.3 Sound Absorption Materials or Structures without Fixed Shapes

Some sound absorptive materials are freely distributed inside the doors, wheelhouses, central console, and pillars. They don't have fixed shapes and are freely placed; or they could have certain shapes, but their installation is relatively simple. Their deformations have little impact on their sound absorption function, and also don't cause any other problems. Even if the fixed installations are loose, they will not induce any abnormal sound. Figure 4.51 shows absorptive materials placed inside an A-pillar, door trim, and inside the rear package tray.

Here, the roof is used as an example to illustrate the effects of absorption materials on sound absorption. Figure 4.52 is a section of a roof structure. It consists of seven layers: the decorative layer, adhesive layer, reinforcement layer, absorption layer, fabric layer, air layer, and steel layer.

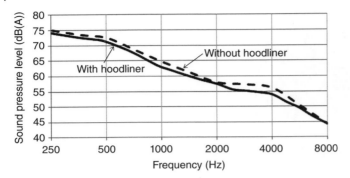

Figure 4.50 Interior noise comparison with and without the hoodliner in the WOT condition.

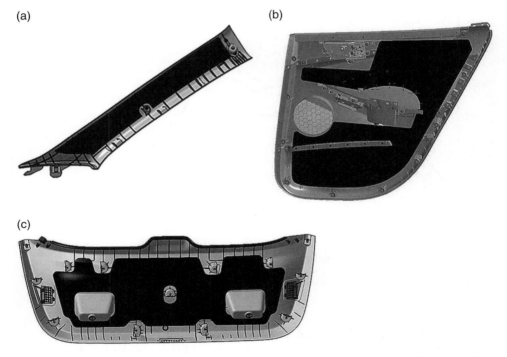

Figure 4.51 Sound absorption structures freely distributed or without fixed shapes: (a) inside A-pillar; (b) inside door trim; (c) inside rear package tray.

The decorative layer is visible, and its main role is cosmetic. The material is usually made of non-woven or woven fabrics. The layer has little function on sound absorption, but it has good air permeability in order to let the absorptive material absorb sound. Foams or sponges are placed on the back of some decoration layers, which increases the thickness and sound absorption performance of the decorative layer.

The function of the adhesive layer is to bond the decoration layer and the reinforcement layer. This layer does not function to absorb sound, but it must have adequate air permeability.

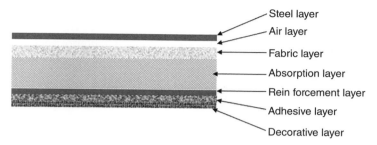

Figure 4.52 A section of a roof structure.

The third layer is the reinforcement layer. The headliner is a heavy and large struc-ture, so it must have sufficient strength. In addition, some small devices could be placed on the top of the headliner, so it must have the capacity to take weight. Like the above layers, the reinforcement layer must have good air permeability as well.

The fourth layer is the absorption layer, which is also called the substrate layer. It is composed of sound absorptive material, and it is the main sound absorption compo-nent of the roof. The first three layers have good air permeability, so the air easily reaches the substrate layer, and a good sound absorption effect can be obtained. Because the roof is close to the passengers, the substrate layer must be made of PU foam materi-als. Occasionally, glass fiber composite materials are used.

The fifth layer is the fabric layer (also called the protection layer), which prevents the absorptive material from moving. If the protection layer is properly designed, it can act as a resonant cavity structure together with the air behind.

The sixth layer is an air layer, on top of which is the steel sheet layer.

The headliner consists of the first five layers. The design of the headliner is very important because it is close to the occupants' ears. Figure 4.53 shows an interior noise

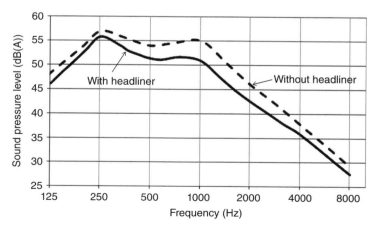

Figure 4.53 Interior noise comparison with and without headliner.

Figure 4.54 Anatomical diagram of the surface and internal acoustic structure of a seat.

comparison with and without the headliner for a passenger car cruising at $120\,km\,h^{-1}$. Across the whole frequency range, the interior sound is reduced by 1–3 dB after installing the headliner.

4.5.1.4 Sound Absorption of Seats

Figure 4.54 shows an anatomical diagram of the surface and internal acoustic structure of a seat. There are three types of surface materials: fabric, real leather, and artificial leather. The internal acoustic structure consists of porous foam materials.

Compared with other sound absorption structures, the seat has two major unique characteristics: a large area and deep thickness. Because its surface area is very large, and its internal structure is filled with porous absorptive materials, its sound absorption capability is very strong. Figure 4.55 shows the percentage of sound absorption for different components of a vehicle body; the seat accounts for nearly half of the total sound absorption. The thickness of the seat foam determines the frequency of the absorbed sound. Only when the thickness is greater than 1/10 of the wavelengths will the sound waves of the corresponding frequencies be absorbed. The seat is much thicker than other sound absorption structures, so it has an advantage over other structures in terms of low frequency sound absorption. Figure 4.56 shows an interior noise comparison with and without seats for a vehicle cruising at $120\,km\,h^{-1}$. Across the whole frequency range, the sound absorption level with seats is about 5 dB higher than without the seats.

The surface material of a seat has a significant influence on its sound absorption capacity. The fabric has good air permeability, and the sound wave easily passes through the surface and enters the foam material, where it is absorbed. The air permeability of leather is very poor because it is difficult for the sound wave to penetrate it, so its sound absorption capacity is much lower than for fabric. Figure 4.57 shows a sound absorption comparison for three different seat surface

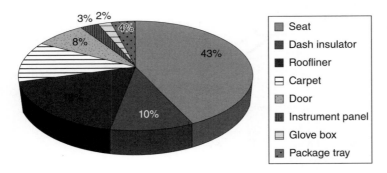

Figure 4.55 Percentage of sound absorption for different components of a vehicle body.

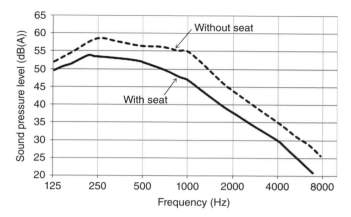

Figure 4.56 Interior noise comparison with and without seats for a vehicle cruising at 120 km h^{-1}.

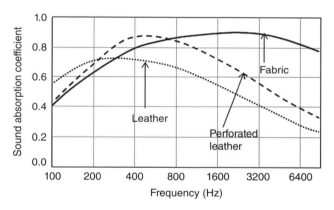

Figure 4.57 Sound absorption comparison for three different surface materials (fabric, leather, and perforated leather) of a seat with the same structure and same internal material.

materials (fabric, leather, and perforated leather) with the same seat structure and the same internal material. The sound absorption coefficient at middle and high frequencies for the fabric seat is much higher than for the leather seat. When perforations are punched into the leather material, its permeability increases, resulting in an increase in its sound absorption coefficient – but it is still far below the fabric's coefficient.

4.5.2 Application of Combination of Sound Insulation Structures and Sound Absorptive Materials

4.5.2.1 Composite Sound Insulation and Absorption Structure
The majority of the sound package used on a vehicle body is a combination of absorptive materials and sound insulation structures, such as the dash mat, carpet, and so

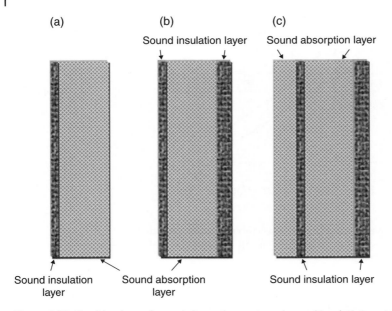

Figure 4.58 Combinations of sound absorption parts and sound insulation parts.

on. Figure 4.58 shows several different combinations of the typical sound absorption parts and sound insulation parts. Figure 4.58a is a combination of a sound absorption layer and a sound insulation layer. Figure 4.58b is a combination of a sound insulation layer, a sound absorption layer, and a sound insulation layer, Figure 4.58c is a combination of multi-layer sound insulation layers and multi-layer sound absorption layers.

Usually, a material or a structure possesses sound absorption and sound insulation capabilities simultaneously. The major function of a sound absorption structure is to absorb the sound wave, but it has a small sound insulation function as well, and vice versa. Figure 4.59 shows the propagation processing of a sound wave in a sound absorption material and a sound insulation material. The sound waves continuously inject, reflect, and transmit, and finally are attenuated, achieving the purpose of reducing the sound energy.

4.5.2.2 Dash Mat Sound Package Structure

Figure 4.60 is a schematic diagram of a dash panel and dash mat. The sound package is composed of a sound absorption layer and a sound insulation layer. Usually, the absorption layer uses PU foam or cotton felt, and the sound insulation layer usually uses EVA. EVA is a heavy material and has good sound insulation performance. If the dash panel is regarded as a sound insulation layer, the combination structure of the sound package is changed into three layers: a sound insulation layer, an absorption layer, and a sound insulation layer. Steel is a very good sound insulation material, but the panel thickness is around 1 mm, so it cannot effectively insulate the engine noise; therefore, additional sound package is needed to improve the sound insulation capability.

Figure 4.61 shows an interior sound pressure comparison with and without the dash mat for a vehicle during acceleration. Without the dash mat, the interior noise increases, especially in the middle and high frequency range; for example, at 2000 Hz, the sound pressure is 3 dB higher without the mat than with the mat.

Due to the high cost of PU foam material, dash mats in some vehicles do not use PU, or even EVA; instead, a double layer of felt is used. Using a double layer increases the absorption capability of felt, but because of its light weight, it has a lower sound insulation capacity than EVA.

The sound insulation of the dash structure not only depends on the thickness of the metal sheet, the sound absorption capability of the absorption layer, and the sound insulation of the insulation layer, but also on the contact status between the sound package and the metal sheet. Figure 4.62 shows two cases of contact between the dash mat and the dash panel. Figure 4.62a shows that the two components are well contacted, so the noise passing through the dash structure is absorbed and insulated by the mat, and the noise transmitted to the vehicle is greatly reduced. Figure 4.62b indicates that their contact is not good, so noise passing through the dash structure directly transmits to the interior through the gap, and the sound insulation capability of the mat is greatly reduced.

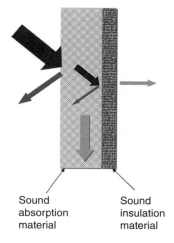

Figure 4.59 Propagation processing of a sound wave in a sound absorption material and a sound insulation material.

4.5.2.3 Carpet Sound Package Structure

Figure 4.63 is a typical carpet sound package structure, consisting of a carpet lining, a sound insulation layer, and a fabric layer. The carpet lining material is foam or felt, which absorbs sound. The main function of the fabric is decoration; however, some fabric has a small capability for sound insulation and absorp-

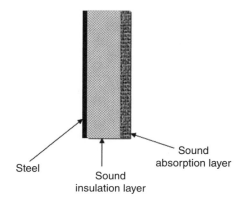

Figure 4.60 Schematic diagram of a dash panel and dash mat.

tion. If the floor is regarded as a sound insulation layer, the floor–liner–sound insulation layer constitutes a double sound insulation system. Its sound insulation capability depends on the surface density of the insulation layer and the thickness of the floor sheet. Figure 4.64 shows an interior noise comparison with and without the floor sound package for a vehicle during acceleration. The interior noise with the carpet sound package is lower than that without the sound package, especially in the middle- and high-frequency range, where the carpet makes the interior noise 2–8 dB lower.

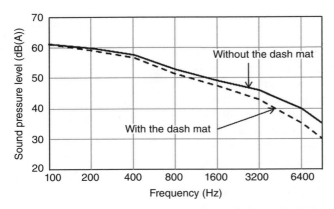

Figure 4.61 Interior sound pressure comparison with and without dash mat for a vehicle during acceleration.

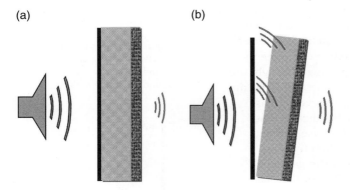

Figure 4.62 Contact between a dash mat and dash panel: (a) good contact; (b) poor contact.

Figure 4.63 A typical carpet sound package structure.

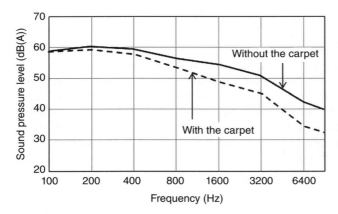

Figure 4.64 Interior noise comparison with and without floor sound package for a vehicle during acceleration.

4.5.2.4 Laminated Glass

The glass used on a body includes the front windshield, rear windshield, and side windows. In recent years, panoramic sunroofs made of a large piece of glass have become more common. The coincidence frequencies of steel body panels are above 10 000 Hz, which is usually far away from the frequencies of the excitation sources, so the coincidence phenomenon seldom happens for metal body panels. However, the coincidence frequency of glass is mostly between 2500 and 3500 Hz, which falls into the frequency range of many external excitation sources, so its sound insulation drops in the coincidence frequency range. Figure 4.65 shows the STL curve of two pieces of glass with thicknesses of 4 and 5 mm, respectively. Within the 3000–4000 Hz range, the sound insulation is reduced due to the coincidence effect.

Recently, in order to compensate for the drop in the STL caused by the coincidence effect, laminated glass structures have been used in automobiles. A thin polyvinyl butyral (PVB) film is sandwiched between two layers of glass to form laminated glass, as shown in Figure 4.66. Figure 4.67 shows a comparison of the STL between laminated glass and ordinary glass with the same thickness. In the frequency range of 3000–4000 Hz, the STL of the ordinary glass is lower than that of the laminated glass. There are three reasons for the increase of STL in the laminated glass. First, the coincidence effect is avoided. Second, the laminated glass forms a double layer of sound

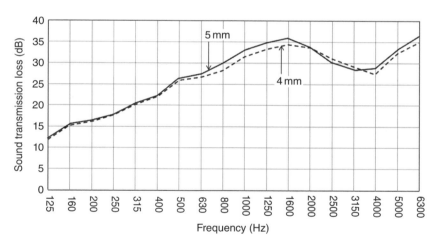

Figure 4.65 STL of two pieces of glass with thicknesses of 4 and 5 mm.

Figure 4.66 A PVB laminated glass structure.

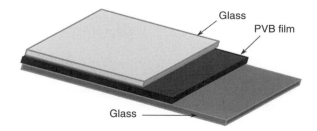

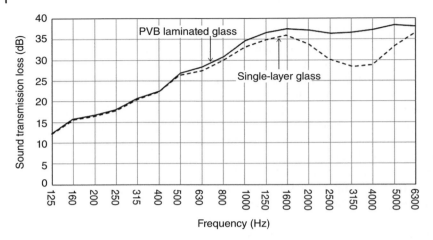

Figure 4.67 Comparison of STL between pieces of PVB laminated glass and ordinary glass with the same thickness.

insulation structures, so the sound transmission characteristics of the single layer of glass are changed. Third, the PVB material has a damping effect.

Laminated glass has been widely used in front windshields, especially in mid-range and luxury vehicles. In some sedans, laminated glass is even used on side door windows.

4.5.2.5 The Laminated Steel Sheet

In Chapter 3, the damping effect of laminated steel plate is introduced in detail. Compared with a conventional single plate, a composite plate not only reduces the vibration, but also improves the sound insulation. Figure 3.60 shows a comparison of two dash panels. Figure 3.60a is a traditional structure, consisting of four layers. Figure 3.60b is a composite structure, consisting of two layers, i.e. a dash panel and a laminated steel sheet.

4.5.3 Application of Sound Baffle Material

Figure 4.68a shows a tube. The fluid (water, air, and so on.) can freely flow through the tube. A piece of material is used to block the tube, as shown in Figure 4.68b, so the fluid cannot flow through it.

When the tube is blocked using a thick baffle material, most of the incident sound wave is reflected back, and a few sound waves pass through the barrier, as shown in Figure 4.69. Thus, it is very difficult for the sound wave to continuously propagate in the tube, achieving the purpose of sound insulation. The baffle material is a special material designed to block the tube.

Many structures on a vehicle body are similar to the tube, such as beams and pillars; these are hollow tubes, as shown in Figure 4.70. The sound wave can freely transmit inside the beams and pillars. There are many openings on the beams and pillars, including function holes and manufacturing process holes. Some holes communicate with the body cavity, such as the mounting hole of the safety belt on the B-pillar. The airflow or

(a)

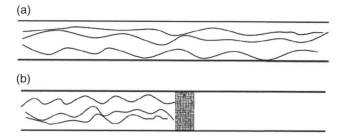

(b)

Figure 4.68 Tube and fluid: (a) unblocked tube; (b) blocked tube.

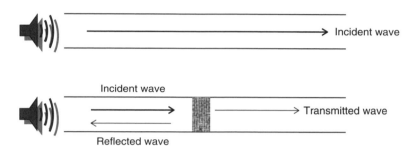

→ Incident wave

Incident wave

→ Transmitted wave

Reflected wave

Figure 4.69 A sound wave propagates in tubes.

(a)   (b)

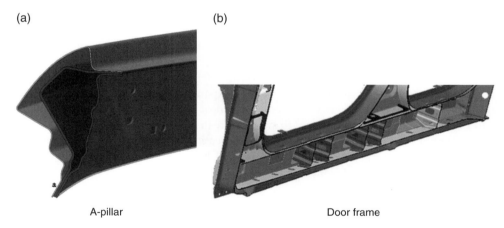

A-pillar   Door frame

Figure 4.70 Hollow tubes of pillars and beams: (a) A-pillar; (b) door frame.

sound wave from outside the vehicle can pass through the beams and pillars, and transmit through the holes into the interior.

In order to prevent the air flowing freely inside the beams and pillars, it is necessary to block the internal channels. Figure 4.71 shows that several locations inside the beams and pillars are blocked by baffles. The airflow cannot be transmitted inside the channels, so the sound transmission is blocked as well.

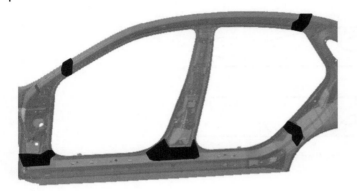

Figure 4.71 Several locations inside beams and pillars are blocked by baffles.

(a)   (b)

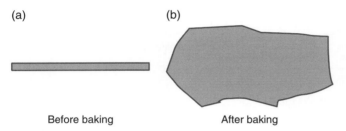

Before baking   After baking

Figure 4.72 Volume comparison of a baffle (a) before and (b) after baking.

The baffling material is a foaming material. Its original size is small, but after baking in a high temperature environment, its volume expands by several times, even dozens of times. Figure 4.72 shows a volume comparison of a baffle before and after baking. After baking, the baffle expands and firmly fills the inside of the tube.

Baffling material can be placed at a designed location inside a beam or pillar using a temporary installation bracket, such as a small hook. Figure 4.73 shows a baffle supported by a bracket that is placed at the lower end of a B-pillar.

4.6 Statistical Energy Analysis and Its Application

Statistical Energy Analysis (SEA) appeared in the 1960s. In order to study the vibration and noise problems of the aviation transmitter generated by the broadband random excitations, a group of US scientists introduced the heat transfer and room acoustics analysis methods. Later, R.H. Lyon summarized the work and provided the SEA method. In the 1980s, this method was used to analyze vehicles' high-frequency noise. Since the 1990s, SEA has been widely used in automotive engineering, and advanced commercial software has been developed.

Figure 4.73 A baffle supported by a bracket is placed at the lower end of a B-pillar.

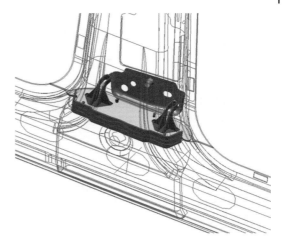

4.6.1 Concepts of Statistical Energy Analysis

SEA is used to analyze vibration and noise problems from two perspectives, "statistics" and "energy," which are explained below.

4.6.1.1 Concept of Statistics

Figure 4.74 shows a typical response spectrum of a vehicle's interior noise and vibration. At low frequencies, the response (such as acceleration, sound pressure, and so on.) curve is very clear, and the magnitude and frequency of each peak can be clearly read out. In the middle frequency range, the peak density increases, and it is hard to identify some peaks. In the high-frequency region, due to the high density of peak responses, it is impossible to distinguish the magnitude and frequency of each peak.

Because it is possible to clearly distinguish the low frequency responses, traditional analytical methods can be employed in the low frequency range to analyze specific velocities, sound pressure levels, and so on. In the high frequency region, due to high density of the modes, it is impossible to obtain the response for each mode. Therefore, the statistical method has to be used to describe average response values within a certain frequency band. For the high density vibration and noise signals, statistics is a good analysis method.

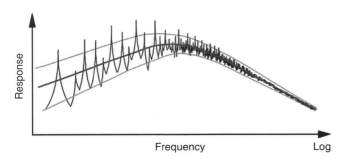

Figure 4.74 A typical response spectrum of a vehicle's interior noise and vibration.

4.6.1.2 Concept of Energy

When the modal density of a vibration system is very high, its kinetic energy and potential energy are almost the same. The total energy of the system, E_V, is expressed as

$$E_V = M\bar{u}^2, \tag{4.51}$$

where M is the system mass, and $\bar{u}$ is the average velocity.

Equation (4.51) can be rewritten as

$$\bar{u} = \sqrt{E_V / M}. \tag{4.52}$$

Equation (4.52) establishes the relationship between the velocity representing the structure vibration and the system energy.

For an acoustic cavity system, when its modal density is large enough, the acoustic energy (E_S) is proportional to the square of the average sound pressure, and expressed as

$$E_S = \frac{\bar{p}^2}{\rho_0 c} V_i, \tag{4.53}$$

where V_i is the volume of the cavity, ρ_0 is the air density, and c is the sound speed.

Equation (4.53) can be rewritten as

$$\bar{p}^2 = \frac{E_S}{V_i} \rho_0 c. \tag{4.54}$$

Equation (4.54) establishes the relationship between the sound pressure representing the sound signal and the system energy.

According to Eqs. (4.52) and (4.54), for a high-frequency and high modal density system, the velocity representing the vibration signal and the sound pressure representing the sound signal can be represented by the energy, so the energy method is a good analytical one for the system.

The statistics method and the energy method are combined to analyze high-frequency noise and vibration problems, and are the foundation of SEA.

4.6.2 Theory of Statistical Energy Analysis

4.6.2.1 Balance of Energy

The SEA of a system is based on the energy balance among inflowing and outflowing energy between the subsystems, and the energy loss inside the subsystems. Figure 4.75 shows a coupled system composed of two subsystems.

For the first subsystem, its energy flow consists of four parts.

1) The power input from the outside: the work done outside the system is noted as $\varPi_1$.
2) The power input to the first system from the second system, which is the lost energy of the second subsystem because of partial energy flows to the outside, and which is expressed as follows:

$$\varPi_{21} = \omega \eta_{21} E_2, \tag{4.55}$$

where η_{21} is the coupling loss factor of the second subsystem to the first subsystem; E_2 is the energy of the second system, and ω is the natural frequency of the subsystem.

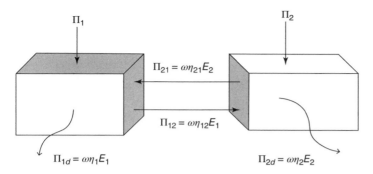

Figure 4.75 A coupled system consisting of two subsystems.

3) Internal power loss: Because of the presence of damping inside the system, part of the energy will be consumed inside the system and converted into other forms of energy. For viscous damping, the consumed power is

$$\Pi_{1d} = c_1 \dot{x}^2 = \omega \eta_1 E_1,$$ (4.56)

where c_1, η_1, and E_1 are the damping coefficient, internal loss factor, and energy, respectively, and $\dot{x}$ is the vibration velocity.

4) The power output to the second system from the first system, which is the lost energy of the first subsystem because of partial energy flows to the outside, and which is expressed as follows:

$$\Pi_{12} = \omega \eta_{12} E_1,$$ (4.57)

where η_{12} is the coupling loss factor of the first subsystem to the second subsystem.

For a conservative system, the inflowing energy equals the internal consumed energy plus the outflowing energy. For the first subsystem, the inflowing power includes the power from outside and the power flowing from the second system. Therefore, this subsystem energy balance equation can be expressed as

$$\Pi_1 + \Pi_{21} = \Pi_{1d} + \Pi_{12}.$$ (4.58)

Substitute Eqs. (4.55–4.57) into Eq. (4.58), obtaining

$$\Pi_1 = \omega \eta_1 E_1 + \omega \eta_{12} E_1 - \omega \eta_{21} E_2.$$ (4.59)

4.6.2.2 Energy Balance Equations of System

In a similar way to the first subsystem, the power balance equation of the second subsystem shown in Figure 4.75 can be expressed as follows:

$$\Pi_2 = \omega \eta_2 E_2 + \omega \eta_{21} E_2 - \omega \eta_{12} E_1,$$ (4.60)

where Π_2 is the input power to the second subsystem from the outside and η_2 is the internal loss factor of the second subsystem.

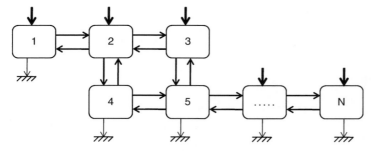

Figure 4.76 A big system consisting of N subsystems.

Put Eqs. (4.59) and (4.60) together, and the power flow equations of the system consisting of the two subsystems are expressed as

$$\omega \begin{bmatrix} \eta_1 + \eta_{12} & -\eta_{21} \\ -\eta_{12} & \eta_2 + \eta_{21} \end{bmatrix} \begin{Bmatrix} E_1 \\ E_2 \end{Bmatrix} = \begin{Bmatrix} \Pi_1 \\ \Pi_2 \end{Bmatrix}. \tag{4.61}$$

This analysis method of the system's power flow is extended to a big system consisting of N subsystems, as shown in Figure 4.76. For each system, the input energy is equal to the internal consumed energy plus the energy flowing to the outside.

For the big system including N sub-systems, the power flow equations are expressed as follows:

$$\omega \begin{bmatrix} \eta_1 + \sum\limits_{j \neq 1} \eta_{1j} & -\eta_{21} & \cdots & -\eta_{N1} \\ -\eta_{12} & \eta_2 + \sum\limits_{j \neq 2} \eta_{2j} & \cdots & -\eta_{N2} \\ \vdots & \vdots & & \vdots \\ -\eta_{1N} & \cdots & \cdots & \eta_N + \sum\limits_{j \neq N} \eta_{Nj} \end{bmatrix} \begin{Bmatrix} E_1 \\ E_2 \\ \vdots \\ E_N \end{Bmatrix} = \begin{Bmatrix} \Pi_1 \\ \Pi_2 \\ \vdots \\ \Pi_N \end{Bmatrix}. \tag{4.62}$$

Equation (4.62) can be also written as

$$\omega[\eta]\{E\} = \{\Pi\}. \tag{4.63}$$

According to Eq. (4.62) or (4.63), after the input power and the loss factor of the system have been given, the system energy will be obtained. After obtaining the energy of the system, according to Eq. (4.52) or (4.54), the vibration velocity or the sound pressure of the system will be obtained.

4.6.3 Assumptions and Applications of Statistical Energy Analysis

The above analyzed system is a conservative one in which the system's energy is balanced, and it also follows statistical law. SEA can be used only under certain conditions and includes a number of assumptions.

4.6.3.1 Assumptions of Statistical Energy Analysis

Four assumptions – a linear system, a conservative system, modal energy balance, and the reciprocity principle – are included in SEA.

1) Linear system: the system is linear, and the coupling between the subsystems is linear.
2) Conservative system: the system's energy is balanced, that is, the energy flows among the subsystems.
3) Modal energy balance: within a frequency band, the energy for all modes is equal.
4) Reciprocity principle: between the two coupling systems, within a frequency band, the products of loss factor and mode number for each subsystem are equal.

4.6.3.2 Application Scope of Statistical Energy Analysis

Before discussing the application scope of the SEA method, some concepts, such as modal density, equivalent frequency, and so on, are introduced below.

1) Modal density: within a frequency range (ω_1, ω_2), the ratio of the number of modes (N) to the bandwidth is defined as modal density, namely

$$n(\omega) = \frac{N}{\omega_2 - \omega_1}. \tag{4.64}$$

2) Equivalent frequency: within a frequency band, there are many modes. According to the principle of modal energy balance, the energy for all modes is equal. Therefore, the power of a single degree of freedom damping system can be equivalent to the power of all the modes within the frequency band. The frequency range corresponding to the single degree of freedom equivalent power is defined as the equivalent power frequency (Δe), and expressed as

$$\Delta e = \frac{\pi}{2}\omega\eta. \tag{4.65}$$

3) Modal overlap: the number of modes within a frequency range is called the modal overlap, denoted by $\bar{D}$, and is expressed as

$$\bar{D} = \Delta e * n(\omega) = \frac{\pi}{2} n(\omega)\omega\eta. \tag{4.66}$$

When the modal overlap is much less than 1, the modes can be clearly identified. When the modal overlap is much larger than 1, it is difficult to distinguish the modes.

Statistical methods should only be resorted to when the modes are difficult to distinguish. The modal overlap can be used as a criterion to judge the application scope of SEA. SEA should be used only when the modal overlap is much larger than 1.

4.6.4 Loss Factor

4.6.4.1 Classification of Loss Factor

According to Eq. (4.61), the loss factor of a system comprises an internal loss factor (η_i) and a coupling loss factor (η_{ij}) between the subsystems.

The internal loss factor refers to the ratio of energy consumed by damping in one cycle to the total mechanical vibration energy. The analysis and measurement of the internal loss factor is described in detail in Chapter 3. Generally, the loss factor of most materials tends to be small, but is very large for a damping structure. A damping structure converts vibration energy to heat to attenuate structural vibration and sound radiation.

The coupling loss factor represents the loss factor between the subsystems. It determines the amount of the energy flowing from one subsystem to another one. The coupling loss factor can refer to the coupling loss factor between solid subsystems, the coupling loss factor between fluid subsystems, and the coupling loss factor between a solid subsystem and a fluid subsystem.

There are three forms of coupling between solid subsystems: point coupling, line coupling, and surface coupling, as shown in Figure 4.77. The coupling among the rods is point coupling, the coupling between a plate and another plate is line coupling, and the coupling between two solid surfaces is surface coupling.

4.6.4.2 Acquisition of Loss Factor

The coupling loss factors can be obtained by analytical methods, testing methods, and numerical calculation methods.

4.6.4.2.1 Analytical Methods

It is very complex to analyze loss factors using the analytical methods that can be applied to some simple structures, such as the coupling between a rod and a plate. Even for simple structures, the mathematical expressions are very complex: for example, for the coupling between a rod and a plate shown in Figure 4.78, the coupling loss factor is expressed as

$$
\eta_{r,p} = \frac{1}{\omega \dfrac{l}{2\pi}\sqrt{\dfrac{\rho_r}{E}}} \left[\frac{1}{\Delta\omega} \int_{\omega-\Delta\omega/2}^{\omega+\Delta\omega/2} \frac{4\mathrm{Re}\left(\dfrac{1}{2A\sqrt{E\rho_r}}\right)\mathrm{Re}\left(\dfrac{1}{8\sqrt{Dh\rho_p}}\right)}{\left|\dfrac{1}{2A\sqrt{E\rho_r}} + \dfrac{1}{8\sqrt{Dh\rho_p}}\right|^2} \, d\omega \right]. \tag{4.67}
$$

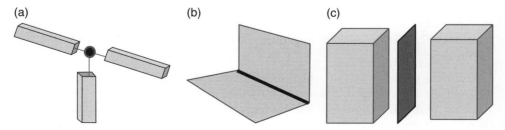

Figure 4.77 Solid couplings: (a) point coupling, (b) line coupling, (c) surface coupling.

Figure 4.78 Coupling between a rod and a plate.

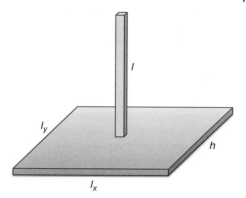

For a vehicle body, coupling between a solid subsystem and fluid subsystem mainly refers to the coupling between a body panel and the air. The coupling loss factor of the panel to the air is expressed as

$$\eta_{pa} = \frac{\rho_0 c \sigma}{\omega \rho_s},$$
(4.68)

where ρ_0 is the density of the air, ρ_s is the surface density of the panel, and σ is the sound radiation coefficient.

There are coupling losses between sound fields, such as the coupling between the body interior cavity and the trunk cavity. The coupling loss factor between two coupling sound fields is expressed as

$$\eta_{12} = \frac{cS}{4\pi V_1},$$
(4.69)

where S is the area of the coupling window and V_1 is the volume of the first cavity.

4.6.4.2.2 Testing Methods
Structural loss factor measurement methods include the steady energy flow method and the transient decaying method.

The loss factor can be written as a function of the input power to the system (Π_{in}) and the system energy (E), and expressed as follows:

$$\eta = \frac{\Pi_{in}}{\omega E}.$$
(4.70)

If the input power and the system energy are measured, the loss factor can be calculated by Eq. (4.70). In the steady energy flow method, a stable power needs to be input into the system. The system's response (velocity, $\bar{u}$, or sound pressure, $\bar{P}$) is then measured, and the system's energy can be calculated. Finally, the loss factor can be calculated by Eq. (4.70).

The purpose of SEA is to obtain the system's response if the power and the loss factor are provided. However, the function of Eq. (4.70) is to calculate the system's loss factor when the input power and the system response are given. Thus, the steady-state testing

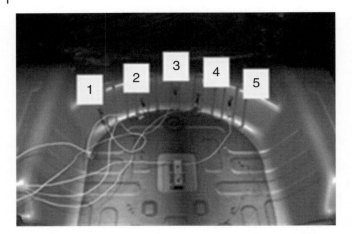

Figure 4.79 Test layout to measure the loss factor of the spare tire well surface.

method can be described as the inverse calculation to the SEA. Because stable input power is need, this method is more complicated.

The transient decaying method uses the decaying characteristics of the vibration or acoustic signals to determine the internal loss factor through the decaying time of the tested signals. The transient decaying method is relatively simple, and is suitable for quick estimation of the loss factor of a structure or a cavity. The method in Chapter 3 for measuring the component loss factor with the SAE J1637 standard is the transient decaying method. This method is widely used to test the loss factor of a vehicle body; a hammer is used to excite the body, and the acceleration responses are measured. Figure 4.79 shows a test layout to measure the loss factor of the spare tire well surface. Five acceleration sensors are placed on the surface, and the hammer is used to strike a point, then the frequency response functions between the responses to the input are obtained. The loss factor can be obtained from the decaying signals. Figure 4.80 provides the loss factors of the five tested points and their average loss factor.

The reason for energy loss inside a cavity is that the sound energy is absorbed, so the sound energy attenuation can be used to measure the loss factor of an acoustic cavity. In a similar way to the measurement of the reverberation time inside a reverberation chamber, the reverberation time of the cavity T_{60} is measured, then its internal loss factor can be calculated, and is expressed as follows:

$$\eta = \frac{2.2}{T_{60}f}. \tag{4.71}$$

For measurement of the loss factor of the body cavity, the volume sound source is used as an excitation, and the sound pressures are tested. Several microphones and the volume sound source are placed inside the vehicle. After the decaying characteristics of the sound signals have been measured, the loss factor can be obtained.

4.6.4.2.3 Numerical Calculation Method

Analytical methods for calculating loss factor are very complex, and it is impossible to obtain an analytical solution for complicated structures. Whereas testing methods are

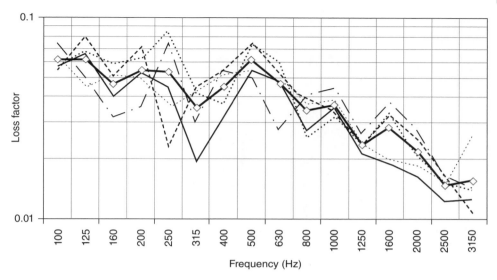

Figure 4.80 Loss factors of five tested points and average loss factor.

difficult and time-consuming, the numerical method is relatively simple and widely used. Finite element analysis can be used to calculate the internal loss factor for each subsystem and the coupling loss factors between the subsystems of a complex system.

4.6.4.3 Reciprocal Principle of Coupling Loss Factor

The loss factors (η_{ij} and η_{ji}) between two coupled systems (system i and system j) are closely related to the modal density. The reciprocal principle can be applied on the two systems, namely

$$\eta_{ij}N_i = \eta_{ji}N_j, \tag{4.72}$$

where N_i and N_j are mode numbers for system i and system j, respectively.

For two coupling systems, if the coupling loss factor from system i to system j is given, the coupling loss factor from system j to system i will be obtained.

4.6.5 Input Power

To calculate the system's energy using the statistical energy method, in addition to the loss factor, the input power must be given. For the vibrating system generated by the external force excitation, its power is

$$\Pi = \frac{1}{2}\mathrm{Re}\left\{F^*u\right\}, \tag{4.73}$$

where F and u are the input force and velocity response, respectively, and $*$ represents a complex conjugate.

By measuring the input force and velocity response, the power input to the system can be calculated. The body structure is subjected to a variety of external vibration excitations, such as from the engine, road, driveline, and so on. The excitation characteristics are introduced in detail in Chapter 5.

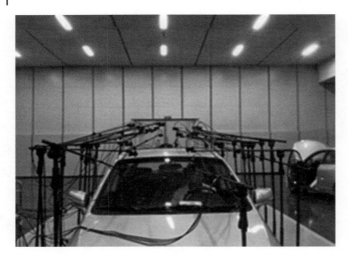

Figure 4.81 Sound pressure testing outside a body.

In addition to being subjected to force excitations, the body is also subjected to sound pressure excitation. The input sound power is expressed as

$$\Pi = IS = \frac{p^2}{\rho_0 c} S, \qquad (4.74)$$

where I is the sound intensity and S is the area.

The sound pressure outside the body can be obtained by measurement or computational fluid dynamics (CFD) calculation. Figure 4.81 shows a picture of the sound pressure testing outside the body.

The body is subjected to excitations of the sound pressure and the airflow field pressure, such as from the engine sound, exhaust tailpipe sound, and so on. Such excitations are introduced in Chapter 5 and Chapter 6.

4.6.6 Application of Statistical Energy Analysis on Vehicle Body

The application of SEA is essentially based on the existing commercial software. With this software, the sound insulation of a body, system, or component can be calculated and analyzed.

4.6.6.1 Acoustic Analysis of Vehicle Body

To analyze the body vibration and noise problems using SEA, structural and acoustic models for the subsystems must be established. All of the available commercial software has functions to build SEA models.

The main elements of the body subsystem include flat plates, single curvature plates and double curvature plates. The subsystem structural models can be obtained by converting computer aided design (CAD) data or finite element models. Figure 4.82 shows an SEA model of a body subsystem structure. The material and geometrical parameters of the models can be obtained by directly converting the finite element models.

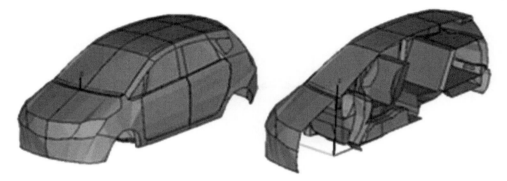

Figure 4.82 SEA model of a body subsystem structure.

The body cavity model is divided into an interior model and an exterior model. The function of the exterior body model (Figure 4.83a) is to input sound pressures. The function of the body interior model (Figure 4.83b) is to calculate the interior sound pressures. The interior cavities include the passenger compartment cavity and the A-, B-, and C-pillar cavities.

After the subsystem models have been established, first the loss factor, geometrical, and physical parameters of the interior materials are added, then the external excitations, including vibration and pressure, are added. Finally, the desired statistic energy parameters can be calculated and obtained using the software.

The applications of SEA on the vehicle body include two aspects, namely calculation of the interior sound pressure and analysis of the body sound package. After the body SEA model has been established, the external sound pressures are obtained by testing, as shown in Figure 4.81, or by CFD calculation, and are then applied to the body surface. Then the interior sound pressure can be calculated. Figure 4.84 shows an interior noise comparison between the measured value and SEA predicted value for a vehicle at a cruising speed of $100 \, \text{km} \, \text{h}^{-1}$. Below 500 Hz, the predicted value and the measured value have a 5–10 dB difference; however, above 500 Hz they are very close, which indicates that SEA is very useful for predicting middle- and high-frequency noise.

(a)   (b)

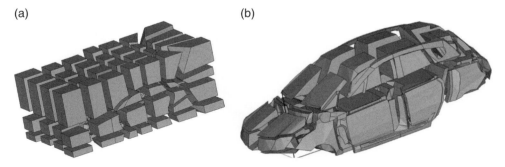

Figure 4.83 An exterior model (a) and interior model (b) of a body cavity model.

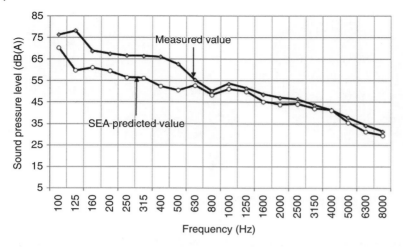

Figure 4.84 Interior noise comparison between measured value and SEA predicted value.

4.6.6.2 Sound Insulation Analysis of Systems and Components

SEA can be used to predict and compare the sound insulation effectiveness of the body systems and components, such as the dash panel, doors, and so on. In a similar way to the vehicle body model, the structural sub-models and acoustic sub-models of the systems or components must be established. In addition, models of the reverberation chamber and anechoic chamber in which the system or component models are placed should be built.

Figure 4.85 shows a cavity model of a dash panel. The dash panel models are placed between the reverberation chamber model and the anechoic chamber model. The noise reduction of the dash panel can be calculated by SEA analysis in a similar way to the method of physical testing.

Figure 4.86 shows an STL comparison between the SEA calculated value and measured value for a dash panel. Above 300 Hz, the calculated and measured values are very close. By using SEA analysis, the sound insulation effectiveness of different designs, such as different materials, different installations, and different structures, can be predicted and compared. The openings on a dash panel have a significant influence on the sound insulation, and these openings can be analyzed by the SEA method as well. Figure 4.86 shows that after a small hole is opened on the dash panel, the STL reduces by 8 dB in the high-frequency range.

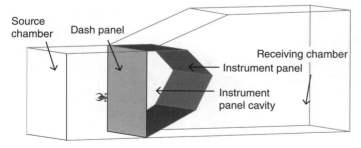

Figure 4.85 Cavity model of a dash panel.

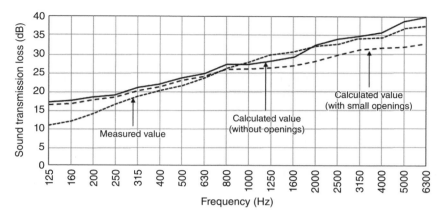

Figure 4.86 STL comparison between the SEA calculated value and measured value for a dash panel.

Bibliography

Alexander J., Reed D., Gerdes R. Random Incidence Absorption and Transmission Loss Testing and Modeling of Microperforated Composites. SAE Paper 2011-01-1626; 2011.

ASTM. *Standard Test Method for Sound Absorption and Sound Absorption Coefficients by the Reverberation Room Method*. ASTM International; 2009.

Beranek L.L. *Acoustics*. Acoustical Society of America; 1996.

Bertolini C., Guj L. Numerical Simulation of the Measurement of the Diffuse Field Absorption Coefficient in Small Reverberation Rooms. SAE Paper 2011-01-1641; 2011.

Bertolini C., Falk T. On Some Important Practical Aspects Related to the Measurement of the Diffuse Field Absorption Coefficient in Small Reverberation Rooms. SAE Paper 2013-01-1972; 2013.

B&K. Fundamentals of the Two-microphone Method. 2009.

Bolduc M., Atalla N., Wareing A. Measurement of SEA Damping Loss Factor for Complex Structures. SAE Paper 2005-01-2327; 2005.

Calçada M., Parrett A. Enhanced Acoustic Performance using Key Design Parameters of Headliners. SAE Paper 2015-01-2339; 2015.

Chae K.S., Lim S.H., Yoo J.W., Hong S.G. Study on Sound Insulation Performance of Vehicle Dash Reinforcements. SAE Paper 2014-01-2085; 2014.

Cherng J.G., Xi Q., Mohanty P., Ebbitt G. A Comparative Study on Sound Transmission Loss and Absorption Coefficient of Acoustical Materials. SAE Paper 2011-01-1625; 2011.

Cherng J.G., Xing S., Wu W. Acoustics Characterization of Nano Enhanced Open Cell Foams. SAE Paper 2015-01-2205; 2015.

Chevillotte F., Panneton R., Wojtowicki J.L., Chaut C. Characterization of the Bulk Elastic Properties of Expanding Foams from Impedance Tube Absorption Tests. SAE Paper 2007-01-2191; 2007.

Chuang C.H., Shu K.T., Liu W., Qian P. A CAE Optimization Process for Vehicle High Frequency NVH Applications. SAE Paper 2005-01-2422; 2005.

Connelly T., Knittel J.D., Krishnan R., Huang L. The Use of in Vehicle STL Testing to Correlate Subsystem Level SEA Models. SAE Paper 2003-01-1564; 2003.

Cordioli J.A., Calçada M., Rocha T., Cotoni V., Shorter P. Application of the Hybrid FE-SEA Method to Predict Sound Transmission Through Complex Sealing Systems. SAE Paper 2011-01-1708; 2011.

Crocker, M.J. (2007). *Handbook of Noise and Vibration Control.* John Wiley & Sons.

Dinsmore M., Bliton R., Perz S. Density Optimization of Underhood Sound Absorber Applications. SAE Paper 2011-01-1634; 2011.

Dinsmore M. Measurement Dynamic Range Considerations for Sound Transmission Loss Testing. SAE Paper 2011-01-1650; 2011.

Dow Automotive. 3D-Cavity Sealing with BETAFOAM™ AFI. 2007.

Duval A., Rondeau J.F., Dejaeger L., Lhuillier F., Monet-Descombey J. Generalized Light-Weight Concepts: A New Insulator 3D Optimization Procedure. SAE Paper 2013-01-1947; 2013.

Ebbitt G., Hansen M. Mass Law - Calculations and Measurements. SAE Paper 2007-01-2201; 2007.

Ebbitt G., Remtema T., Scheick J. Sound Absorbers in Small Cavities. SAE Paper 2013-01-1945; 2013.

Everest, F.A. (2001). *The Master Handbook of Acoustics.* McGraw-Hill.

Färm A., Boij S., Glav R. On Sound Absorbing Characteristics and Suitable Measurement Methods. SAE Paper 2012-01-1534; 2012.

Fasse M. The Current State of Pre-Formed versus Bulk Cavity Filler Technologies in the Automotive, Medium Duty and Heavy Truck Markets. SAE Paper 2013-01-1946; 2013.

Fasse M., Brichet N. Innovative Thermoplastic Cavity Filler Design Solutions. SAE Paper 2013-01-1939; 2013.

Ferreira P.E., Melo F.M.D., Silva M.G.D. Sound Package Evaluation of Dash Panel Focusing on Weight Reduction using Particle Velocity Measurements. SAE Paper 2012-36-0627; 2012.

Filho N.G.R.D.M., Morais M.V.G.D., Ferreira A.C., Carvalho M.O.M.D. Experimental Modal Identification of Vibro-Acoustic Cavities with Calibrated Acoustic Source. SAE Paper 2012-36-0619; 2012.

Fredö C.R. A Modification of the SEA Equations: A Proposal of How to Model Damped Car Body Systems with SEA. SAE Paper 2005-01-2436; 2005.

Fung K.K.H., Li X., Huang W., Wentzel R.E., Zhu K. Measurement of Sound Transmission Loss Properties in Single & Multi-layered Systems – A Comparative Study between Two-room and Standing Wave Tube Techniques. SAE Paper 2011-01-1653; 2011.

Gish L.E., Sanderson M.A. Use of Layered Media for Noise Abatement in Automotive Interiors: A Balanced Approach. SAE Paper 2001-01-1456; 2001.

Green E.R. Improving Tube Sound Transmission Loss Measurements Using the Transfer Matrix Technique to Remove the Effect of Area Changes. SAE Paper 2015-01-2310; 2015.

Green E.R., Frey A. The Effect of Edge Constraints on the Measurement of Automotive Sound Package Materials Using a Tube Apparatus. SAE Paper 2013-01-1942; 2013.

Griffen C.T., Khambete S., Sampath A. An Efficient Design & Manufacturing Process for Automotive Dash Insulators with Optimally Tuned Performance. SAE Paper 2003-01-1421; 2003.

Grosso A., Bree H.E.D., Steltenpool S., Tijs E. Scan and Paint for Acoustic Leakage Inside the Car. SAE Paper 2011-01-1673; 2011.

Gur Y., Pan J., Huber J., Wallace J. MMLV: NVH Sound Package Development and Full Vehicle Testing. SAE Paper 2015-01-1615; 2015.

Han J., Herrin D.W., Seybert A.F., Accurate Measurement of Small Absorption Coefficients. SAE Paper 2007-01-2224; 2007.

Hirabayashi T., Rusch P., McCaa D., Saha P., Rebandt R. Application of Noise Control and Heat Insulation Materials and Devices in the Automotive Industry. SAE Paper 951375; 1995.

Hua X., Herrin D.W. Reducing the Uncertainty of Sound Absorption Measurements Using the Impedance Tube Method. SAE Paper 2013-01-1965; 2013a.

Hua X., Herrin D.W., Jackson P. Enhancing the Performance of Microperforated Panel Absorbers by Designing Custom Backings. SAE Paper 2013-01-1937; 2013.

Hua X., Herrin D.W. Reducing the Uncertainty of Sound Absorption Measurements Using the Impedance Tube Method. SAE Paper 2013-01-1965; 2013b.

Jain D.B., Saha P. Predicting the Acoustical Performance of Weak Paths in a Sound Package System. SAE Paper 2005-01-2520; 2005.

Jain S.K., Shravage P., Joshi M., Karanth N.V. Acoustical Design of Vehicle Dash Insulator. SAE Paper 2011-26-0022; 2011.

Kim N.N., Lee S., Bolton J.S., Hollands S., Yoo T. Structural Damping by the Use of Fibrous Materials. SAE Paper 2015-01-2239; 2015.

Kobayashi N., Tachibana H. A SEA-Based Optimizing Approach for Sound Package Design. SAE Paper 2003-01-1556; 2003.

Kolano R. Design and Construction of an Innovative Sound Transmission Loss Testing Fixture. SAE Paper 2013-01-1963; 2013.

Laux P.C., Veen J.R., Unglenieks R.J., Dinsmore M.R. Novel Design of a Multi-Function Acoustics Laboratory for the Testing and Evaluation of Automotive Acoustics Systems and Components. SAE Paper 2001-01-1489; 2001.

Lee J.W., Lee W., Lee S.N., Park S.W. A Study for Improving the Acoustic Performance of Dash Isolation Pad Using Hollow Fiber. SAE Paper 2013-01-0101; 2013.

Liu J., Herrin D.W. Effect of Contamination on Acoustic Performance of Microperforated Panels. SAE Paper 2011-01-1627; 2011.

Liu J., Herrin D.W., Seybert A.F. Application of Micro-Perforated Panels to Attenuate Noise in a Duct. SAE Paper 2007-01-2196; 2007.

Liu W., Tao D., Kathawate G. Use of Statistical Energy Analysis Method to Predict Sound Transmission Loss of Sound Barrier Assemblies. SAE Paper 1999-01-1707; 1999.

Liu Z., Fard M., Jazar R. Development of an Acoustic Material Database for Vehicle Interior Trims. SAE Paper 2015-01-0046; 2015.

Luo Z., Schneider M., Cherng J.G. Investigation and Validation of Transmission Loss for Vehicle Components with a Large Aperture. SAE Paper 2001-01-1621; 2001.

Mendonca G., Connelly T., Bonthu S., Shorter P. CAE-Based Prediction of Aero-Vibro-Acoustic Interior Noise Transmission for a Simple Test Vehicle. SAE Paper 2014-01-0592; 2014.

Misaji K., Tada H., Yamashita T., Mantovani M., Gaudino C., Falk T. Prediction of the SEA Input Parameters for the Sound Package. SAE Paper 2003-01-1022; 2003.

Musser C., Manning J., Peng G.C. Prediction of Vehicle Interior Sound Pressure Distribution with SEA. SAE Paper 2011-01-1705; 2011.

Nelisse H., Onsay T., Atalla N. Structure Borne Insertion Loss of Sound Package Components. SAE Paper 2003-01-1549; 2003.

Noguchi Y., Doi T., Tada H., Misaji K. Development of a Lightweight Sound Package for 2006 Brand-New Vehicle Categorized as C. SAE Paper 2006-01-0710; 2006.

Pan J., Saha P., Veen J.R. Random Incidence Sound Absorption Measurement of Automotive Seats in Small Size Reverberation Rooms. SAE Paper 2007-01-2194; 2007.

Pan J., Semeniuk B., Ahlquist J., Caprioli D. Optimal Sound Package Design Using Statistical Energy Analysis. SAE Paper 2003-01-1544; 2003.

Pang, J., Sheng, G., and He, H. (2006). *Automotive Noise and Vibration – Theory and Application*. Beijing Institute of Technology Press.

Parrett A.V., Wang C., Zeng X., Nielubowicz D., Snowden M. Application of Micro-Perforated Composite Acoustic Material to a Vehicle Dash Mat. SAE Paper 2011-01-1623; 2011.

Patching M., Taylor J., Khan H. Carpets & Floor Surface Technologies: An Overview Considering Global Vehicle Design & Performance. SAE Paper 2007-01-2193; 2007.

Pieper R.M., Alexander J.H., Bolton J.S., Yoo T. Assessment of Absorbers in Normal-Incidence Four-Microphone Transmission-Loss Systems to Measure Effectiveness of Materials in Lateral-Flow Configurations of Filled or Partially Filled Cavities. SAE Paper 2007-01-2190; 2007.

Pietrzyk A. Prediction of Airborne Sound Transmission into the Passenger Compartment. SAE Paper 2015-01-2266; 2015.

Prasanth B, Wagh S., Hudson D. Evaluation of Acoustic Performance of Expandable Foam Baffles and Correlation with Incab Noise. SAE Paper 2011-01-1624; 2011.

Qian Y., Vanbuskirk J., Gorzelski T. Sound Package Weight Reduction: An Analysis Through Tests and SEA Models. SAE Paper 1999-01-1696; 1999.

Raveendra S.T., Zhang W. Vibro-acoustic Analysis Using a Hybrid Energy Finite Element/Boundary Element Method. SAE Paper 2007-01-2177; 2007.

SAE International Surface Vehicle Recommended Practice. Laboratory Measurements of the Airborne Sound Barrier Performance of Automotive Materials and Assemblies. SAE standard J1400, Rev. 2017.

Saha P. Application of Noise Control Materials to Trucks and Buses. SAE Paper 2002-01-3063; 2002.

Saha P. The Thought Process for Developing Sound Package Treatments for a Vehicle. SAE Paper 2011-01-1679; 2011.

Shkreli V., Vandenbrink K.A. The Use of Subjective Jury Evaluations for Interior Acoustic Packaging. SAE Paper 2003-01-1506; 2003.

Siavoshani S., Frost J. ACOUSTOMIZE™ A Method to Evaluate Cavity Fillers NVH & Sealing Performance. SAE Paper 2011-01-1672; 2011.

Stotera D., Connelly T., Gardner B., Seifferlein E., Alvarez R.D.A. Testing and Simulation of Anti-Flutter Foam and High Damping Foam in a Vehicle Roof Structure. SAE Paper 2013-01-1944; 2013.

Suresh S., Kastner J., Lim T. Transmission Loss Analysis through Porous Laminated Glass using Transfer Matrices. SAE Paper 2011-01-1629; 2011.

Tanaka H, Miyama Y., Murakami H., Enomoto T. Balance Weight-Saving with Performance of Acoustic Isolation Using Hybrid SEA Model. SAE Paper 2012-01-0216; 2012.

Tao K., Parrett A., Nielubowicz D. Headliner Absorption Parameter Prediction and Modeling. SAE Paper 2015-01-2303; 2015.

Tathavadekar P., Alvarez R.O.D.A., Sanderson M., Hadjit R. Hybrid FEA-SEA Modeling Approach for Vehicle Transfer Function. SAE Paper 2015-01-2236; 2015.

Tracey B.H., Huang L. Transmission Loss for Vehicle Sound Packages with Foam Layers. SAE 1999-01-1670; 1999.

Veen J.R. Material Construction, Manufacturing Processes and Acoustic Performance Characteristics of Fibrous Absorber Materials Used in the Automotive Industry. SAE Paper 2005-01-2381; 2005.

Unglenieks R.J. Dependence of Sound Package Item Sensitivities on Initial Conditions. SAE Paper 2005-01-2423; 2005.

Ver, I.L. and Beranek, L.L. (2006). *Noise and Vibration Control Engineering: Principles and Applications.* John Wiley & Sons.

Visconte G., Fasana A. Transmission Loss Prediction of Multilayered Components – A New Impedance Formulation to Take in Account Finite Structural Connection. SAE Paper 2011-01-1715; 2011.

Völker C., Thesing U. Acoustic Optimization of a Blow Molded Resonant Absorber. SAE Paper 2003-01-1568; 2003.

Wang C., Tao K., Parrett A. Performance Equivalent Thickness of a Sound Insulation System. SAE Paper 2013-01-1981; 2013.

Wang C., Parrett A. Damping Mass Effects on Panel Sound Transmission Loss. SAE 2011-01-1633; 2011.

Wentzel R.E., Aubert A.C. An Interactive Approach to the Design of an Acoustically Balanced Vehicle Sound Package. SAE Paper 2007-01-2314; 2007.

Woodcock R., Ebbitt G. Modeling the Vibro-Acoustical Behavior of Composite Multi-layered Systems. SAE Paper 2001-01-1413; 2001.

Wyerman B.R., Reed D.B. The Role of a Fiber Decoupler on the Acoustical Performance of Automotive Floor Systems. SAE Paper 2007-01-2185; 2007.

Wyerman B.R., Reed D.B., Krumnow T. The Impact of Sample Size on Sound Absorption Measured in a Small Reverberation Room. SAE Paper 2013-01-1967; 2013.

Yamamoto T., Maruyama S., Shimada H. Weight Reduction of Damping Materials on Vehicle Body Panels by using an Optimization with Sound Pressure Constraints. SAE Paper 2012-01-0220; 2012.

Zhang Q., Parrett A., Wang C., Wang D., Huang M. SEA Modeling of A Vehicle Door System. SAE Paper 2005-01-2427; 2005.

Zhou G., Tian X., Zhu K., Huang W., Wentzel R.E., Care M.J. Study on the Influence of Material Parameters to Acoustic Performance. SAE Paper 2015-01-2200; 2015.

Zhu J., Hammelef D., Wood M. Power-Based Noise Reduction Concept and Measurement Techniques. SAE Paper 2005-01-2401; 2005.

5

Vehicle Body Sensitivity Analysis and Control

5.1 Introduction

5.1.1 System and Transfer Function

A system is a collection of interacting or interdependent components forming an integrated whole, and has a special function. Any object can be regarded as a system, such as a vehicle body, an engine, etc. A dinner table can be also regarded as a system.

If an input is applied on a system, an output will be generated. The input could be force, temperature, pressure, etc. the output could be force, displacement, speed, etc. Figure 5.1 shows the relationship between the input and the output of a system. The input is also called excitation and the output is called response.

For a dynamic structural system or an acoustic system, both the input and the output signals are functions of frequency and are expressed by $X(\omega)$ and $Y(\omega)$, respectively. The relationship between the input and the output signals can be characterized by a transfer function. The transfer function is defined as the ratio of the output signal to the input signal, as shown in Eq. (1.4). In order to maintain the integrity of this chapter, Eq. (1.4) is rewritten as follows:

$$H(\omega) = \frac{Y(\omega)}{X(\omega)} \tag{5.1}$$

The transfer function is also a function of frequency, which represents the inherent characteristics of a dynamic system, that is, each dynamic system has a unique transfer function. When the transfer function of the system is independent of the input and the output, the system is called a linear system, which means that no matter how the input changes, the system's transfer function does not change, that is, the transfer function of the linear system is fixed. For example, for a single degree of freedom (SDOF) system shown in Figure 3.45, the elastic force is proportional to the displacement and the damping force is proportional to the velocity. For this linear system, the transfer function between the displacement response and the force input is fixed, that is, the input force could be 10 or 1000 N, but the transfer function does not change, as shown in Figure 5.2.

The superposition principle is suitable for the linear system. For the linear system, if the input is $X_1(\omega)$, the output $Y_1(\omega)$ is generated, and if the input is $X_2(\omega)$, the output $Y_2(\omega)$ is generated, which are expressed as follows:

Noise and Vibration Control in Automotive Bodies, First Edition. Jian Pang.
© 2019 China Machine Press. All rights reserved. Published 2019 by John Wiley & Sons Ltd.

| Input $X(\omega)$ | → | System $H(\omega)$ | → | Output $Y(\omega)$ |

Figure 5.1 Relationship between input and output of a system.

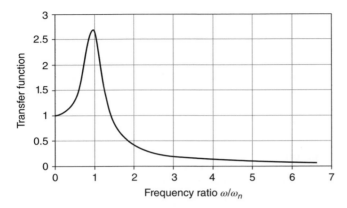

Figure 5.2 Transfer function of a single degree of freedom system.

$$Y_1(\omega) = H(\omega)X_1(\omega) \tag{5.2a}$$

$$Y_2(\omega) = H(\omega)X_2(\omega) \tag{5.2b}$$

When the input is $\varepsilon_1 X_1(\omega) + \varepsilon_2 X_2(\omega)$, the output will be $\varepsilon_1 Y_1(\omega) + \varepsilon_2 Y_2(\omega)$,

$$\varepsilon_1 Y_1(\omega) + \varepsilon_2 Y_2(\omega) = H(\omega)\left[\varepsilon_1 X_1(\omega) + \varepsilon_2 X_2(\omega)\right] \tag{5.3}$$

where $\varepsilon_1, \varepsilon_1$ are the coefficients related to the inputs.

According to the superposition principle, the total input can be decomposed into a number of sub-inputs. If the output for each sub-input is obtained, the total output response will be obtained by superposition of all sub-outputs. For a linear system, the relationship between the input and the output for each subsystem can be analyzed independently, which brings great convenience to analyze a system having a complex input.

A nonlinear system, in contrast to a linear system, is a system that does not satisfy the superposition principle; instead, its output is not directly proportional to the input. The transfer function of the nonlinear system is not fixed, that is, when the input changes, not only the output changes, but also the transfer function changes.

The automotive seat is a typical nonlinear system. Figure 3.104 shows the vibration transfer function between the seat cushion response and the platform excitation. The transfer function changes with the input change. The transfer function curves have two distinct nonlinear characteristics. Both the natural frequencies and magnitudes of the transfer functions decrease with the floor acceleration increases.

In the real world, there are few linear systems, but many systems can be regarded as linear systems. Strictly speaking, the vehicle body is a nonlinear system, but from the perspective of dynamic analysis, the body structure can be regarded as a linear system.

5.1.2 Vibration and Sound Excitation Points on Vehicle Body

Various vibration and noise excitations are exerted on the body, and then transferred to the passengers. After the vibration excitations are exerted on the body, the vibrational waves propagate inside the body frames and panels, and then transmit to the passengers' hands and bodies, so they perceive the vibration. At the same time, the excited panels radiate sound into the interior, so they hear the noise. The interior noise generated by the external vibration and transmitted through the body structures is called structural-borne noise. Outside noise passing through the body insulation into the interior and heard by the occupants is called airborne noise.

The occupant responses depend not only on the magnitudes and characteristics of the excitation sources but also on the transfer characteristics of the vehicle body structures. The excitation locations, that is, the points the vibration and noise applied on the body, are very important for analysis of the body transfer functions. The following materials will introduce these excitation points.

5.1.2.1 Vibration Excitation Points on Body

Many systems, such as power plant, exhaust, intake, driveline, suspension, steering system, electrical system, etc., are installed on the body directly. When these systems are operating, their exciting forces will be transmitted to the body through the corresponding attached points with the body. The main attached points on the body are listed as follows:

1) The attached points between the power plant and the body
2) The attached points between the exhaust hangers and the body
3) The attached points between the suspension and the body
4) The attached points between the driveline and the body
5) The attached points between the cooling system and the body
6) The attached points between the electrical system and the body
7) The attached points between the pipes and the body
8) The attached points between the accessories and the body

Figure 5.3 shows the main attached points on the body.

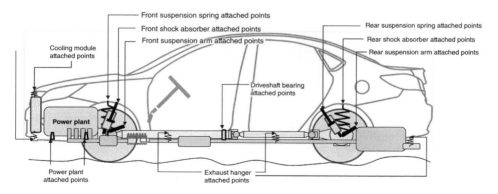

Figure 5.3 Main attached points on a body.

The connection between the body and the attached systems can be divided into two categories: flexible connection and rigid connection.

The flexible connection means that a flexible connecting element, such as vibration isolator, rubber pad, etc., is placed between an attached system and the body, as shown in Figure 5.4. According to the size and function of the flexible elements, the flexible connection can be divided into four categories:

1) *Engine mount.* The engine mount is a flexible element to take high engine load. Its size is relatively large, and it has high dynamic stiffness. The mount includes rubber mount, hydraulic mount, active mount, and semi-active mount. The engine mount is used to connect the power plant and the body through brackets, as shown (1) in Figure 5.5.

2) *Bushing.* The bushing is a flexible element used to carry load in suspension system, driveline system, etc., for example, the bushings used to connect the suspension arms and the body, to connect the subframe and the body, and to connect the driveline and the body, etc. Similar to the engine mount, the bushings also include rubber bushing, hydraulic bushing, active bushing, and semi-active bushing. One difference of the bushing from the engine mount is that it is usually connect with the body directly, as shown (2) in Figure 5.5; that is, there are no brackets between them.

3) *Pad.* The pad is a small size and thin rubber element, and its dynamic stiffness is low, such as the pads to isolate pipes, cooling module, resonator, etc as shown (3) in Figure 5.5.

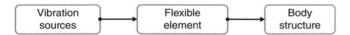

Figure 5.4 External excitation – flexible element – body.

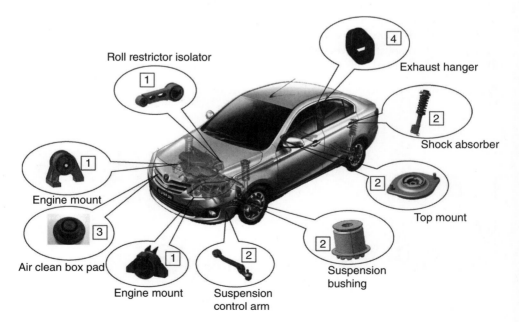

Figure 5.5 Flexible elements used to connect other systems and the body.

4) *Hanger isolator.* The hanger isolator is a rubber element used to isolate the exhaust system from the body, as shown (4) in Figure 5.5. Usually, the small brackets called exhaust hangers are used between the two systems.

The flexible elements attenuate the forces and vibration transmitted to the body from the attached systems, so the flexible element design is very important for vibration and structural-borne noise control.

The rigid connection refers to the direct connection between an attached system and the body, such as the rigid connections between the subframe and the body, between the steering shaft and the body, etc. so the attached systems transfer the excitations to the body directly.

5.1.2.2 Sound Excitation Points on Body

The vibration excitations directly apply on the body structure, but the noise excitation sources are not; instead, most of them are outside the body. When the noise sources are transmitted to the body, partial sound energy is attenuated by the body sound insulation structure and the rest passes through the body, entering the interior. The main noise source points include:

1) Engine radiation noise excitation points
2) Intake orifice noise excitation point
3) Exhaust orifice noise excitation point
4) Transmission radiation noise excitation points
5) Driveline radiation noise excitation points
6) Road noise excitation points
7) Wind noise excitation points
8) Electrical devices noise excitation points
9) Body panel (excited by wind, engine, road) radiation noise points
10) Accessories (such as side mirror excited by the airflow) noise excitation points
11) Other shells (intake resonator, muffler, etc.) Radiation noise excitation points

Figure 5.6 shows the main noise source excitation points.

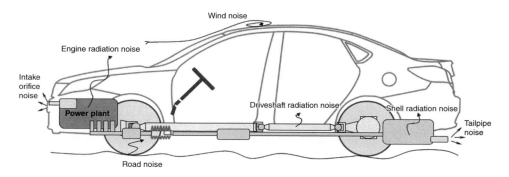

Figure 5.6 Main noise source excitation points.

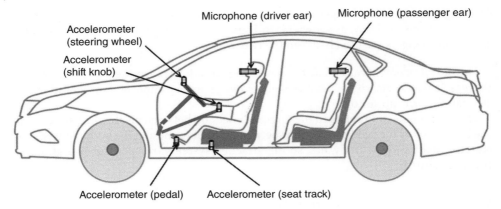

Figure 5.7 Human body response points.

5.1.3 Response Points

The occupants perceive noise by ears and vibration by hands, feet, and bodies, so the ears, hands, feet, and bodies are the response points, as shown in Figure 5.7.

Noise response points:

- Ears

Vibration response points:

- Points on steering wheel
- Points on the shift knob
- Points on floor
- Points on seat track
- Points on the seat cushion

5.1.4 Body Sensitivity

The outside source inputs include noise excitations and vibration excitations. The human body responses include the heard sound and perceived vibration. The ratio of a response to an excitation is defined as a vehicle body transfer function. The transfer function can be divided into three categories:

1) Transfer functions between the human body vibration responses and outside force (vibration) excitations
2) Transfer functions between the human body sound responses and outside force (vibration) excitations
3) Transfer functions between the human body sound responses and outside sound excitations.

In the automobile noise, vibration, and harshness (NVH) world, the engineers usually classify the transfer function as sensitivity. The sensitivity indicates the strength of a system response to an outside excitation. For a fixed excitation, the bigger the system response, the more sensitive the system, and vice versa. Corresponding to the transfer functions, the vehicle body sensitivities can be divided into three categories:

1) Vibration–vibration sensitivity
2) Sound–vibration sensitivity
3) Sound–sound sensitivity

In the above sensitivity descriptions, the first word represents response and the second word symbolizes excitation. For example, in "sound–vibration sensitivity," the *sound* represents the interior noise response and the *vibration* means the external force excitation. The three categories of sensitivity will be briefly described as follows.

In this book, both "transfer function" and "sensitivity" are used, and their meanings are the same. Which one is preferred to be used depends on different engineers' habits.

5.1.4.1 Vibration–Vibration Sensitivity (Vibration–Vibration Transfer Function)

Vibration–vibration sensitivity refers to the ratio of the interior vibration response to the outside force excitation. The interior vibration response is represented by $V(\omega)$, and the outside force excitation is represented by $F(\omega)$, so the vibration–vibration sensitivity, H^{VF}, is expressed as

$$H^{VF}(\omega) = \frac{V(\omega)}{F(\omega)} \tag{5.4}$$

5.1.4.2 Sound–Vibration Sensitivity (Sound–Vibration Transfer Function)

Sound–vibration sensitivity refers to the ratio of the interior sound response to the outside force excitation. The interior sound response is represented by $P(\omega)$, and the outside force excitation is represented by $F(\omega)$, so the sound–vibration sensitivity, H^{SB}, is expressed as

$$H^{SB}(\omega) = \frac{P(\omega)}{F(\omega)} \tag{5.5}$$

The superscript SB in the sensitivity expression represents structural-borne noise. The first letters in the two words *structural* and *borne* are used to symbolize SB.

5.1.4.3 Sound–Sound Sensitivity (Sound–Sound Transfer Function)

Sound–sound sensitivity refers to the ratio of the interior sound response to the outside sound excitation. The outside sound excitation is represented by $P_E(\omega)$, so the sound–sound sensitivity, H^{AB}, is expressed as

$$H^{AB}(\omega) = \frac{P(\omega)}{P_E(\omega)} \tag{5.6}$$

The superscript AB in the sensitivity expression represents airborne noise. The first letters in the two words *air* and *borne* are used to symbolize AB.

5.2   Source–Transfer Path–Response Model for Vehicle Body

5.2.1   Source–Transfer Path–Response Model

The transmission process of noise and vibration can be described as an Excitation–Transfer Path–Occupant model, as shown in Figure 5.8; i.e. the outside "noise and vibration sources" are transmitted into the interior through the "transfer paths" and then perceived by the "occupants".

The transmission process can be abstracted as a source–transfer function–response mathematic and analytical model. The "source" refers to the outside vibration source and noise source. It is fundamental to understand the sources characteristics in order to analyze the responses. The source characteristics that are functions of frequency, engine speed, and order, etc. are very complex. The "transfer function" represents the frequency characteristics of the body structure. The body is a barrier to attenuate noise and vibration transmission, so it is important to understand the frequency characteristics of the transfer function in order to effectively control the inputs. The "response" refers to the occupant's sensations of hearing and touching that are determined by the source and the transfer function. Since the responses include noise response and vibration response, the model is divided into two sub-models, i.e. "source–transfer function–vibration" model and "source–transfer function–noise" model.

5.2.2   Source–Transfer Function–Vibration Model for Vehicle Body

When the vehicle body is subjected to an outside vibration excitation (f_i), the occupants will perceive the vibration. Figure 5.9 shows the vibration transfer processing of a path (path i). The transfer function between the interior vibration and the excitation is expressed by $H_i^{VF}(\omega)$. The interior vibration response can be expressed by the excitation and the transfer function as follows:

$$V_i(\omega) = H_i^{VF}(\omega) f_i(\omega) \tag{5.7}$$

There are many vibration sources applied on the body. Assuming that N forces are applied on different positions or directions, each vibration is transmitted to the body, and then to the occupant, as shown in Figure 5.10. The perceived vibration by the

Figure 5.8 Transmission process of "Excitation–Transfer Path–Occupant."

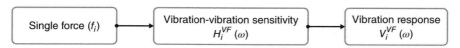

Figure 5.9 Vibration transfer processing of path i.

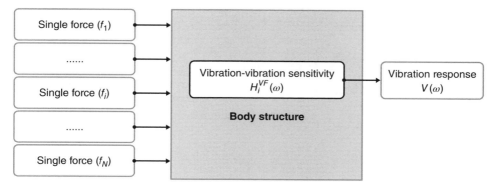

Figure 5.10 Transfer processing of *N* vibration sources to the interior vibration.

occupant is the summation of each response transmitted by the corresponding excitation, and expressed as follows:

$$V(\omega) = \sum_{i=1}^{N} V_i(\omega) = \sum_{i-1}^{N} H_i^{VF}(\omega) f_i(\omega) \tag{5.8}$$

5.2.3 Source–Transfer Function–Noise Model for Vehicle Body

The interior noise comes from two sources: noise generated by outside force excitations and by outside noise excitations. The interior noise is a summation of the structural-borne noise and the airborne noise.

5.2.3.1 Transmission of Structural-Borne Sound

The outside vibration excitations generate not only the interior vibration but also the structural-borne noise. Figure 5.11 shows one transfer path (the *i*th path) of the transfer processing of the outside vibration source to the interior noise, where the transfer function is represented by $H_i^{SB}(\omega)$. The interior noise response can be expressed by the excitation and the transfer function as follows:

$$P_i(\omega) = H_i^{SB}(\omega) f_i(\omega) \tag{5.9}$$

All of the *N* vibration sources are transferred to the body and then generate the structural-borne noise, as shown in Figure 5.12. The heard sound by the occupant is the summation of each response transmitted by the corresponding excitation, and it is expressed as follows:

$$P^{SB}(\omega) = \sum_{i-1}^{N} H_i^{SB}(\omega) f_i(\omega) \tag{5.10}$$

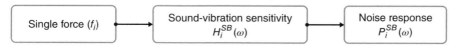

Figure 5.11 Transfer processing of the *i*th vibration source to the interior noise.

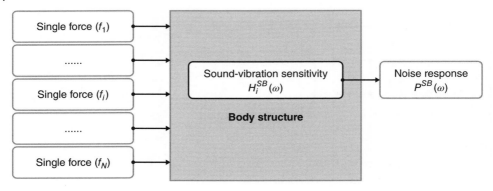

Figure 5.12 Transfer processing of the *N* vibration sources to the interior noise.

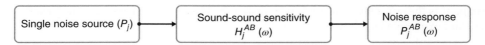

Figure 5.13 Transfer processing of the *j*th noise source to the interior.

5.2.3.2 Transfer of Airborne Sound

Figure 5.13 shows the transfer processing of the *j*th noise source to the interior. The outside noise penetrates the body and then reaches to the occupant's ears. The transfer function is expressed by $H_j^{AB}(\omega)$. The interior noise response can be expressed by the excitation and the transfer function as follows:

$$P_j^{AB}(\omega) = H_j^{AB}(\omega)P_j(\omega) \tag{5.11}$$

All of the *M* outside noise sources penetrate the body and transfer to the occupant's ears, as shown in Figure 5.14. The heard sound is the summation of each sound response contributed by each noise source, and is expressed as follows:

$$P^{AB}(\omega) = \sum_{j=1}^{M} H_j^{AB}(\omega)P_j(\omega) \tag{5.12}$$

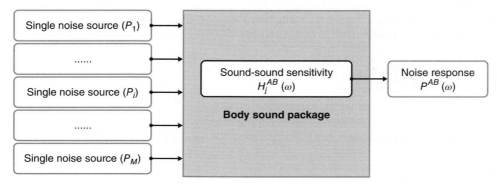

Figure 5.14 Transfer processing of the *M* noise sources to the interior.

5.2.3.3 Summation of Interior Noise

The vibration exerting on the body generates the structural-borne noise, and the outside noise induces the airborne noise. The overall interior noise is the summation of the structural-borne noise and the airborne noise, as shown in Figure 5.15.

The overall interior noise is expressed as

$$P(\omega) = P^{SB}(\omega) + P^{AB}(\omega) = \sum_{i=1}^{N} H_i^{SB}(\omega) f_i(\omega) + \sum_{j=1}^{M} H_j^{AB}(\omega) P_j(\omega) \tag{5.13}$$

Eq. (5.8) integrates the overall interior vibration contributed by all the vibration sources and the corresponding paths. Eq. (5.13) provides the overall interior noise contributed by all the structural-borne noise and the airborne noise and their corresponding paths. If the overall interior noise and the contributions by each structural-borne and airborne path are plotted on one graph, it will be clear to recognize each path's contribution. Figure 5.16 shows the overall interior sound pressure, the overall

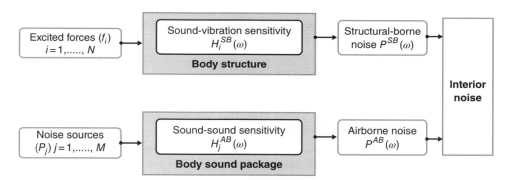

Figure 5.15 Overall interior noise is summation of the structural-borne noise and the airborne noise.

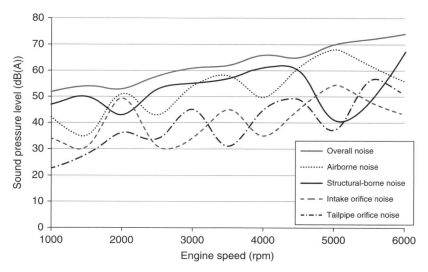

Figure 5.16 Overall interior sound pressure, overall structural-borne noise, overall airborne noise, and the noise contribution by each path for a vehicle at WOT condition.

structural-borne noise, the overall airborne noise, and the noise contribution by each path for a vehicle at wide-open throttle (WOT) condition. From the graph, it is clear to see the contributions by the structural-borne noise, the airborne noise, and each path at every engine speed. Similarly, if the overall interior vibration and the contribution by each vibration transfer path are plotted on one graph, the contribution by each path will be clearly identified. The method to put the overall noise and vibration and the contribution by each path together is called transfer path analysis (TPA).

5.3   Characteristics and Analysis of Noise and Vibration Sources

The studied object of this chapter is the sensitivity or transfer function that is independent of the input and the output for a linear system. However, understanding the sources and their characteristics is helpful for the body sensitivity and response analysis. In addition, it is necessary to understand the excitation sources characteristics for reading other chapters, such as the excitations for panel structures (Chapter 3) or noise sources for the sound package (Chapter 4). In this chapter, the major excitation sources applying on the body will be introduced.

For some systems, due to the vibration and noise coming from the same source, their vibration and noise characteristics are the same or similar; for example, engine noise characteristics and engine mount vibration characteristics are very similar, so they are cataloged as the same group to be analyzed. For other systems, their sources characteristics are different, such as engine noise and wind noise, so they are analyzed separately.

In this section, the source characteristics of some typical systems are analyzed, and they can be summarized into the following categories:

1) Excitation source characteristics of engine and related systems
2) Excitation source characteristics of drivetrain system
3) Excitation source characteristics of tires
4) Excitation source characteristics of rotary machines
5) Excitation source characteristics of random or impulse inputs
6) Excitation source characteristics of wind noise

In Chapter 6, wind noise characteristics will be introduced, so it will be not covered in this chapter. The first five categories will be described in the following materials.

5.3.1   Excitation Characteristics of Engine and Related Systems

The engine is the main source of noise and vibration in an automobile. The systems relating to the engine, including intake system, exhaust system, power plant mounting system, and drivetrain system, have the same noise and vibration characteristics as the engine.

Most automobile engines are four-stroke engines, where the pistons complete four separate strokes (suction, compression, combustion, and exhaust) during one working cycle. The crankshaft rotates two cycles or 720° to complete one working cycle. When the engine works, huge combustion pressure and inertia force are generated, forming four different excitations: gas combustion torque, inertia torque, unbalanced inertia force, and unbalanced moment.

The engine excitations are related to its speed. The idle speed is generally between 600 and 1000 rpm, and the acceleration speed is usually between 1000 and 6000 rpm. Under different rotating speeds and loads, the engine combustion is not the same, which generates noise and vibration with different magnitudes and characteristics.

The engine noise and vibration are related not only with frequency but also with the engine speed and firing order, so they are different from other noise and vibration problems. The specific features of the engine noise and vibration are summarized as follows:

- Frequency feature
- Speed feature
- Order feature

The combined sounds by different frequency, speed, and order have different sound quality. Similar to the engine noise and vibration features, the noise and vibration characteristics of the intake system and the exhaust system, and the vibration features of the mounting system are also related with frequency, speed, and order. The relation among frequency (f), speed, and order can be expressed as follows,

$$f = \frac{rpm}{60} * order = \frac{rpm}{60} * \frac{number\ of\ cylinders}{2} \tag{5.14}$$

The firing order for a four-cylinder engine is the second order, and the engine also has harmonic firing orders that are integer times of the second order, i.e. fourth order, sixth order, eighth order, etc. In addition to the firing order and harmonic firing orders, there exist other integer orders such as first order, third order, fifth order, etc. The "half orders" are very important ones influencing sound quality, including 0.5th order, 1.5th order, 2.5th order, etc. Table 5.1 lists relationship among engine frequency, speed, and order of the first four orders for a four-cylinder engine.

Table 5.1 Relationship among engine speed, order, and frequency of first four orders for a four-cylinder engine.

Engine speed (rpm)\order	0. 5	1	1. 5	2	2. 5	3	3. 5	4
600	5.0	10.0	15.0	20.0	25.0	30.0	35.0	40.0
700	5.8	11.7	17.5	23.3	29.2	35.0	40.8	46.7
800	6.7	13.3	20.0	26.7	33.3	40.0	46.7	53.3
900	7.5	15.0	22.5	30.0	37.5	45.0	52.5	60.0
1000	8.3	16.7	25.0	33.3	41.7	50.0	58.3	66.7
2000	16.7	33.3	50.0	66.7	83.3	100.0	116.7	133.3
3000	25.0	50.0	75.0	100.0	125.0	150.0	175.0	200.0
4000	33.3	66.7	100.0	133.3	166.7	200.0	233.3	266.7
5000	41.7	83.3	125.0	166.7	208.3	250.0	291.7	333.3
6000	50.0	100.0	150.0	200.0	250.0	300.0	350.0	400.0

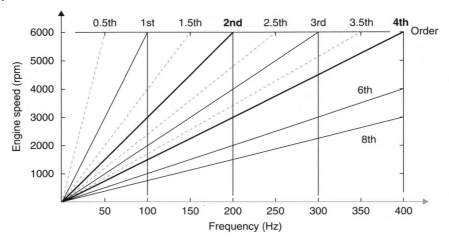

Figure 5.17 Relationship among speed, frequency, and order.

If an idle speed is determined, the corresponding frequency can be determined as well; for example, for 750 rpm idle speed, the frequency corresponding to the firing order is 25 Hz.

During acceleration, the engine speed varies from 1000 to 6000 rpm. The relationship among frequency, speed, and order listed in Table 5.1 is plotted on a graph, as shown in Figure 5.17. The horizontal coordinate is frequency, and the longitudinal coordinate is engine speed. The scattering lines represent orders. From this graph, it is very intuitive and clear to identify the relationship among the noise and vibration and frequencies and orders.

In addition to the relationship among the speed, order, and frequency, there is another excitation feature for the engine and related systems: resonance. The engine, the intake system, and the exhaust system could generate interior booming noise due to loud noise sources. The engine and exhaust system may also generate a structural resonance, or their excitations are resonant with the mounting brackets or exhaust hangers, which induce the body to be resonant, generating structural booming noise. Figure 5.18 shows that an engine excitation is resonant with a mounting bracket at 280 Hz, resulting in 280 Hz interior booming.

Other systems, such as the drivetrain, are related with the engine excitations and also have the same feature relating to the order and speed, but the drivetrain has its own characteristics as well, so the system will be described separately.

5.3.2 Excitation Characteristics of Drivetrain System

The drivetrain system includes the power plant internal driveshaft system consisting of engine, clutch, and transmission, as shown in Figure 5.19, and the external driveshaft system comprising front and rear half shafts, driveshaft for rear-wheel drive (RWD) or four-wheel drive (4WD) vehicle, rear axle, as shown in Figure 5.20. The characteristics of the drivetrain system are closely related to the engine speed, order, frequency, and resonance. In addition, it has its unique characteristics, mainly in the gear meshing.

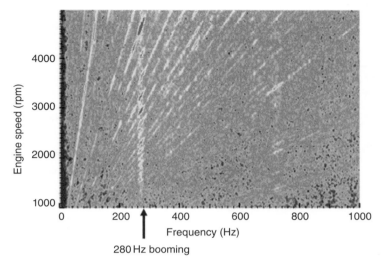

Figure 5.18 Interior booming generated by resonance between an engine excitation and a mounting bracket at 280 Hz.

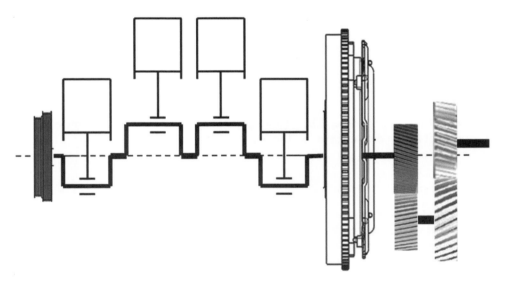

Figure 5.19 A power plant internal driveshaft system.

5.3.2.1 Meshing Orders of Drivetrain Systems

The rotational speeds of the shafts and gears inside a drivetrain system are proportional to the engine speed, so the orders of the drivetrain components can be determined by the engine orders. Figure 5.21 shows a schematic diagram of an engine and a transmission, and there are two pairs of gears inside the transmission where the corresponding teeth numbers are z_1 and z_2, z_3 and z_4, respectively. In the first pair of gears, the first gear is directly connected to the engine through a shaft, so its meshing order is the tooth

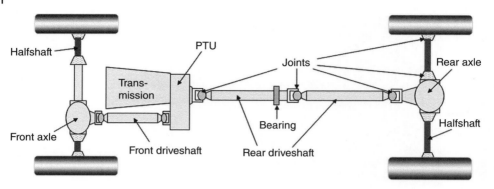

Figure 5.20 An external driveshaft system.

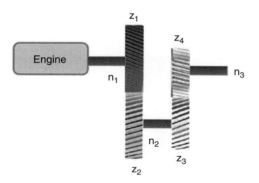

Figure 5.21 Schematic diagram of an engine and a transmission.

number of the first gear, i.e. z_1. The engine's power and torque will be transmitted to the second shaft by the meshing pair, so the second shaft speed is:

$$n_2 = \frac{z_1}{z_2} n_1 \qquad (5.15)$$

where n_1 and n_2 are rotational speeds of the first shaft and the second shaft, respectively.

The third gear is driven by the second shaft connecting to the second gear. The meshing order for the second pair of gears is determined by the rotational speed of the second shaft and tooth number of the third gear. Using n_1 as a reference, the meshing order of the second pair of gears is obtained, and is $\frac{z_1}{z_2} z_3$.

The fourth gear meshes with the third gear, so rotational speed for the third shaft is

$$n_3 = \frac{z_3}{z_4} n_2 = \frac{z_1}{z_2} \frac{z_3}{z_4} n_1 \qquad (5.16)$$

An example is used to illustrate the relationship between the orders and the gears. Assume the tooth numbers for two pairs of gears are $z_1 = 29$ and $z_2 = 38$, $z_3 = 17$ and $z_4 = 73$, respectively. The gear ratio for the two pairs of gears are $z_2/z_1 = 1.31$ and $z_4/z_3 = 4.294$, respectively. The overall gear ratio is $(z_2/z_1)(z_4/z_3) = 5.626$. The meshing order for the first pair of gear is the 29th order, so the meshing order for the second pair of gear is 12.97th (=29*17/38) order.

5.3.2.2 Noise and Vibration Characteristics of the Drivetrain System

In the drivetrain system, the noise and vibration include four phenomena: whine, rattle, second-order vibration, and resonance.

When a pair of gears is engaged, the slip at the contacted area results in different linear speeds for the two gears at the engaged point, which could induce a high-frequency sound. The meshing speed difference can be represented by transmission errors. Because of the relatively high frequency, the noise sounds like screaming, so it is called whine or whistle. The whine has a very obvious order feature. According to relationship between teeth number of the meshing gear and the engine speed, the meshing order can be calculated. The whine could penetrate the transmission or axle shells and then the vehicle body, and enter into the interior, forming the airborne noise. The gear meshing could induce the internal structures and the transmission shell to vibrate, and the vibration will be transmitted to the interior through the shifter cable, engine bracket, etc., forming the structural-borne noise. Figure 5.22 shows a color map of a vehicle interior noise. In the high-order region, there are two distinct straight lines that represent the transmission whine generated by the gear meshing. The two lines represent 19.63th order and 31th order gear meshing whines, respectively.

Compared with the engine firing order and harmonic order sounds, the sound pressure level of the whine is much lower, but due to its high frequency, it is easily perceived.

There are many gears inside the transmission, but only one pair of gears is engaged at an operation condition, while the other gears rotate freely, driven by the shafts. There exist backlashes between a pair of freely rotating gears, as shown in Figure 5.23. If their rotary speeds are different, they will impact each other, which generates the rattle sound.

Figure 5.22 A color map of a vehicle interior noise and two gear-meshing whines.

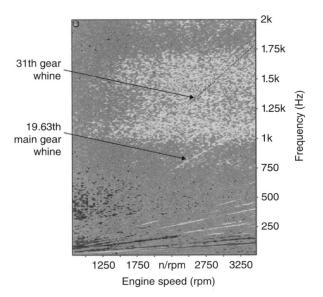

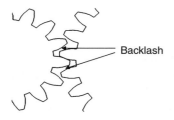

Figure 5.23 Backlash between a pair of freely rotating gears.

Different from the whine that is strongly related to the orders, the rattle is a broadband noise, as shown in Figure 5.24. Its frequency range is usually from 300 to 5000 Hz. The rattle noise can be transmitted into the interior by both airborne and structural-borne paths.

For RWD or AWD vehicles, the driveshaft usually comprises two or three shafts. The shafts are usually connected by Cardan joints, as shown in Figure 5.25. The axes of the two shafts do not coincide; instead, there is an angle between them. The driving shaft rotates one cycle at a constant speed, but the rotational speed of the driven shaft changes twice, from fast to slow, and then from slow to fast, as shown in Figure 5.26, which results in the second-order vibration that will be transmitted into the interior as the structural-borne noise.

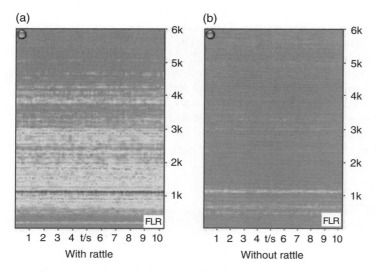

Figure 5.24 Color map of rattle noise: (a) with rattle (b) without rattle.

Figure 5.25 A Cardan joint.

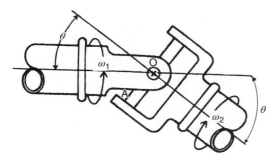

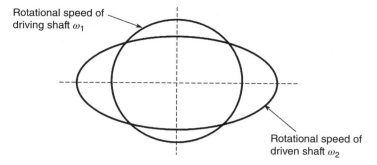

Figure 5.26 Diagram of second-order motion.

5.3.3 Excitation Characteristics of Tires

Tire noise generated by interaction between tires and road surface is one of the most important noise sources of a vehicle. The noise and vibration related to the tires can be divided as three categories: broadband noise generated by the interaction between the tire and the road surface, narrow band cavity noise induced by the tire's cavity modes, and low-frequency vibration produced by tire dynamic imbalance.

5.3.3.1 Mechanism of Tire Noise

The tire surface has treads, grooves, etc. When a vehicle moves forward, the tread are impacted, which generates impact noise. When the tire is squeezed, the treads, the grooves, and the road surface form small, closed spaces. After the airbags inside the closed spaces break, the cavity noise will be generated. At the same time, the tire and the road could slip against each other, generating the friction noise. The impact noise, cavity noise, and friction noise have broadband frequency. The interior noise on smooth roads is lower than that on rough roads, as shown in Figure 5.27.

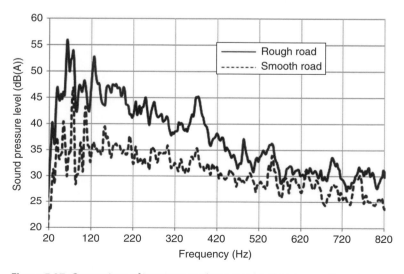

Figure 5.27 Comparison of interior sound pressure level on a smooth road and rough road.

5.3.3.2 Tire Cavity Mode and Resonance

The air inside the tire forms an enclosed cavity. The modal frequency of the first-order acoustic mode can be estimated by following equation:

$$f = \frac{c}{2\pi r} \tag{5.17}$$

where r is radius of the tire central line and c is the sound speed.

The modal frequency of the tire first-order cavity acoustic mode for the passenger vehicles is usually between 210 and 240 Hz. When the modal frequency is consistent with the suspension frequency, a resonance could happen between the tire cavity mode and the suspension mode, and the resonated vibration will be transmitted to the vehicle body. Figure 5.28 shows the interior noise of a sedan cruising at a speed of $45\,\mathrm{km\,h^{-1}}$. The frequency of the tire cavity acoustic mode is 232 Hz, and the modal frequency of the suspension arm is 230 Hz, so the resonance happens and the 232 Hz interior booming is generated.

5.3.3.3 Tire Imbalanced Excitation

The tire is a circular structure. Due to radial stiffness difference on each direction, manufacturing error, non-concentricity between the axes of the tire and the rotating shaft, etc., the tire could be eccentric, so dynamic imbalance and a centrifugal force are generated when the vehicle is running. The excitation force causes the suspension structure to vibrate, which is transmitted to the body. The frequency of the dynamic imbalance force of the tire is expressed as follows:

$$f = \frac{nu}{2\pi r} \tag{5.18}$$

where u is the vehicle speed, r is the tire radius, $n = 1, 2, 3 \ldots$

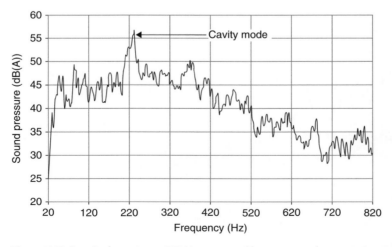

Figure 5.28 Interior booming at 232 Hz generated by resonance between tire cavity acoustic mode and suspension mode.

Table 5.2 Excitation frequencies of first and second orders of tire's dynamic imbalance at different vehicle speeds.

Vehicle speed (km h^{-1})	First order (Hz)	Second order (Hz)
80	12.7	25.4
85	13.4	26.8
90	14.1	28.2
95	14.8	29.6
100	15.5	31
110	17.3	34.6
120	18.7	37.4
125	19.4	38.8

The tire dynamic imbalance is the major reason to cause a nibble problem of a steering wheel. For example, for a passenger vehicle installed with 175/65R14 tires traveling at speeds from 80 to 120 km h^{-1}, the first-order excitation frequency is from 12.7 to 18.7 Hz. The steering wheel nibble is easily generated by the low-frequency excitation. Table 5.2 lists the excitation frequencies of first and second orders of the tire dynamic imbalance at different vehicle speeds.

5.3.4 Excitation Characteristics of Rotary Machines

The rotary machines mentioned in this book refer to other "rotating" devices excluding engine, transmission and tire, such as generator, water pump, oil pump, compressor, blower, and fan. They are directly or indirectly connected to the body, and are important noise and vibration sources. For example, the cooling fan induces the front-end cooling module to vibrate, which is transmitted to the interior as structural-borne noise. At the same time, the noise generated by the fan is transmitted directly to the interior.

The excitations of these rotary machines are related to the rotation speeds, so the generated noise and vibration should be represented by the sound levels, frequency, and order. The excitation frequency is calculated as follows:

$$f = \frac{N}{60} n \tag{5.19}$$

where N is the rotating speed and n = 1, 2, 3, ... is orders.

For example, the speed of a fan is 2100 rpm, so the fundamental frequency of its excitation is 35 Hz.

The rotary machines can be divided into two categories, one related to the engine speed and the other independent of the engine speed. The rotary machines related to the engine speed include the generator, water pump, oil pump, and compressor. The rotary shafts of these machines are connected with the engine crankshaft by belts or chains or gears, so their excitation orders can be converted to the engine orders. Table 5.3 lists the gear ratio of the rotary machines to the engine crankshaft for an engine.

Table 5.3 Gear ratio of rotary machines to engine crankshaft for an engine.

Name of Rotary Machines	Transmission Ratio of Rotary Machines to Engine
Crankshaft gear	18
Camshaft gear	36
Generator	29.76
Water pump	7.8
Oil pump	22
Compressor	11.8

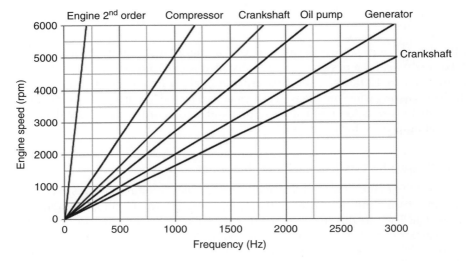

Figure 5.29 Orders of rotary machines on an engine.

The orders of the rotary machines are related not only to the gear ratios but also to their structures. The ratio between the rotary machines and the engine crankshaft, the blades, bearings, and grooves inside the water pump, generator, and compressor, determine the rotational machines' orders. Based on the engine speed and order, the orders and frequencies of these rotational machines can be plotted, as shown in Figure 5.29. From this plot, it is clear to see their order and frequency characteristics, which can help to quickly identify the sources of noise and vibration.

The rotary machines that are independent of the engine speed include the fuel pump, fan, blower, etc. Similar to Table 2.3, their speed and frequency characteristics can be listed, which can help to quickly identify the sources of noise and vibration.

5.3.5 Excitation Characteristics of Random or Impulse Inputs

In addition to the excitations described above and wind excitation, the body is also subjected to some random or pulse inputs.

Many pipes and cables are installed on a vehicle body, such as air-conditioning pipe, fuel pipe, selection cable, shifter cable, etc. Despite small size of the pipes and cables, their contribution to the interior noise and vibration cannot be ignored. For example, the thin and long fuel pipes are connected to the underbody by fasteners; however, improper connection could increase the interior idle noise 1–3 dB(A).

The excitation source of the pipes is mainly from the fluid pulsation inside the pipe. The fuel is continuously injected by the pulsation that excites the pipe wall and transmits the vibration to the body. Therefore, the vibration isolation between the body and the pipe should be effective in order to reduce the pulsation transmission, such as using flexible pipe wall, soft vibration isolation pad, etc.

The power plant vibration could be transferred to the dash panel and the floor by selection cable and shifter cable. Simultaneously, the cables could radiate noise directly. So, it is necessary to reduce the noise and vibration transmission contributed by the cables, such as increasing the vibration isolation between the cable and the front panel, adding mass on the cable, etc.

The excitations of the pipes and cables have the same characteristics as the pulsation or random inputs.

5.4 Dynamic Stiffness and Input Point Inertance

5.4.1 Mechanical Impedance and Mobility

In order to facilitate the analysis of vibration problems, several concepts related to transfer function and sensitivity are introduced: mechanical impedance and mechanical mobility.

The mechanical impedance is a measure of how much a structure resists motion when subjected to a harmonic force. It is defined as a ratio of the excitation to the response of a system subjected to a harmonic excitation and noted as Z. The vibration responses include displacement, velocity, and acceleration, so the corresponding impedances have displacement impedance (Z^X), velocity impedance (Z^V), and acceleration impedance (Z^A).

For a viscoelastic system with a single degree of freedom (DOF), its dynamic equation and the response solution are described in Chapter 3. For the completeness of this chapter, Eq. (3.54) is written again as follows,

$$m\ddot{x} + c\dot{x} + kx = f \tag{5.20}$$

The force applied on the mass is $f = Fe^{j\omega t}$, and the displacement response is $x = X_0 e^{j(\omega t - \varphi)}$, so the corresponding velocity and acceleration responses are

$$\dot{x} = j\omega X_0 e^{j(\omega t - \varphi)} = j\omega x \tag{5.21a}$$

$$\ddot{x} = -\omega^2 X_0 e^{j(\omega t - \varphi)} = -\omega^2 x = j\omega\dot{x} \tag{5.21b}$$

Substituting the displacement and the force into Eq. (5.20), the displacement impedance is obtained:

$$Z^X = \frac{f}{x} = \left(k - m\omega^2\right) + jc\omega \tag{5.22}$$

Magnitude of the displacement impedance is

$$\left|Z^X\right| = \sqrt{\left(k - m\omega^2\right)^2 + \left(c\omega\right)^2} \qquad (5.23)$$

Substituting the velocity and the force into Eq. (5.20), the velocity impedance is obtained:

$$Z^V = \frac{f}{\dot{x}} = \frac{\left(k - m\omega^2\right) + jc\omega}{j\omega} \qquad (5.24)$$

Magnitude of the velocity impedance is

$$\left|Z^V\right| = \frac{1}{\omega}\sqrt{\left(k - m\omega^2\right)^2 + \left(c\omega\right)^2} \qquad (5.25)$$

Substituting the acceleration and the force into Eq. (5.20), the acceleration impedance is obtained:

$$Z^A = \frac{f}{\ddot{x}} = \frac{\left(k - m\omega^2\right) + jc\omega}{-\omega^2} \qquad (5.26)$$

Magnitude of the acceleration impedance is

$$\left|Z^A\right| = \frac{1}{\omega^2}\sqrt{\left(k - m\omega^2\right)^2 + \left(c\omega\right)^2} \qquad (5.27)$$

The mechanical mobility is a measure of how easily a structure moves when subjected to a harmonic force. It is defined as a ratio of the response to the excitation of a system subjected to a harmonic excitation, and noted as Y. The mechanical mobility is the reciprocal of the mechanical impedance.

The vibration responses include displacement, velocity, and acceleration, so corresponding motilities have displacement mobility (Y^X), velocity mobility (Y^V), and acceleration mobility (Y^A). The relationships between the mobility magnitudes and the impedance magnitudes can be expressed as follows:

$$\left|Y^X\right| = 1/\left|Z^X\right| \qquad (5.28a)$$

$$\left|Y^V\right| = 1/\left|Z^V\right| \qquad (5.28b)$$

$$\left|Y^A\right| = 1/\left|Z^A\right| \qquad (5.28c)$$

5.4.2 Driving Point Dynamic Stiffness

5.4.2.1 Concept of Dynamic Stiffness

Stiffness represents the rigidity of an object. The stiffness of a body is a measure of the resistance offered by an elastic body to deformation. The stiffness is the intrinsic characteristic of a linear system that is independent of input and output. After a static force is applied on the system, the displacement can be measured. The stiffness is a ratio of the force to the response, which is called static stiffness. The stiffness k in Eq. (5.20) is a typical static stiffness of a single DOF system.

Substituting the velocity and the acceleration in Eq. (5.21) into Eq. (5.20), the system stiffness becomes

$$k_d = \frac{f}{x} = \left(k - m\omega^2\right) + jc\omega \qquad (5.29)$$

The stiffness in Eq. (5.29) is a function of the excitation frequency (ω), that is, the stiffness is not a fixed value; instead, it changes with the frequency. The stiffness in Eq. (5.29) is called *dynamic stiffness*, and its magnitude is expressed as follows:

$$\left|k_d(\omega)\right| = \sqrt{\left(k - m\omega^2\right)^2 + (c\omega)^2} \qquad (5.30)$$

The dynamic stiffness depends on mass, damping, and static stiffness of the system. Figure 5.30 shows a dynamic stiffness curve for the single DOF. The mass, damping, and static stiffness values used in this curve are $m = 2\,\text{kg}$, $c = 10\,\text{Ns}\,\text{m}^{-1}$, $k = 2000\,\text{N}\,\text{m}^{-1}$, respectively.

For a special case where the system is subjected to a static load, instead of a dynamic load, i.e. $\omega = 0$, substitute $\omega = 0$ into Eqs. (5.29) or (5.30); then the dynamic stiffness becomes static stiffness, i.e. $k_d = k$. So, the static stiffness is a special case of the dynamic stiffness.

Comparing Eq. (5.30) and Eq. (5.23), we can see that the two expressions are the same. It can thus be concluded that the dynamic stiffness is the displacement impedance.

$$\left|k_d\right| = \left|Z^X\right| = \sqrt{\left(k - m\omega^2\right)^2 + (c\omega)^2} \qquad (5.31)$$

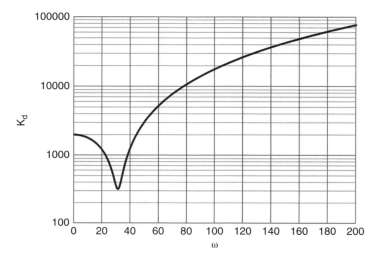

Figure 5.30 A dynamic stiffness curve for a single DOF.

5.4.2.2 Driving Point Dynamic Stiffness

In a system, if the excitation and the response are at the same point that is called *driving point*, the dynamic stiffness obtained is called *driving point dynamic stiffness*. For a single DOF system shown in Figure 3.45, the excitation point and the response point is the same one; therefore, the dynamic stiffness expressed in Eq. (5.29) is the driving point dynamic stiffness.

In a multi-DOF system or a continuous system, there are many excitation points and response points, as shown in Figure 5.31.

The modal analysis theory and the transfer function between single output and single input for a multi-DOF system are introduced in Chapter 2. Eq. (2.55) provides the transfer function expression between the response point (*l* point) and the excitation point (*p* point), expressed as follows:

$$H_{lp}(\omega) = \frac{x_l(\omega)}{f_p(\omega)} = \sum_{r=1}^{N} \frac{\phi_{lr}\phi_{pr}}{\left(K_r - \omega^2 M_r + j\omega C_r\right)} \tag{5.32}$$

Equation (5.32) represents the displacement mobility of the system under single point excitation.

When the response point and the excitation point are the same one, assume that the point is *l*, so Eq. (5.32) changes to

$$H_{ll}(\omega) = \frac{x_l(\omega)}{f_l(\omega)} = \sum_{r=1}^{N} \frac{\phi_{lr}\phi_{lr}}{K_r - \omega^2 M_r + j\omega C_r} \tag{5.33}$$

Equation (5.33) is the displacement mobility at the same point or driving point mobility. Since the mobility and the dynamic stiffness (or impedance) are reciprocal, the driving point dynamic stiffness is the reciprocal of the driving point mobility. Based on Eq. (5.33), the driving point dynamic stiffness can be obtained.

5.4.3 IPI and Driving Point Dynamic Stiffness

Equation (5.29) is rewritten as follows:

$$x = \frac{f}{k_d} \tag{5.34}$$

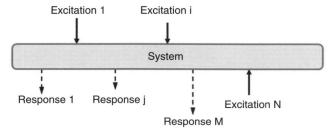

Figure 5.31 Excitation points and response points for a multi-DOF system or a continuous system.

The expression shows that the larger the driving point dynamic stiffness, the smaller the displacement response for a fixed excitation. The driving point dynamic stiffness is the displacement impedance, so, in other words, the larger the impedance, the smaller the displacement response. According to Eqs. (5.24) and (5.26), the same conclusions can be obtained for the velocity and acceleration responses, i.e. the larger the velocity imped-ance, the smaller the velocity response; and the larger the acceleration impedance, the smaller the acceleration response. Therefore, it is very important to control the driving point dynamic stiffness in order to reduce the vibration transmitting to the body.

Engineers and scientists used to use acceleration to characterize dynamic response of a system. Compared with the displacement and velocity, acceleration is much easier to measure. Therefore, it is necessary to establish the relationship between the accelera-tion and the dynamic stiffness.

Substituting Eq. (5.21b) into (5.26), we get

$$Z^A = \frac{f}{-\omega^2 x} = -\frac{1}{\omega^2} k_d \tag{5.35}$$

Eq. (5.35) establishes the relationship between the acceleration impedance and the dynamic stiffness. The relationship between their magnitudes can be expressed as

$$\left|Z^A\right| = \frac{1}{\omega^2}\left|k_d\right| \tag{5.36}$$

In vehicle body NVH analysis, the static stiffness and dynamic stiffness are usually used as the design targets because they are directly related with structural characteris-tics. Usually, the dynamic stiffness is set to a fixed value, that is, it is a constant for all frequencies. The driving point dynamic stiffness lower than $10^6 N/m$ indicates that the local structure is too weak and easily excited by external excitations. If the stiffness is higher than $10^8 N/m$, the local structure is so strong that the external excitations have no impact on the structure. The higher the stiffness, the better. However, increasing the dynamic stiffness means that the body weight and cost will increase simultaneously. According to the engineering experience, when the driving point dynamic stiffness reaches $10^7 N/m$ or $10,000 N/mm$, the local structure is regarded as strong enough to resist any external input, and the external vibration transmitted to the body can be effectively controlled. Therefore, $10^7 N/m$ is used as the target for body driving point dynamic stiffness. In testing or analysis of the body dynamic stiffness, the above three values are often put together and used as the references, as shown in Figure 5.32.

Corresponding to the values in Figure 5.32 and according to Eq. (5.36), the accelera-tion impedance can be drawn, as shown in Figure 5.33. The horizontal reference lines in Figure 5.32 become decaying curves with the frequency increase.

The measured accelerations and forces can be converted into acceleration imped-ances, or driving point dynamic stiffness, which are plotted together with the reference lines, as shown in Figure 5.34. From the plot, the dynamic stiffness at each frequency can be clearly identified. The higher the curve, the bigger the dynamic stiffness, and vice versa. At some frequencies, we can see some downward peaks where the dynamic stiffness has the lowest values. The downward peaks are the "problem" peaks that indicate the structures are weak.

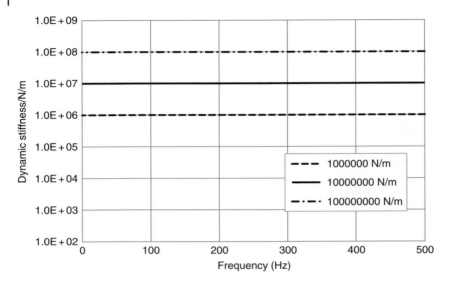

Figure 5.32 Reference values for body driving point dynamic stiffness.

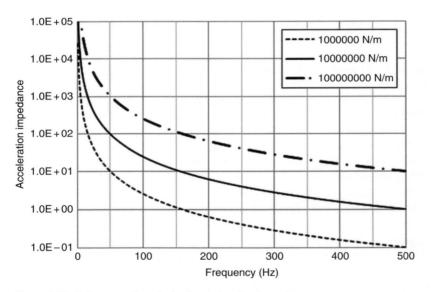

Figure 5.33 Reference values for body acceleration impedance.

Engineers prefer to see the upward peaks and judge them as "resonant" peaks or "problem" peaks. Instead, they are not used to see the downward "problem" peaks in Figure 5.34. In order to let the problem peaks change from downward to upward, the curves in Figure 5.34 must be reversed. To reach this purpose, a new concept, input point inertance (IPI), is introduced.

The word *inertance* means inertia. Described by mechanical terminology, the inertance is mobility. IPI is the acceleration mobility, i.e. the transfer function between the

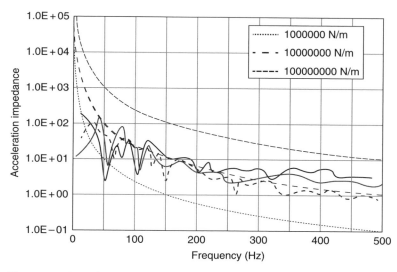

Figure 5.34 Tested acceleration impedances and reference lines of dynamic stiffness.

acceleration response and the input force. The main concern for IPI analysis is its magnitude, and usually its phase is neglected, so the IPI can be expressed as

$$IPI = \left| \frac{\ddot{x}}{f} \right| = \frac{\omega^2}{\sqrt{\left(k - m\omega^2\right)^2 + (c\omega)^2}} \tag{5.37}$$

Eq. (5.37) shows that the IPI and the acceleration impedance are reciprocal. Plot the three referenced curves in Figure 5.34 in an IPI plot, as shown in Figure 5.35. The curves' values increase with the frequency increase. The previous "downward" peaks become "upward" peaks, which means that the "problem" peaks become upward. The higher the peak value, the lower the driving point dynamic stiffness. Thus, the tendency of the peaks is consistent with the traditional "resonance" peaks, which will be convenient for engineers to identify the problems.

5.4.4 Control of Driving Point Dynamic Stiffness

Control of the driving point dynamic stiffness can be implemented from three aspects: driving point local structure, the bracket connected with the driving point, and input energy.

5.4.4.1 Driving Point Local Structure
First, the overall structure stiffness will affect the local structure stiffness. Take a power plant installed on the body front side frames as example to illustrate the influence. The frames stiffness not only determines the body overall stiffness, but also affects the local stiffness of the connecting points between the power plant and the body. Figure 5.36 shows two beams, one with opening section, and the other one with a closed section. As described in Chapter 2, the stiffness of the closed section beam is much higher than

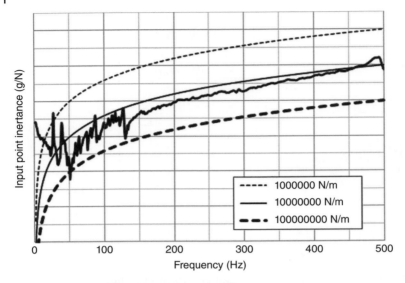

Figure 5.35 IPI and driving point dynamic stiffness.

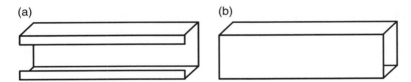

Figure 5.36 Two beams: (a) opening section, (b) closed section.

that of the opening section beam. Therefore, the opening section beam in the vehicle body structure should be avoided.

Second, optimization of beam section. The beam stiffness is determined by the moment of inertia of its cross section, so the optimal design of the beam section is very important for the driving point dynamic stiffness. For example, some locations on body frames and pillars have to be designed as open sections, but the open-section areas must be optimized in order to avoid decrease of the local stiffness. Figure 5.37 shows that a supporting bracket is added on an open-section beam. The IPIs for the original open-section beam and the reinforced beam are compared, as shown in Figure 5.38. After the open-section beam is reinforced, the overall stiffness increases and its resonant frequency shifts higher.

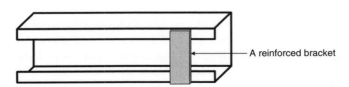

Figure 5.37 A reinforced bracket is added to an open-section beam.

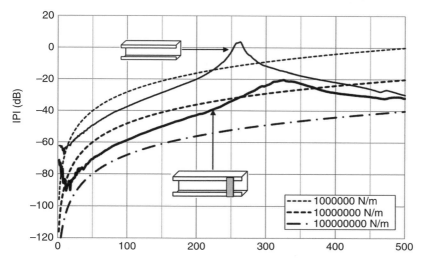

Figure 5.38 IPI comparison for open-section beam and reinforced beam.

Third, reinforcement of driving points. In some cases where the beam cross section cannot be changed, the driving point dynamic stiffness can be increased by reinforcing the local structures, such as punching ribs, adding brackets.

Fourth, increasing structure thickness or changing material. Increasing the thickness will increase the stiffness, but the weight and cost increase simultaneously. If material changes, the young modulus change, so the stiffness will change.

5.4.4.2 Bracket Connecting with Driving Point

Usually, the brackets are used to connect the excitation systems and the body, such as the engine brackets shown in Figure 3.88. The excitation source is not directly applied on the body, but on the brackets, so they can be regarded as the body extended parts. The IPIs depend not only on the driving point stiffness but also on the brackets stiffness.

The body and the bracket can be regarded as a series connection of two springs, as shown in Figure 5.39. Assuming that the stiffness of the body driving point is K_{b1}, the bracket stiffness is K_{b2}, then, the overall stiffness of the body and the bracket is,

$$\frac{1}{K_a} = \frac{1}{K_{b1}} + \frac{1}{K_{b2}} \tag{5.38}$$

Only when the stiffness of the bracket is infinite, that is, $K_{b2} \to \infty$, will the overall stiffness be the stiffness of the body driving point, i.e. $K_a = K_{b1}$. In this case, the influence of the bracket can be neglected.

However, the bracket stiffness cannot be infinite. If the bracket stiffness is the same as the body driving point stiffness, that is, $K_{b1} = K_{b2}$, then the overall stiffness is reduced to half.

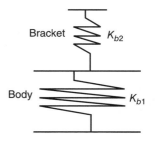

Figure 5.39 A series connection of two springs.

5.4.4.3 Input Energy

When the driving point is subjected to an external excitation, the external energy will be input to the system. The input energy is

$$W = \oint f dx = \int_0^T F \sin \omega t X_0 \cos(\omega t - \varphi) dt = \pi F X_0 \sin \varphi \qquad (5.39)$$

Substitute the relationship between displacement, force and dynamic stiffness $X_0 = \dfrac{F}{K(\omega)}$ into Eq. (5.39). From this, we obtain

$$W = \frac{\pi F^2 \sin \varphi}{K(\omega)} \qquad (5.40)$$

According to Eq. (5.40), it can be concluded that the energy input to the system depends on the magnitude of the excitation force and the driving point dynamic stiffness. The excitation force is from outside, and the bigger excitation force, the more the energy input to the system.

Assuming the external excitation is fixed, the lower the driving point dynamic stiffness, the more the energy will be input to the system, and vice versa. The higher dynamic stiffness increases the impedance, which reduces the magnitudes of the waves propagating in the structure, so the vibration and noise transmitted into the interior is reduced accordingly. Therefore, control of the driving point dynamic stiffness of a system is very important.

5.5 Vibration–Vibration Sensitivity and Sound–Vibration Sensitivity

After the external vibration sources are applied on a vehicle body, the vibration waves propagate inside frames, beams, pillars, and panels, and then reach to the steering wheel and the seats. The vibrational waves transmitted inside the body structures radiate sound into the interior, forming the structural-borne noise.

5.5.1 Transfer Processing of Vibration Sources to Interior Vibration and Vibration–Vibration Sensitivity

5.5.1.1 Transfer Processing of Vibration Excitation to Interior Vibration

Many vibration sources applying on the vehicle body are listed in section one in this chapter. The locations the occupants can perceive the vibration include the steering wheel, the floor or seat track, seat cushion, and shift knob. Take one excitation source (e.g. engine excitation at a power plant mounting point) and one response location (e.g. steering wheel) as an example to illustrate the transfer processing of the vibration source to the interior, as shown in Figure 1.31.

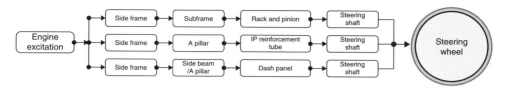

Figure 5.40 Transfer of vibrational flow from engine excitation to steering wheel by three paths.

The engine excitation is input to the mounting point on the body, and then transmitted into the side frame, and finally, the vibration reaches to the steering wheel through three paths, as shown in Figure 5.40:

1) *Path one.* The vibrational waves start from the excitation points and propagate in the side frames. At the joints between the frames and the subframe, partial waves are attenuated by the subframe bushings, and partial waves continue to travel in the subframe. Because the rack and pinion steering linkage is installed on the subframe, the vibrational waves are transmitted into the steering linkage, and then reach to the steering wheel through the steering shaft.
2) *Path two.* The vibrational waves inside the side frames are transmitted to the A pillars. Because the cross-car beam (CCB) or called IP reinforcement tube is installed on the A pillars, the waves pass through the tube, and then reach the steering wheel. Finally, the driver perceives the vibration.
3) *Path three.* The waves propagating inside the side frames and A pillars will excite the dash panel. The steering shaft is connected to the panel by brackets, so the vibrational waves will reach to the steering wheel through the shaft and the tube.

Some vibration sources have many transfer paths to reach the occupants, such as cooling fan excitation; while some sources have only one path, such as exhaust hanger.

There are many (N) vibration sources applying on the body, and each source has three directions (x, y, and z). The vibration for each source in one direction will generate three direction responses at one perceived location. For example, the vibration at one engine mounting in each direction (x, y, and z) will induce three direction responses on the steering wheel, as shown in Figure 5.41.

Among the N sources, the excitation force for ith source in j direction (x, y, z) is noted as $F_{i,j}$. The velocity responses at the steering wheel in x-, y-, z-directions are represented

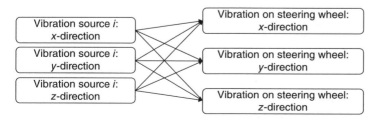

Figure 5.41 Vibration at one source generating responses on steering wheel in three directions.

by $V_{i,j,x}^{str}$, $V_{i,j,y}^{str}$ and $V_{i,j,z}^{str}$. The vibration–vibration sensitivity between the steering wheel response and the excitation force is

$$H_{i,j,x}^{str} = \frac{V_{i,j,x}^{str}}{F_{i,j}} \tag{5.41}$$

The corresponding velocity on the steering wheel in x-direction can be expressed as

$$V_{i,j,x}^{str} = H_{i,j,x}^{str} F_{i,j} \tag{5.42}$$

The velocity on the steering wheel in x-direction generated by all N sources is

$$V_x^{str} = \sum_{i=1}^{N} \sum_{j=1}^{3} V_{i,j,x}^{str} \tag{5.43}$$

Similarly, the velocity responses on the steering wheel in y- and z-directions can be obtained as follows:

$$V_y^{str} = \sum_{i=1}^{N} \sum_{j=1}^{3} V_{i,j,y}^{str} \tag{5.44}$$

$$V_z^{str} = \sum_{i=1}^{N} \sum_{j=1}^{3} V_{i,j,z}^{str} \tag{5.45}$$

The overall velocity on the steering wheel is the vector summation of the responses in the three directions, expressed as

$$V^{str} = \sqrt{\left(V_x^{str}\right)^2 + \left(V_y^{str}\right)^2 + \left(V_z^{str}\right)^2} \tag{5.46}$$

Similarly, the velocity responses for other locations (such as floor, seat, and shift knob) excited by all sources can be obtained. The corresponding sensitivities can be simultaneously acquired.

5.5.1.2 Testing and Analysis of Vibration–Vibration Sensitivity

The body sensitivity can be acquired either by testing or by analysis (such as finite element analysis).

The accelerometers are placed on the steering wheel, seat track, seat cushion, shift knob, and other locations where the occupants could touch, and a hammer is used to excite the points where the external excitations are applied on the body, such as the engine mounting points, as shown in Figure 5.42. Both the accelerations and forces are tested simultaneously, and then the vibration–vibration sensitivity can be obtained by processing the tested data.

The tested object could be a trimmed body or a whole vehicle. The trimmed body or the whole vehicle can be hanged by the soft ropes or supported by the elastic cushions

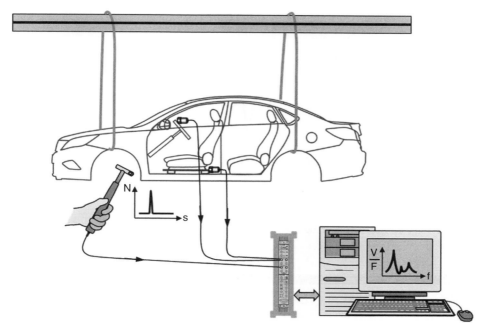

Figure 5.42 Diagrammatic sketch of vibration–vibration sensitivity testing.

so as to form the "free-free" boundary. If the whole vehicle is tested and the hammer cannot directly reach some excitation points where they are occupied by other systems, for example, the engine mounting locations are occupied by the mounts, the locations closest to the excitation points can be chosen as the impact points. Sometimes, the special accessory tools must be used.

Analysis using finite element method is similar to the test. After the excitations and responses are obtained, the transfer functions or vibration–vibration sensitivity can be calculated.

Graphic display for the vibration–vibration sensitivity has two forms: curve graph and color map.

Figure 5.43 shows a tested vibration–vibration sensitivity curve. The horizontal coordinate is frequency, and the vertical coordinate is the sensitivity value $(m\,s^{-1}\,N^{-1})$. From this curve, it is clear to identify the sensitivity value at each frequency.

Figure 5.44 shows a color-strip map for a vibration–vibration sensitivity path. The horizontal direction represents frequency, and the color intensity symbolizes the sensitivity value. The magnitudes of the sensitivity at each frequency can be clearly recognized, for example, the highest value is at 480 Hz.

In order to identify each path's contribution to the interior vibration, all vibration–vibration sensitivities can be plotted together. Figure 5.45 shows a color map of the vibration–vibration sensitivities of all sources to the steering wheel for a vehicle. From the color map, it is clear to present the most sensitive paths for each frequency. For example, the paths for the engine right mount in x- and y-directions are the most sensitive paths between 330 and 350 Hz. If the interior vibration is high at the frequency range, it will be benefit to reduce the interior vibration by attenuating the two path sensitivities.

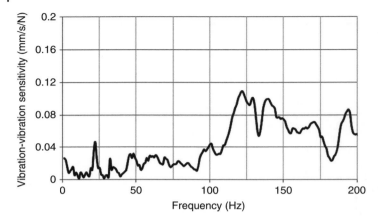

Figure 5.43 A tested vibration–vibration sensitivity curve.

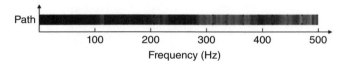

Figure 5.44 A color-strip map for a vibration–vibration sensitivity path.

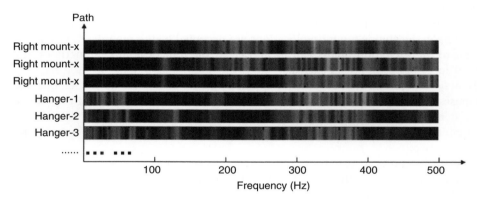

Figure 5.45 A color map of vibration–vibration sensitivities.

5.5.2 Transfer Processing of Vibration Sources to Interior Noise and Sound–Vibration Sensitivity

5.5.2.1 Transfer Processing of Vibration Excitation to Interior Noise

The power plant is used as an example to illustrate the vibration transfer processing to the interior noise. The vibrational waves generated by the engine excitation propagate in the side frames and cross members and pillars, and finally excite the panels. The excited panels radiate sound to the interior, as shown in Figure 5.46.

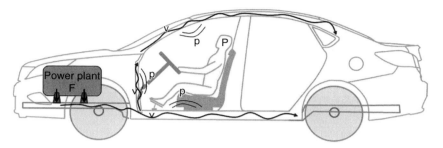

Figure 5.46 Transfer processing of engine mount vibration to interior noise.

Among the N sources, the excitation force for ith source in j direction (x, y, z) is noted as $F_{i,j}$. The interior sound pressure is represented by $P_{i,j}^{SR}$. The sound-vibration sensitivity for this path is

$$H_{i,j}^{SB} = \frac{P_{i,j}^{SB}}{F_{i,j}} \qquad (5.47)$$

Equation (5.47) can be rewritten to express the sound pressure as follows:

$$P_{i,j}^{SB} = H_{i,j}^{SB} F_{i,j} \qquad (5.48)$$

The overall interior sound pressure for all N excitation sources is

$$P^{SB} = \sum_{i=1}^{N} \sum_{j=1}^{3} P_{i,j}^{SB} \qquad (5.49)$$

5.5.2.2 Testing and Analysis of Sound–Vibration Sensitivity
Similar to the vibration–vibration sensitivity, the sound–vibration sensitivity can be obtained either by testing or by analysis (such as finite element analysis).

The excitation method to measure the sound–vibration sensitivity is the same as that of the vibration–vibration sensitivity, that is, a hammer is used to impact the excitation points. Microphones are placed in the occupant ears' locations to measure the sound response, as shown in Figure 5.47. The sound–vibration sensitivity can be acquired by processing the measured sound and vibration signals.

In engineering practices, a hammer is used to impact the excitation points, and the vibration and sound the occupants perceive are measured simultaneously, so the vibration–vibration sensitivities and sound–vibration sensitivities can be obtained simultaneously.

The sound–vibration sensitivity is usually expressed as a curve varying with frequency, as shown in Figure 5.48. Sometimes, it can also be presented by a color map. From the curve, the sensitivity magnitudes for each frequency can be identified. For example, the sensitivity magnitude at 180 Hz is 60 dB N^{-1} that is higher than the acceptable value.

In order to compare the sensitivity magnitudes for each transfer path, all the sensitivities can be plotted on one graph, as shown in Figure 5.49. From the graph, the main

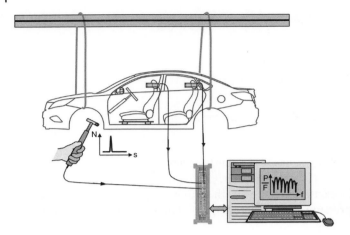

Figure 5.47 Diagrammatic sketch of the sound–vibration sensitivity testing.

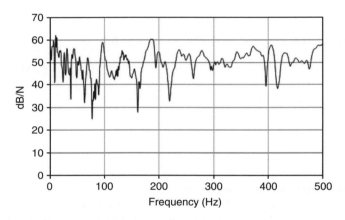

Figure 5.48 Sound–vibration sensitivity curve.

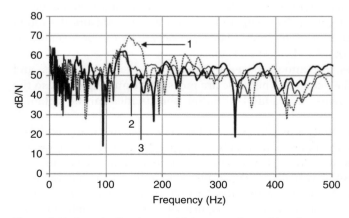

Figure 5.49 Sound–vibration sensitivity curves for multi-paths.

transfer paths of the structural-borne noise can be identified. For example, Figure 5.49 shows the sensitivity curves of three paths. The sensitivity for path one has a peak value $70\,\mathrm{dB\,N^{-1}}$ at 140 Hz, and the sensitivities for paths two and three have peak values $62\,\mathrm{dB\,N^{-1}}$ at 130 Hz. The values are too high to be accepted, which illustrates that these paths could contribute a lot to the interior noise at these frequencies. Therefore, it is extremely important to reduce the sensitivities for these paths at corresponding frequencies.

5.5.3 Sensitivity Control

The process of transferring the vibration sources to the interior sound and vibration can be cascaded to several moduli: excitation and driving point modulus, beam (pillar) module, vibration isolation module, steering module, and panel module. The control of the sensitivities can be implemented from the five modules.

5.5.3.1 Control of Driving Point

Eqs. (5.22–5.27) indicate that for the same external excitation, the higher the driving point impedance (displacement impedance, velocity impedance, and acceleration impedance) or the dynamic stiffness, the smaller the response (displacement, velocity, and acceleration). Eq. (5.40) shows that for the same external excitation, the higher the driving point dynamic stiffness, the less the energy input to the system. The energy consumption for higher impedance component is more than that for lower impedance component. Thus, the first step to controlling the energy input is to increase the driving point dynamic stiffness.

If a body structure cannot be changed, the external excitation points must be exerted on the body locations with higher dynamic stiffness. Figure 5.50 shows a vehicle cross section. Four points (A, B, C, and D) on the underbody are chosen as the external excitation points. Point A is on the side frame, point B is near the frame, point C is away from the frame, and point D is on the middle of the underbody. The same excitation is applied to the four points, and the corresponding sound–vibration sensitivities for points A, B, C, and D are 53, 55, 63, and $72\,\mathrm{dB\,N^{-1}}$, respectively. The sensitivity

Figure 5.50 Four points on the underbody of a vehicle cross section.

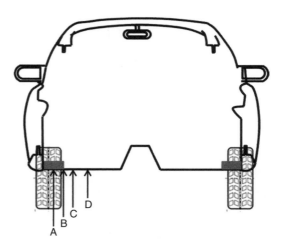

magnitudes increase with the distance away from the frame increase. Among the four points, the frame stiffness is the highest and the stiffness at the middle underbody is the lowest. Therefore, the excitation point should be chosen on or near to the frame.

Modal shape distributions can be used to help control the energy input to the body. If the excitation point is selected at a mode node, the input energy for the corresponding modal frequency is zero. Figure 5.51 shows the first bending mode of a body and three excitation points E, F, and G. E is at the node, F is close to the node, and G is at the point of the maximum modal amplitude. The sound-vibration sensitivities for the three points are 55, 58, and 70 dB N^{-1}, respectively. The closer the excitation point to the node, the less the energy input to the system. Therefore, the external excitation source should be placed on or close to the modal nodes.

5.5.3.2 Control of Beam Structure

Compared with the thin and large surface body panels, the frames (beams) and pillars are much stronger, so they can be regarded as solid structures rather than plates. Vibration in the solid structure is mainly transmitted in the form of bending waves.

Pipe section change results in sound impedance change, so partial sound waves traveling inside the pipe will be reflected back at the changed section, and partial sound waves will continue to propagate. Propagation of the structural bending wave inside the beams has the similar phenomenon. Figure 5.52 shows a constant section beam (a) and a beam with smaller section in the middle (b). The dimensions of the two beams are the same. The same excitations are applied at one end (point A) and responses are obtained on the other end (point B). Figure 5.53a and b shows the comparisons of time-domain and frequency-domain responses of the two beams at point B, respectively. The response of the beam with the smaller section in the middle is much lower than that of the constant section beam, and its frequencies shift as well.

A similar result can be obtained for the case where the beam section increases. Therefore, the structural wave can be attenuated by changing the beam section, further, the transmitted energy is attenuated.

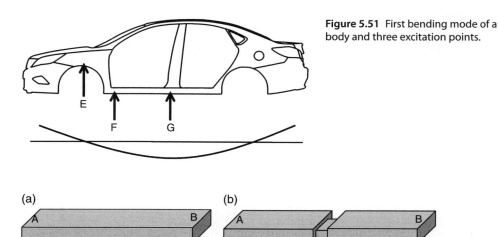

Figure 5.51 First bending mode of a body and three excitation points.

(a) (b)

Figure 5.52 (a) A constant section beam; (b) a beam with smaller section in the middle.

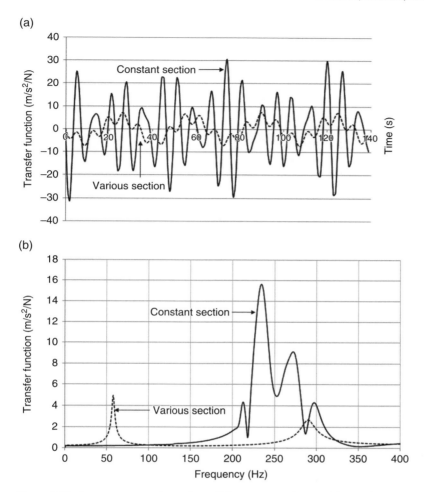

Figure 5.53 Response comparison of two different section beams: (a) time-domain, (b) frequency-domain.

5.5.3.3 Control by Isolation Elements

If isolation elements are used in the transfer paths, the structural wave transmission will be greatly attenuated. Figure 5.54 shows two connections between the subframes and the body. One is that the subframe is solidly connected with the body (a), and it is called rigid-connected subframe. The other one is that a rubber is inserted between the subframe and the body, and it is called flexible-connected subframe. Figure 5.55 shows the sound–vibration sensitivities for the two structures subjected to the same excitation. The structural-borne noise transmission is greatly reduced for the flexible-connected subframe.

Because the structural waves are attenuated in the transfer paths, the vibration transmitted to the body panels and to the occupants will be significantly reduced, which is a common method to isolate the paths to reduce the vibration–vibration and sound–vibration sensitivities. However, for sporty vehicles, the users enjoy driving funs and handling performances, so the subframes and the bodies are usually rigidly connected.

(a)   (b)

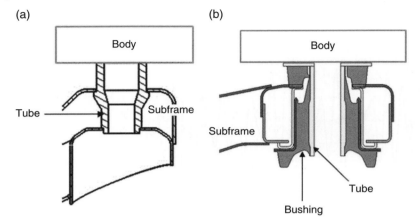

Figure 5.54 Connections between subframes and body: (a) rigid-connected subframe; (b) flexible-connected subframe.

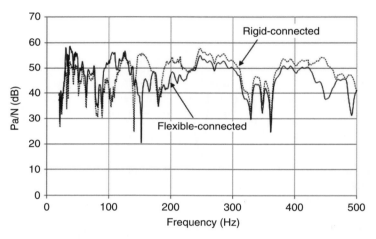

Figure 5.55 Comparison of sound-vibration sensitivities for two subframes (rigid-connected subframe, and flexible-connected subframe) subjected to the same excitation.

5.5.3.4   Control of Steering System Modes

The purpose to control the steering system modes is to separate the external excitation frequencies and the system modal frequencies and to reduce the vibration–vibration sensitivity. Readers can refer to the materials on the modal control of the steering system in Chapter 3.

5.5.3.5   Control of Panel Structures

When the structural waves are transmitted to the panels, they will be excited to vibrate and radiate sound, and the occupants can perceive the vibration and sound. The panel structure design is very important to reduce the vibration–vibration and sound–vibration sensitivities. The panel control methods include stiffness control, damping processing, mass control, dynamic absorber control, etc. Please refer to Chapter 3 for the detailed materials.

5.5.4 Sensitivity Targets

5.5.4.1 Vibration–Vibration Sensitivity Target

The vibration responses are represented by velocity or by acceleration. In the sensitivity analysis, the ratio of the velocity response to the force excitation is commonly used, and corresponding unit is $\mathrm{m\,s^{-1}\,N^{-1}}$, or $\mathrm{mm\,s^{-1}\,N^{-1}}$.

The relationship between the human body sensitivity to the frequency has been described in Chapter 3. The most sensitive frequency in vertical direction is between 4 and 8 Hz, as shown in Figure 3.102. Above 8 Hz, the acceleration linearly increases with the frequency increase, i.e. the acceleration lines are straight ones, and the relationship can be expressed as follows:

$$\ddot{x} = B\omega \tag{5.50}$$

where B is a constant and ω is frequency.

Integrate the acceleration, the velocity is obtained as follows:

$$\dot{x} = \frac{\ddot{x}}{j\omega} = \frac{B}{j} \tag{5.51}$$

The velocity magnitude is

$$|\dot{x}| = B \tag{5.52}$$

The velocity is a constant, so it is a flat, straight line, i.e. the velocity doesn't vary with the frequency change. Because frequencies of almost all the excitations and the body modes are above 8 Hz, so it is very convenient and easy to use the flat velocity line to represent the vibration target. Correspondingly, the ratio of velocity response to force excitation is adopted as the vibration-vibration sensitivity, and the straight target lines can be set up, as shown in Figure 5.56. For the passenger vehicles, usually the sensitivity target at idling condition is set up as $0.1\,\mathrm{mm\,s^{-1}\,N^{-1}}$.

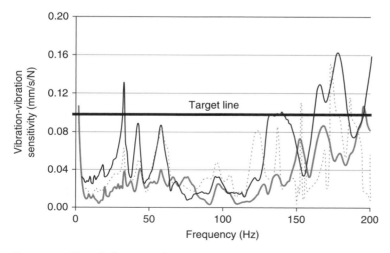

Figure 5.56 Tested vibration–vibration sensitivity and target line at idling.

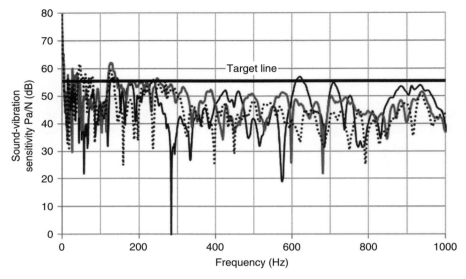

Figure 5.57 Tested sound–vibration sensitivity curves and the target line.

5.5.4.2   Sound–Vibration Sensitivity Target

The sound–vibration sensitivity is the ratio of the interior sound to the applied force, so its unit is $\mathrm{Pa\,N^{-1}}$, or $\mathrm{dB\,N^{-1}}$. The sensitivity target is usually set up as $0.01\,\mathrm{Pa\,N^{-1}}$, or $55\,\mathrm{dB\,N^{-1}}$. Figure 5.57 shows several tested sensitivity curves and the target line.

5.6   Sound–Sound Sensitivity and Control

5.6.1   Sound Transmission from Outside Body to Interior

Figure 5.6 provides the main noise sources that are transmitted to the interior as the airborne noise. Taking the engine radiation noise as an example to illustrate the transfer processing of the external noise to the interior, when the radiated noise encounters the body, partial energy is reflected back, partial energy is absorbed by the sound package, and the rest energy passes through the body openings and penetrates the body panels to the interior.

Generally speaking, the source sound and the interior noise are nondirectional sounds. Assume that the sound pressures at the ith noise source and at the occupant's ear are P_i and P_i^{AB}, respectively, the corresponding sound–sound sensitivity is expressed as

$$H_i^{AB} = \frac{P_i^{AB}}{P_i} \tag{5.53}$$

Eq. (5.53) can be rewritten to represent the interior noise generated by the ith source, as follows:

$$P_i^{AB} = H_i^{AB} P_i \tag{5.54}$$

The overall interior noise contributed by M sources is summation of the interior noise contributed by each source, expressed as follows:

$$P^{AB} = \sum_{i=1}^{M} P_i^{AB} \tag{5.55}$$

5.6.2 Expression of Sound–Sound Sensitivity

Eqs. (5.53–5.55) only describe the concept of sound–sound sensitivity. In engineering analysis and application, three quantitative methods are usually used to represent the sensitivity: Noise Reduction, Noise Transfer Function, and Power Based Noise Reduction.

5.6.2.1 Noise Reduction

Two microphones are placed inside and outside the body, respectively, as shown in Figure 5.58, and the sound pressures are measured. The noise reduction (NR) is defined the difference between the outside sound pressure level and the interior sound pressure level, and expressed as

$$NR = SPL_{out} - SPL_{in} \tag{5.56}$$

where SPL_{out} and SPL_{in} are the outside sound pressure level and the interior sound pressure level, respectively.

A specific environment is required to measure the body noise reduction in order to eliminate the interference of the environmental noise. The specific environment refers to the reverberation chamber and the anechoic chamber. The methods to measure the body noise reduction include the reverberation chamber method and the anechoic chamber method. Figure 5.58 shows the NR measurement in a reverberation chamber.

A vehicle or a body is placed in a reverberation chamber. A sound source is placed outside the vehicle and a plurality of microphones are arranged inside and outside of

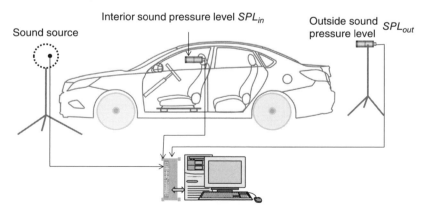

Figure 5.58 Diagram of noise reduction measurement in a reverberation chamber.

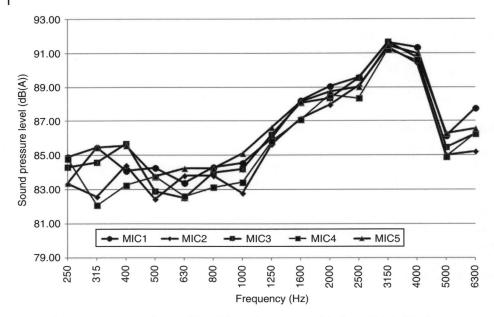

Figure 5.59 Sound pressure levels of five different locations outside the vehicle inside the reverberation chamber.

the vehicle. The sound source generates sound, and the sound pressures at different locations are measured. Figure 5.59 shows the sound pressure levels of five different locations outside the vehicle. The sound pressure levels of these different points are very close, which indicates uniform distribution of the sound field in the reverberation room.

Because of the uniformity of the sound field in the reverberation chamber, the sound pressure level at any point outside the vehicle can represent the sound pressure level of the source. The body's noise reduction can be obtained by subtracting the interior sound pressure level from the outside sound pressure level. If the microphones are positioned in different locations inside the body, such as dash panel, door, roof, the noise reductions for these components can be obtained, as shown in Figure 5.60.

The overall noise reduction can be obtained by the reverberation room method, but the noise reduction contributed by each sound source cannot be acquired because of the uniformity of the outside sound sources. In order to overcome the drawback, the measurement inside the anechoic room is needed. A special sound simulator is placed at a source location to generate white noise. For example, an engine sound simulator is a specially made device where several speakers with different sizes are installed on a box and are used to simulate the engine noise. The engine sound simulator is placed inside the engine compartment, and the vehicle is placed in the anechoic chamber, as shown in Figure 5.61. The sound pressures at six surfaces of the simulator and inside the body are measured. Subtracting the interior sound pressure level from the simulator sound pressure levels, the corresponding noise reductions between the six surfaces and the interior will be obtained, as shown in Figure 5.62. The figure shows that the noise reductions for different engine surfaces are different.

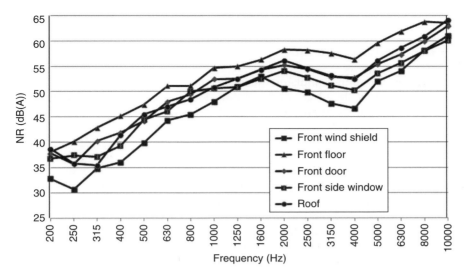

Figure 5.60 Noise reductions at different locations inside the body.

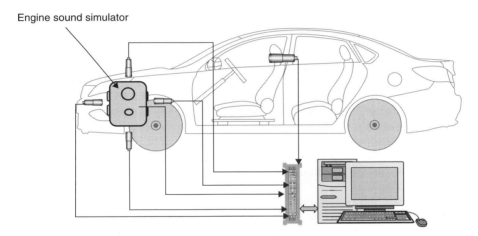

Figure 5.61 Noise reduction measurement inside the anechoic room.

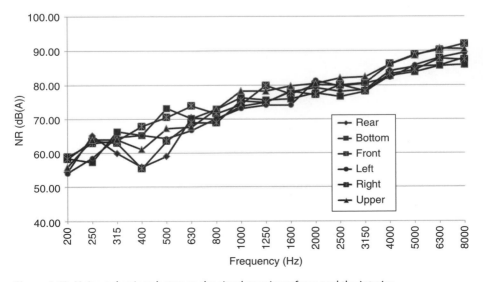

Figure 5.62 Noise reductions between the simulator six surfaces and the interior.

Similar to the engine sound simulator, the intake and exhaust sound simulator or the tire sound simulator is placed at corresponding locations so the noise reductions between these noise sources and the interior can be measured.

5.6.2.2 Acoustic Transfer Function

The sound simulators are complex and expensive, and different systems use different simulators. In engineering, in addition to using the simulators to measure the sound-sound sensitivity, the volume velocity source is widely used as sound source, as shown in Figure 5.63. It is placed at a noise source location, such as engine compartment, exhaust tailpipe orifice, etc., and the volume velocity or acceleration at sound source and the interior sound pressure are measured.

Acoustic transfer function (AFT) is defined as a ratio of the sound pressure at a response point to the volume acceleration at the sound source point, and expressed as

$$H^{PQ} = \frac{p}{Q_a} \tag{5.57}$$

where Q_a is the volume acceleration.

Figure 5.64 shows the tested acoustic transfer functions of a car at two locations. Smaller value indicates less noise transmitted from the outside source to the interior. Thus, the lower the curves, the better the sound insulation. The trend of ATF values is just the opposite of the NR. The bigger the NR value, the better. The ATF values decrease with frequency increase, which indicates that the vehicle body has better sound insulation at high frequencies.

According to the reciprocal principle, the sound source can be placed inside the body. The interior volume acceleration and the outside sound pressure are measured, and then the same acoustic transfer function will be obtained.

5.6.2.3 Power-Based Noise Reduction

Power-based noise reduction (PBNR) is defined as the ratio of sound power at a source point to the mean squared sound pressure at a response point, and is expressed as

$$PBNR = 10\lg\left(\frac{\Pi}{p \cdot p^*} \frac{1}{\text{Ref}}\right) \tag{5.58}$$

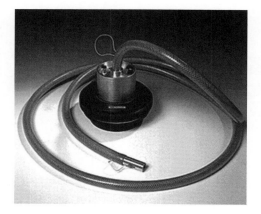

Figure 5.63 Volume velocity source.

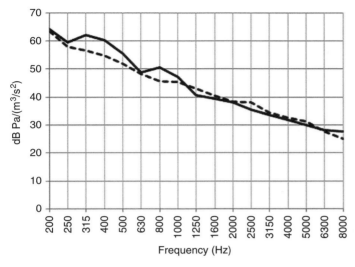

Figure 5.64 Tested acoustic transfer functions of a car.

where Π is the sound power at source point in a free field; p is the sound pressure at the response point, p^* is conjugate of the sound pressure p. Ref is a reference value and is defined as ratio of the reference sound power to square of the reference sound pressure, and expressed as

$$\mathrm{Ref} = \frac{\Pi_{ref}}{p_{ref}^2} = \frac{1}{\rho c} \tag{5.59}$$

where Π_{ref} and p_{ref} are the reference sound power and the reference sound pressure, respectively, and the corresponding air impedance is $400\,\mathrm{kg}\,(\mathrm{m}^{-2}\,\mathrm{s}^{-1})$.

In the free field, the sound power at the source point is

$$\Pi = \frac{\rho Q_a Q_a^*}{4\pi c} \tag{5.60}$$

where Q_a is the volume acceleration at the sound source, Q_a^* is conjugate of Q_a. Substituting Eqs. (5.59) and (5.60) into Eq. (5.58) results in the following:

$$PBNR = 10\lg\left(\frac{Q}{p}\left(\frac{Q}{p}\right)^*\right) + 10\lg\frac{\rho^2}{4\pi} = -20\lg\left|\frac{p}{Q_a}\right| - 9.4 \tag{5.61}$$

where p/Q_a is the ATF magnitude. Eq. (5.61) indicates that the PBNR can be calculated by tested acoustic transfer function.

The relationship between PBNR and NR is

$$PBNR = NR + 10\lg\frac{A_1\alpha_1}{4} \tag{5.62}$$

where A_1 and α_1 are sound absorption area and sound absorption coefficient of the sound source, respectively.

$A_1\alpha_1$ represents the effective absorption of the sound source. Eq. (5.62) indicates that the higher the NR, the higher the PBNR. The NR only contains the noise attenuation of the sound source side and the receiving side, while the PBNR not only contains the attenuation but also includes the sound insulation and absorption of both sides. Therefore, PBNR can better characterize the overall sound insulation and sound absorption of the body. The PBNR value with an absorptive pad installed outside the body is higher than that without the pad, and the gap increases with frequency increase, which indicates that PBNR reflects the sound absorption characteristics of the sound sources.

5.6.3   Targets and Control of Sound–Sound Sensitivity

5.6.3.1   Target

The expressions and measurement methods of the vibration-vibration sensitivity and sound–vibration sensitivity are mature. For a long time, the engineers have the clear and consistent targets, for example, the sound–vibration sensitivity target $55\,\text{dB}\,\text{N}^{-1}$ has been ingrained in the minds of the NVH engineers. However, there are several expressions for sound–sound sensitivity, but there is no consensus on the targets. The scientists and engineers are still exploring more effective expressions and values of the sensitivity.

Usually, benchmarking method is used for target setting. Figure 5.65 provides several vehicles' noise reductions between the exhaust orifice noise and the interior noise, and

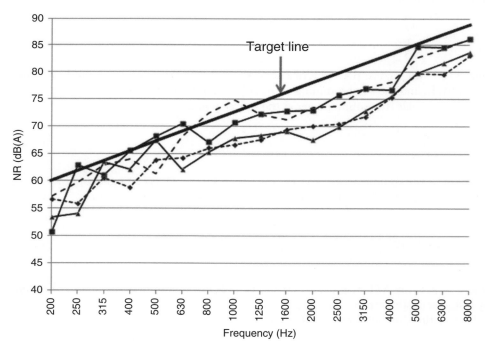

Figure 5.65 Benchmarked noise reductions between the exhaust orifice noise and the interior noise, and the set target.

the set target. The higher the curves, the higher the sound transmission loss, the better the body sound insulation. For example, during early stages of a vehicle development, if the sound package of a vehicle is determined as a leader in its segment-market, its sound insulation target line should be higher than that of the benchmarked vehicles, as shown in Figure 5.65.

5.6.3.2 Control of Sound–Sound Sensitivity

Control of the sound–sound sensitivity depends on the body sound package and the panels' sound radiation capacity. First, good body sealing must be guaranteed, and the body openings must be avoided. Second, good sound insulation should be achieved. Third, the body should have good sound absorption performance. Fourth, the panels' sound radiation must be suppressed, and the panels' modes must be separated from the body cavity modes. Please refer to Chapter 3 and Chapter 4 for the materials on the panel radiation and sound package.

Bibliography

Baik, H.-S., Jung, S.-G., and Kang, Y.-J. (2006). The Study on the Optimization of Attachment Stiffness in Vehicle Body. SAE Paper 2007-01-2346.

Bogema, D., Schuhmacher, A., Newton, G. et al. (2013). High-Frequency Time Domain Source Path Contribution: From Engine Test Bench Data to Cabin Interior Sounds. SAE Paper 2013-01-1957.

Caillet A. and Blanchet, D. (2015). Evolution of Trim Modeling with PEM for Structureborne Noise Prediction in Full Vehicle. SAE Paper 2015-01-2235.

Charpentier, A. Sreedhar, P., and Fukui, K. (2007). Using the Hybrid FE-SEA Method to Predict Structure-borne Noise Transmission in Trimmed Automotive Vehicle. SAE Paper 2007-01-2181.

Cotoni, V., Gardner, B., and Shorter, P. et al. (2005). Demonstration of Hybrid FE-SEA Analysis of Structure-Borne Noise in the Mid Frequency Range. SAE Paper 2005-01-2331.

Daly, M.A. (2003). Influence of Mount Stiffness on Body/Subframe Acoustic Sensitivities. SAE Paper 2003-01-1714.

Dubbaka, K.R., Zweng, F.J. and Haq, S.U. (2003). Application of Noise Path Target Setting Using the Technique of Transfer Path Analysis. SAE Paper 2003-01-1402.

Dumbacher, S M., Witter, M.C., Brown, D.L. et al. (1999). Evaluation of Sensors for Noise Path Analysis Testing. SAE Paper 1999-01-1859.

Eisele, G., Wolff, K., Alt, N. et al. (2005). Application of Vehicle Interior Noise Simulation (VINS) for NVH Analysis of a Passenger Car. SAE Paper 2005-01-2514.

Everest, A.F. (2001). *The Master Handbook of Acoustics*. McGraw-Hill.

Ewins, D.J. (1984). *Modal Testing: Theory and Practice*. Research Press Press.

Guedes, R. and Gonçalves, P.J.P. (2003). Investigation of Sub-System Contribution to a Pickup Truck Boom Noise Using a Hybrid Method Based on Noise Path Analysis to Simulate Interior Noise. SAE Paper 2003-01-3677.

Gwaltney, G.D. and Blough, J.R. (1999). Evaluation of Off-Highway Vehicle Cab Noise and Vibration Using Inverse Matrix Techniques. SAE Paper 1999-01-2815.

Hashioka, M. and Kido I. (2007). An Application Technique of Transfer Path Analysis for Automotive Body Vibration. SAE Paper 2007-01-2334.

Haste, F. and Nachimuthu, A. (1999). Calculating Partial Contribution Using Component Sensitivity Values: A Different Approach to Transfer Path Analysis. SAE Paper 1999-01-1693.

Hong, S.B. and Vlahopoulos, N. (2005). Application of a Hybrid Finite Element Formulation for Analyzing the Structure-Borne Noise in a Body-in-White. SAE Paper 2005-01-2421.

Kalsule, D., Hudson, D., Yeola, Y. et al. (2011). Structure Borne Noise and Vibration Reduction of a Sports Utility Vehicle by Body-Mount Dynamic Stiffness Optimization. SAE Paper 2011-01-1599.

Kim, H., Do, J., Oh, S. et al. (2013). Optimization of Body Attachment for Road Noise Performance. SAE Paper 2013-01-0369.

Kim, H.-S. and Yoon, S.-H. (2007). A Design Process using Body Panel Beads for Structure-Borne Noise. SAE Paper 2007-01-1540.

Kim, K. and Choi, I. (2003). Design Optimization Analysis of Body Attachment for NVH Performance Improvements. SAE Paper 003-01-1604.

Knechten, T., Morariu, M.-C., and van der Linden, P.J.G. (2015). Improved Method for FRF Acquisition for Vehicle Body NVH Analysis. SAE Paper 2015-01-2262.

Kousuke, N. and Junji, Y. (2006). Method of Transfer Path Analysis for Vehicle Interior Sound with No Excitation Experiment, F2006D183, FISITA.

Lee, J.H., Oh, K., Park, Y.-S. et al. (2001a). Transfer Path Analysis of Structure-Borne Shock Absorber Noise in a Passenger Car. SAE Paper 2001-01-1441.

Lee, S.-K., Park, K.-S., Lee, M.-S. et al. (2001b). Vibrational Power Flow and Its Application to a Passenger Car for Identification of Vibration Transmission Path. SAE Paper 2001-01-1451.

LMS. (2005). Best Practices / Planning for TPA, LSM International.

LMS. (2013). LMS Test.Lab Transfer Path Analysis, LSM International.

Madjlesi, R., Khajepour, A., and Ismail, F. (2003). Advance Noise Path Analysis, Robust Engine Mount Optimization Tool. SAE Paper 2003-01-3117

Manning, J.E. (2003). SEA Models To Predict Structureborne Noise In Vehicles. SAE Paper 2003-01-1542.

Mazzarella, L., Godano, P., and Horak, J. (2007). Reciprocal Powertrain Structure-borne Transfer Functions Synthesis for Vehicle Benchmarking. SAE Paper 2007-01-2354.

Park, H.-C. and Yoon, S.-H. (2007). Contribution Analysis of Vehicle Interior Noise Using Air-borne Noise Transfer Function. SAE Paper 2007-01-2359.

Parrett, A., Zhang, Q., Wang, C. et al. (2003). SEA in Vehicle Development Part I: Balancing of Path Contribution for Multiple Operating Conditions. SAE Paper 2003-01-1546.

Peng, G.C. (2011). Measurement of Exterior Surface Pressures and Interior Cabin Noise in Response to Vehicle Form Changes. SAE Paper 2011-01-1618.

Plunt, J. (2005). Examples of Using Transfer Path Analysis (TPA) together with CAE-Models to Diagnose and Find Solutions for NVH Problems Late in the Vehicle Development Process. SAE Paper 2005-01-2508.

Rust, A. and Edlinger, I. Active Path Tracking. A Rapid Method for the Identification of Structure Borne Noise Paths in Vehicle Chassis. SAE 2001-01-1470.

Shiozaki, H., Geluk, T., and Daenen, F. et al. (2012). Time-domain Transfer Path Analysis for Transient Phenomena Applied to Tip-in/Tip-out (Shock & Jerk). SAE Paper 2012-01-1545.

Sottek, R., Sellerbeck, P., and Klemenz, M. (2003). An Artificial Head Which Speaks from Its Ears: Investigations on Reciprocal Transfer Path Analysis in Vehicles, Using a Binaural Sound Source. SAE Paper 2003-01-1635.

Tournour, M.A., Kosaka, F., and Shiozaki, H. (2007). Fast Acoustic Trim Modeling Using Transfer Admittance and Finite Element Method. SAE Paper 2007-01-2166.

Tsujiuchi, N., Koizumi, T., and Nagao, T. (2011). Vibration Transmission Analysis of Automotive Body for Reduction of Booming Noise. SAE 2011-01-1691.

Unruh, J.F., Till, P.D., and Farwell, T.J. (2000). Interior Noise Source/Path Identification Technology. SAE Paper 2000-01-1709.

van der Auweraer, H., Mas, P., Dom, S. et al. (2007). Transfer Path Analysis in the Critical Path of Vehicle Refinement: The Role of Fast, Hybrid and Operational Path Analysis. SAE Paper 2007-01-2352.

van der Linden, P.J.G. and Wyckaert, K. (1999). Modular Vehicle Noise and Vibration Development. SAE Paper 1999-01-1689.

Ver, I.L. and Beranek, L.L. (2006). *Noise and Vibration Control Engineering: Principles and Applications*. Wiley.

Yamamoto, S. Kobayashi, N., and Yamaoka, H. (2011). Vehicle Interior Noise and Vibration Reduction Method Using Transfer Function of Body Structure. SAE Paper 2011-01-1692.

Zhu, J., Hammelef, D. and Wood, M. (2005). Power-Based Noise Reduction Concept and Measurement Techniques. SAE Paper 2005-01-2401.

6

Wind Noise

6.1 Introduction

6.1.1 Problems Induced by Wind Noise

At low speed, the vehicle interior noise is dominated by the noise generated by engine and related systems such as intake system, exhaust system, etc. With vehicle speed increase, road noise increases, and could dominate the interior noise at middle speed, such as from 50 to 80 km h^{-1}, especially on rough road. At high speed, such as over 100 km h^{-1}, the wind noise gradually becomes prominent, and even could mask the engine noise and the road noise. Figure 6.1 shows the relationship between vehicle speed and engine noise, road noise, and wind noise.

In recent years, with noise control technology development of the engine, intake, exhaust, tire and suspension, transfer paths, etc. these traditional noise sources are greatly suppressed, which pushes wind noise in the forefront of major noise sources. Wind noise has become one of the top complaint issues by the customers.

Wind noise, or aerodynamic noise, is induced by interaction between a moving vehicle and the airflow. Wind noise could provide passengers with the impression that doors or windows are not perfectly closed or leakages exist on the body. Loud wind noise could influence conversation between passengers; even to the point that they cannot clearly hear each other. Sometimes, the sound level of the wind noise is not high, but some passengers can clearly hear the low-level and high-frequency leakage sound, and they complain about the noise because the high-frequency leakage noise make them uncomfortable.

Today, people spend more and more time on highway and their demands on low wind noise have become a prominent requirement for vehicle noise, vibration, and harshness (NVH) development because the wind noise will make them annoyance, even fatigue. The passengers request that sound level of the wind noise should be low, and there are no leakage noise and no high frequency tonal noise. The target for wind noise control is not only to reduce the noise level but also to improve its sound quality. The ideal wind noise is: low sound level, no leakage noise, no high frequency whistle, no buffeting sound, and no turbulence noise.

Wind noise control includes two aspects: body design control and manufacture process control. This book covers only the body design for wind noise control, including body overall styling, design of local structures and appendages, and body sealing.

Noise and Vibration Control in Automotive Bodies, First Edition. Jian Pang.
© 2019 China Machine Press. All rights reserved. Published 2019 by John Wiley & Sons Ltd.

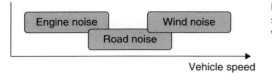

Figure 6.1 Relationship between vehicle speed and engine noise, road noise, and wind noise.

- *Body overall styling.* If the body overall styling is not smooth, such as non-smooth transitions between engine hood and front windshield, protrusions on underbody, etc. the airflow blowing to these locations will induce the turbulence that generates the wind noise.
- *Body local shape and appendage design.* The non-smooth transition shapes or protrusions on the local structures will generate the wind noise. For example, swirl and turbulence will be generated when the airflow blows the A-pillar rain gutter. The large gaps or margins on the body surface will form the cavities where the cavity noise will be generated when they are blown by the airflow.

  Some appendages are mounted outside the body, such as antennas and mirrors. When the airflow blows to the protrusions, vortex and turbulence are generated, inducing the wind noise. If they are properly designed, the noise will be greatly reduced. In addition, installations of these appendages are also very important; for example, the same mirror installed on different locations of the body will generate different wind noise
- *Body sealing.* Small openings on the body will permit outside sound to enter the interior directly, and also induce the whistle noise.

6.1.2 Sound Sources and Classification of Wind Noise

Wind noise can be categorized from two aspects. The first aspect is categorized by the sound being perceived and measured, such as the noise generated by pulsating pressure, the noise generated from the apertures. The second aspect is categorized by physical mechanism of the wind noise, such as monopole noise or dipole noise.

6.1.2.1 Wind Noise Categorized by Sound Being Perceived and Measured

The frictions between the airflow and different locations of the body generate different noise sources that can be grouped into four categories:

1) *Pulsating noise.* The airflow acting on the body generates a vortex, which forms the pressure fluctuation on its surface. The noise generated by the vortex disturbances is called pulsating noise.
2) *Aspiration noise.* The wind noise outside the vehicle could passes through the body apertures and enters into the interior. Even if there are no apertures on the body at vehicle standstill condition, but when it moves, small openings could appear between relatively moving parts such the doors and the body frames. For this case, the outside noise penetrates the openings and enters into the interior. The wind noise penetrating through the apertures or openings is called aspiration noise.
3) *Buffeting noise.* When a sunroof or a side window is opened, the body works like a resonator, and the low-frequency booming noise called buffeting noise will be generated by the interaction between the airflow and the opened body cavity.

4) *Cavity noise.* The gaps always exist on body surface, such as the gap between the A pillar and front door. If the gap is large enough, a small cavity will be formed. When the airflow blows to the small cavity, the airflow oscillates inside the cavity and generates noise that is called the cavity noise.

6.1.2.2   Wind Noise Categorized by Physical Mechanism

In classical acoustics, sound waves are generated by surface vibration of an object. The vibration induces alternating change between compression and expansion of the fluid on the object surface, which generates sound. The simplest unit to generate sound is the point sound source. The combinations of different point sources form different sound generating bodies. The most typical sound sources include monopole sound source, dipole sound source, and quadrupole sound source.

The wind noise belongs to aeroacoustics, and its sound generation mechanism is different from the classical acoustics. However, the sound sources in aeroacoustics can be analogous to monopole source, dipole source, and quadrupole source in classical acoustics. According to this analogy, from the perspective of sound generation mechanism, the wind noise can be divided as monopole noise source, dipole noise source, and quadrupole noise source. Sound sources for different poles are related to vehicle speeds and have different sound radiation capability.

The first wind noise source is the monopole sound source. The monopole sound source is a source of pulsating ball which continuously and periodically expands and contracts with small magnitude, and uniformly radiates spherical waves, as shown in Figure 6.2. The monopole sound source is generated by the movement of the unstable volume airflow. At low Mach number, sound radiation of the monopole sound source is the most efficient. When the fluctuating pressures on the body surface push the outside unstable airflow to move into the vehicle interior, the monopole sound source is generated. The sound power of the monopole sound source is proportional to the fourth power of the airflow velocity, and expressed as

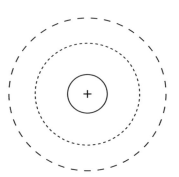

Figure 6.2 Monopole sound source.

$$W \sim \left(\rho_0 l^2 c^3\right)\frac{v^4}{c^4} \tag{6.1}$$

where, ρ_0 is the air density, l is characteristic dimension of an object, v is the airflow velocity, and c is the sound speed.

The sound power level for the monopole sound source is expressed as

$$L_W \sim 40\lg v \tag{6.2}$$

The second wind noise source is dipole sound source. The dipole sound source consists of a pair of monopole sound sources that are very close each other and their phases are opposite, as shown in Figure 6.3. The dipole sound source is generated by

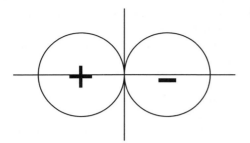

Figure 6.3 Dipole sound source.

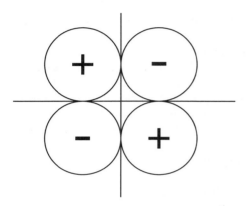

Figure 6.4 Quadrupole sound source.

unstable flow pressure on a rigid body surface or turbulence impact on an object surface. For example, when an antenna is subjected to airflow impact, a Von Carmen vortex is formed and an unstable force is generated, which results in the dipole sound source. The sound power for the dipole sound source is proportional to the sixth power of the airflow velocity, and expressed as

$$W \sim \left(\rho_0 l^2 c^3 \right) \frac{v^6}{c^6} \tag{6.3}$$

The sound power level for the dipole sound source is expressed as

$$L_W \sim 60 \lg v \tag{6.4}$$

The third wind noise source is quadrupole sound source. The quadrupole sound source consists of a pair of dipole sound sources that are very close each other and their phases are opposite, as shown in Figure 6.4. When two fluid elements impact each other, the fluid generates unstable internal stress, forming the quadrupole sound source. The quadrupole sound source exists in an unstable shear turbulent layer. The sound power of the quadrupole sound source is proportional to the eighth power of the airflow velocity, expressed as

$$W \sim \left(\rho_0 l^2 c^3 \right) \frac{v^8}{c^8} \tag{6.5}$$

The sound power level for the quadrupole sound source is expressed as

$$L_W \sim 80 \lg v \tag{6.6}$$

The ratio of the dipole sound source intensity to the monopole sound source intensity is proportional to square of Mach number, while the ratio of the quadrupole sound source intensity to the dipole sound source intensity is proportional to square of Mach number. Thus, for supersonic speeds, the sound intensity of the quadrupole sound source is much higher than that of the dipole sound source, and the sound intensity of the dipole sound source is much higher than that of the monopole sound source. The automobile speed belongs to low Mach number ($M < 0.3$); thus for vehicle wind noise, the source sound intensity of the monopole is much higher than that of the dipole, and the source sound intensity of the dipole is much higher than that of the quadrupole. Since the aspiration noise belongs to the monopole sound source, it dominates the wind noise if an unstable leakage exists. If there is no leakage, the dominated wind noise is contributed by the dipole sound source.

6.2 Mechanism of Wind Noise

According to categorization by the sound being perceived and measured, the wind noise is divided as pulsating noise, aspiration noise, buffeting noise, and cavity noise. According to the categorization by the physical mechanism, the wind noise is divided as monopole sound source, dipole sound source, and quadrupole sound source. In this book, wind noise will be described by the sound being perceived and measured; however, the physical mechanism will be used to explain the causes of the wind noise.

6.2.1 Pulsating Noise

6.2.1.1 Mechanism of Pulsating Noise

Figure 6.5 shows an ideal streamlined object. When the airflow blows it, the airflow perfectly attaches the object surface and the flow field is smooth. The ideal object is a rigid body that has no deformation.

When the airflow blows the object, the pressure fluctuations are generated on its surface, which will induce wind noise, called pulsating noise or wind rush noise. Since the pulsating noise source belongs to the dipole sound source, its sound intensity is proportional to the sixth power of the airflow velocity.

Figure 6.5 An ideal streamlined object.

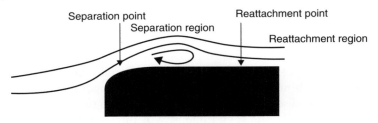

Figure 6.6 A rectangular object with an arc angle.

Figure 6.7 A rectangular object.

Figure 6.6 shows a rectangular object with an arc angle. When the airflow blows it, first, the airflow attaches with its surface; and then leaves it, that is, the airflow and the object are separated; finally, the airflow attaches the object again. The area the airflow separates from the object is called separation region, and the area the airflow reattaches the object is called reattachment region. In the separation region, the airflow forms vortex flow. The sound generated by the vortex flow is much louder than the sound generated by the surface pressure fluctuations, even 10 times bigger. The sound inside the reattachment region is huge as well. The pressure fluctuation in reattachment region is larger than that in the attachment area, so corresponding noise in the reattachment region is higher.

Figure 6.7 shows a rectangular object. When the airflow blows it, the airflow doesn't attach with the object; instead, the airflow and the object are separated, forming a separation region. After a long-distance separation, the airflow reattaches with the object, forming a reattachment region. Compared with the rectangular object with an arc angle, the rectangular object has much longer separation region.

6.2.1.2 Body Pulsating Noise

When the airflow blows the body, unstable pressure fluctuation on its surface is formed to generate the dipole sound source, radiating sound in all directions. Even if the body surface is very smooth, rigid, and with no leakage, the pressure fluctuation always exists because the boundary layer of the airflow attached on its surface is in turbulent condition, which illustrates that no matter how smooth the body surface, the pulsating noise always exists.

Figure 6.8 shows an airflow field distribution of a passenger sedan. There are many attached regions and separation regions. The most typical separation regions include the areas of sideview mirrors, A pillars, and antenna. In the separation regions, there exists strong vortex, and the noise generated by the vortex is 10 times louder than noise generated by the pressure fluctuation in the attached regions. Thus, the separation regions are the main contributors of the pulsating noise.

Figure 6.8 Airflow field distribution of a passenger sedan.

The pulsating noise is the most important component of the body wind noise – that is inevitable. The pulsating noise is generated by the dipole sound source and has relatively wide frequency band. Usually, the pulsating noise is stronger than the aspiration noise and the cavity noise.

6.2.1.3 Control of Body Pulsating Noise
The reason generating the body pulsating noise is the airflow pressure fluctuations on its surface, and the separation of the airflow from the body rapidly increases the noise level. Therefore, control of the pulsating noise should be started from its generation mechanism; that is, the body should possess good streamlined surface and the separation between the airflow and the body should be minimized.

Good streamlined surface lets the airflow perfectly attach the body where the pressure fluctuation will be significantly reduced. In the phase of the body styling, it is extremely important to consider influence of the airflow on the flow resistance and wind noise. In the phase of the prototype, it is almost impossible to modify the body structure.

At interfaces of several body surfaces, transition must be smooth enough in order to avoid mutated area, so, the separation regions can be avoided or significantly reduced. For example, in A-pillar area, front windshield, door, side window, and roof are interfaced together. The interface of these components should be designed to be as smooth as possible.

For the body appendage design, influence of the separation regions generated by the airflow must be considered.

6.2.2 Aspiration Noise

Aspiration noise refers to the noise transferred from the outside into the interior through the body openings. Due to the openings, the outside noise "leaks" into the interior, so the aspiration noise is also called leak noise.

6.2.2.1 Conditions to Generate Aspiration Noise

Aspiration noise appears in two situations. One is due to poor static sealing and the other is due to poor dynamic sealing. Because of design and/or manufacture errors, the static apertures detailed in Chapter 4 could appear.

Small openings could appear for a vehicle with good static sealing when it travels down the road. The openings existing at the vehicle running condition are called dynamic apertures. For example, a door and a body are well sealed, but when the vehicle is driven down the road, the dynamic apertures could emerge because their relative displacements are different. The outside noise will be transmitted into the interior through the openings.

For a moving vehicle, the body surface is subjected to the pressure (p_s) induced by the airflow. The pressure consists of two parts and is expressed by

$$p_s = \bar{p} + p' \tag{6.7}$$

where, $\bar{p}$ is the average pressure, and p' is the fluctuating pressure.

The fluctuating pressure is the main cause that generates the pulsating noise, while the average pressure is the main cause of aspiration noise.

The door presses the body frame by pressure, p_0, called door sealing pressure. When the vehicle moves, the airflow speed is very fast and the average pressure on the body surface rapidly reduces, which results in a pressure difference between inside pressure (p_2) and outside pressure. When the pressure difference is higher than the door sealing pressure, i.e.

$$|p_2 - \bar{p}| > p_0 \tag{6.8}$$

the door is pushed and the dynamic opening between the door and the body appears, resulting in the aspiration noise.

6.2.2.2 Categories of Aspiration Noise

The aspiration noise can be categorized as three groups.

The first category is the noise transmitted into the interior through the openings/apertures. Outside noise such as engine noise and tire noise is directly transferred into the interior through the static openings and dynamic openings.

The second category is the noise generated by the airflow disturbance at outside margins and transferred into the interior. Figure 6.9 shows an aperture on the body and the interaction between the outside airflow and the edge of the margin. The interaction results in vortex or turbulence that is the unstable mass flow, generating sound that radiates in all directions. One direction is toward to the aperture, so it is a monopole noise source.

The mass flow moves into the aperture. As long as it leaves the aperture, it emits into the interior and rapidly expands, forming a new unstable flow, even vortex, and generating dipole noise source. The free turbulence entering the interior generates quadrupole sound source. So, in this category of aspiration, monopole, dipole, and quadrupole sound sources coexist. The mass flow speed is much slower than the Mach number, thus, the monopole source has much stronger sound radiation capacity than the dipole sound source, so the contribution by the dipole sound source can be neglected. Therefore, the monopole sound source dominates the aspiration noise.

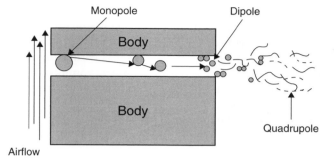

Airflow

Figure 6.9 Aspiration noise development procedure by outside airflow moving into interior through an aperture.

Sound intensity of the aspiration noise is lower than the pulsating noise, even much lower, but its energy concentrates on a single frequency or narrow-band frequencies that are close to the most sensitive frequency range of human hearing, so it will make people more uncomfortable than the pulsating noise.

The third category is the noise caused by impact between the inside airflow and the outside airflow, and then it is transmitted into the interior. The inside pressure is much higher than the outside pressure, i.e.

$$p_2 \gg p_0 \tag{6.9}$$

The inside airflow moves toward the outside, as shown in Figure 6.10. When it reaches to the outside edge of the aperture, it collides with the outside airflow, so the outside turbulence is rapidly intensified and noise at the edge is significantly increased.

6.2.2.3   Location and Control of Aspiration Noise
The aspiration noise can be identified in the areas of both static apertures and dynamic apertures. The major locations the aspiration noise could be found include:

- The apertures between the door and the body
- The apertures between the mirror base and the door
- The apertures between the movable side glass and the seal
- The apertures between the fixed glass and the seal
- The apertures between the door handle and the door
- The apertures between the trunk lid and the body

Figure 6.10 Inside airflow moving toward to outside.

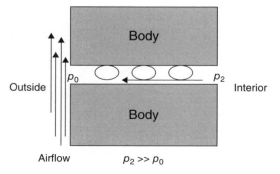

Aspiration noise is dominated by the monopole sound source and consists of high frequency tone signals varying with time. The occupants are usually sensitive to the high-frequency tones that make them annoying. Even if the sound level for the aspiration noise is not high, it is easily and clearly perceived because it is within the most sensitive hearing frequency range. Aspiration noise is one of the most prominent noise problems that customers complain about, so the first task of wind noise control is to control the aspiration noise.

The keys to control the aspiration noise is to eliminate static and dynamic apertures on the vehicle body, i.e. to achieve satisfied static sealing and dynamic sealing.

6.2.3 Buffeting Noise

6.2.3.1 Mechanism of Buffeting Noise

When sunroof or side windows are opened during cruising, a low frequency and loud booming can be perceived. This special noise is called buffeting noise.

Figure 6.11 shows a moving vehicle with an open sunroof. There is an unstable shear layer flow on the body surface. When the shear flow encounters the front edge of the sunroof (point A), the vortex sheds from the body and moves back with the shear layer flow. When the vortex meets the rear edge of the sunroof (point B), the vortex is broken and the scattering pressure waves are generated. Partial pressure waves enter into the body cavity, partial waves radiate to outside, and the rest are reflected to the front edge (point A) and form a new vortex. The new vortex moves back again and meets the rear edge (point B). After the new vortex is broken, new scattering pressure waves are formed. The processing "vortex movement – broken – reflection – new vortex – broken" repeats with a certain frequency, which is the source of the sunroof buffeting.

When the frequency of the processing "vortex movement – broken – reflection – new vortex – broken" is consistent with the body cavity frequency, body resonance happens. The cavity resonant frequency, or the buffeting frequency, depends on the vehicle speed, cavity volume, opening shape and area, etc. The buffeting frequency is very low, even lower than 20 Hz. The passengers couldn't hear the sound, but can perceive that they are assaulted by the pulse waves. The pulse energy is huge and the sound level could be higher than 100 dB. The buffeting noise makes people very uncomfortable.

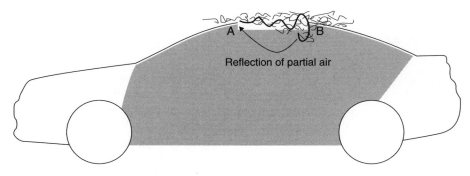

Figure 6.11 A moving vehicle with an opening sunroof.

(a)   (b)

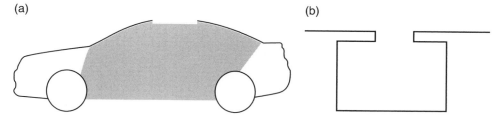

Figure 6.12 Analogy between a body cavity (a) and a Helmholtz resonator (b).

The vehicle compartment is an empty space. When the sunroof is opened, the space and the sunroof opening form a Helmholtz resonator, as shown in Figure 6.12. The space is the Helmholtz's volume and the opening can be regarded as a connecting tube of the resonator. The opening area and height difference between opening inside and outside can be regarded as the tube's section area and length, respectively.

6.2.3.2 Calculation of Buffeting Noise

The body with the sunroof opening can be regarded as a Helmholtz resonator, as shown in Figure 6.12. The Helmholtz resonator is composed of a cavity and a tube. When the outside pressure pushes the air inside the tube to move, the air inside the cavity is compressed and then expanded and pushes the air inside the tube to move in the opposite direction. The Helmholtz resonator is similar to a single degree of freedom (SDOF) mass-spring system. The air in the tube is analogous to the mass, and the air in the cavity is analogous to the spring, as shown in Figure 6.13.

The frequency of the Helmholtz resonator is expressed as

$$f = \frac{c}{2\pi}\sqrt{\frac{A}{Vl}} \tag{6.10}$$

where V is the cavity volume, A and l are section area and length of the tube, respectively. Equation (6.10) can be used to calculate the buffeting frequency. V is the interior volume, A and l are the area and the height of the opening, respectively.

6.2.3.3 Control of Buffeting Noise

The buffeting is induced during the processing of "vortex movement – broken – reflection – new vortex – broken." As long as the processing cycle is broken, the buffeting is eliminated.

(a)   (b)   (c)

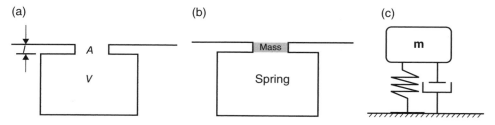

Figure 6.13 Analogy between a Helmholtz resonator and a mass-spring system (a) Helmholtz resonator; (b) Equivalent mass-spring system; (c) single degree of freedom mass-spring system.

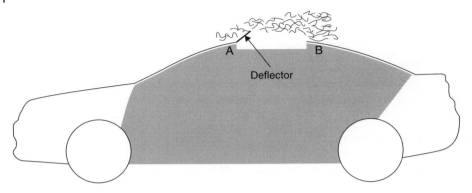

Figure 6.14 A deflector added at front edge of the sunroof.

Figure 6.14 shows that a deflector is added at front edge of the sunroof. The airflow moves along with the deflector. When the flow sheds from the deflector, it moves on top of the cavity and doesn't enter into the cavity. After the airflow passes over the rear edge, it reattaches the body. Because there exists no buffeting cycle and no airflow enters into the cavity, the buffeting disappears.

6.2.4   Cavity Noise

Gaps or margins exist at the interface areas on the body surface. The outside gap and aperture are two different concepts. The aperture means the small openings between the body outside and inside, where the airflow and sound can be transmitted between the both sides. The gap refers to the body surface openings at the interface among several component surfaces, where the airflow cannot be communicated between inside and outside.

In some areas, the outside gaps form small cavities on the body surface, for example, the gap between A pillar and the front door, the gap between the side mirror and the surrounding parts. Figure 6.15 shows a cross section of B pillar region. There are seals between the front door and the B pillar, and between the rear door and the B pillar. There exists a gap between the front and the rear doors. Therefore, a small cavity is formed in area of the doors, the B pillar and the seals. The small cavity and the body interior are not communicated. When the airflow blows the cavity, a special noise, called cavity noise, is generated.

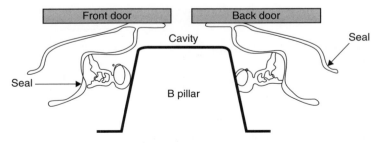

Figure 6.15 A small cavity in area of the doors, the B pillar and the seals.

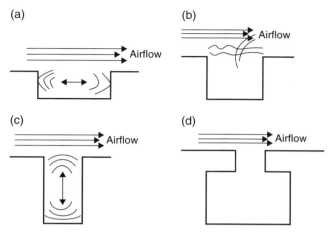

Figure 6.16 Types of small cavities on body surface (a) wide opening and shallow depth cavity, (b) wide opening and deep depth cavity, (c) narrow opening and deep depth cavity, (d) Helmholtz resonator.

The small cavities can be classified into several types, such as wide opening and shallow depth cavity (a), wide opening and deep depth cavity (b), narrow opening and deep depth cavity (c), Helmholtz resonator (d), etc., as shown in Figure 6.16. For the wide opening and shallow depth cavity and the Helmholtz resonator, the airflow movement is the same as the buffeting processing. For the wide opening and deep depth cavity, the sound wave is transmitted horizontally. For the narrow opening and deep depth cavity, the sound wave is transmitted in depth direction. When the airflow blows to the small cavities, the generated noise is a tonal sound. Unlike the low-frequency buffeting, the cavity noise frequencies are much higher due to its small volume. Sometimes for a tiny cavity, the noise becomes a whistle.

The method to eliminate or reduce the cavity noise is to eliminate the cavities. In body design, the cavities should be avoided. In some areas where cavities cannot be avoided, some actions must be implemented to cover the cavities. For example, the cavity shown in Figure 6.15 can be filled using a rubber cake, as shown in Figure 6.17a, or covered by a piece of tape, as shown in Figure 6.17b.

6.3 Control Strategy for Wind Noise

Wind noise sources include pulsating noise source, aspiration noise source, cavity noise source, and buffeting noise source. Transfer paths include airborne path and structural borne path. The wind noise control should be implemented by both source control and path control.

6.3.1 Transfer Paths of Wind Noise

The transfer paths for wind noise are the same as others introduced in previous chapters, including the airborne paths and the structural borne paths. The unstable airflow

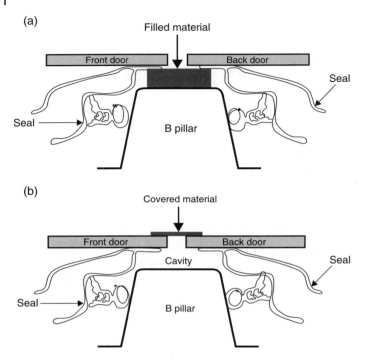

Figure 6.17 Control of the cavity in B-pillar region: (a) filed, (b) covered.

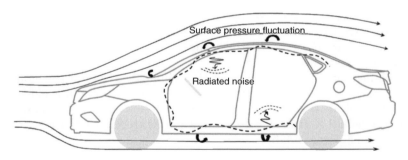

Figure 6.18 Structural borne paths of wind noise.

pressure fluctuation induces the body panels to vibrate, and then the vibrated panels radiate sound into the body interior. This noise transfer path is a typical structural borne path, as shown in Figure 6.18.

The airborne path relates to the body structure and the sealing. Partial outside sound is insulated by the body structure and partial sound is transmitted through the body into the interior, as shown in Figure 6.19. The apertures permit the outside sound to pass through the body directly and enter the interior. Even though there are no apertures on the body at vehicle standstill condition, small opening could emerge on the body when it runs down the road. Dynamic sealing is more important for wind noise control, which will be introduced in this chapter.

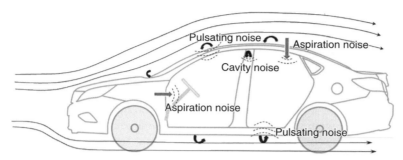

Figure 6.19 Airborne paths of wind noise.

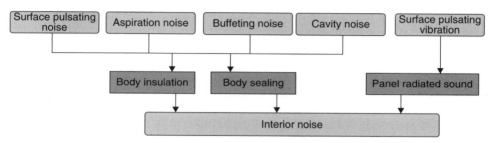

Figure 6.20 Noise sources and transfer paths of wind noise.

Figure 6.20 describes the noise sources and the transfer paths of the wind noise. The pulsating noise source and the cavity noise source are outside the body, so the method to reduce the noise transferred into the interior is to control the body's sound insulation performance. The aspiration noise passes through body apertures and enters into the interior, so the method to reduce it is to increase body sealing performance, especially the dynamic sealing. The buffeting is induced by opening the sunroof, so the method to reduce the noise is to control the design of sunroof and its surrounding structures.

6.3.2 Control Strategy of Wind Noise

According to the characteristics of the noise sources and the transfer paths, the wind noise can be controlled from two aspects: source control and transfer path control. The source control is to optimize body design based on analysis of interaction between the airflow and the body. The transfer path control is to control dynamic sealing, body panel vibration, and its sound radiation.

6.3.2.1 Control of Wind Noise Sources
The source control for the wind noise is to reduce the noise source by modifying the body structure based on analyzing interaction between the airflow and the body, and mechanism of the wind noise, including the pulsating noise source, the aspiration noise source, the cavity noise source, and the buffeting noise source.

Control of the pulsating noise source should be started from overall body styling and local structure design. The objective is to reduce the interaction effectiveness between the airflow and the body, and further, to reduce the pressure fluctuation on the body surface.

Transitions among the panel surfaces should be as smooth as possible. The separation regions and reattachment regions should be avoided or reduced. The local structures should be designed to guide the airflow to move smoothly on their surface in order to reduce the surface pressure fluctuation.

To achieve the goals, the body and local structures can be designed from two aspects. First, the interfaces between the body and the attached components, such as the side mirror, should be benefit for smooth airflow. The unsmooth transition and narrow distance between the mirror and the body could induce huge local vortex. Second, the local structures or components should have rounded and smooth surfaces. For example, the mirror should have a streamlined shape in order to guide the airflow to move around its surface smoothly. Another example, an antenna design should prevent generating the vortex; otherwise, an annoying tonal noise will be generated.

Control of the aspiration noise focuses on the dynamic sealing. First, the body should have a good static sealing. Second, relative displacement for neighbor components and seal characteristics should be analyzed for a moving vehicle. Only if the body has both good static sealing and dynamic sealing can the aspiration noise be avoided or reduced.

Control of the cavity noise focuses on reducing the outside gaps (margins). Big gaps and cavities at interfaces among the panels should be avoided to prevent the airflow to blow them.

The buffeting control targets the interaction between the airflow and the body cavity. A deflector should be properly designed in order to guide the airflow to move over the cavity and reattach the body behind the rear edge of the cavity opening.

Mechanisms for the four categories of wind noise are different, but from the perspective of sound pole sources, they contain some common characteristics. To effectively analyze the wind noise sources, their generation mechanisms and relating sound poles should be considered simultaneously.

6.3.2.2 Transfer Path Control Strategy of Wind Noise

The structural borne path of the wind noise is the body panels. The pressure fluctuation of the airflow excites the panels, and then the panels radiate sound into the interior. Thus, the structural borne transmission of the wind noise is determined by the panel structures. The body panels are subjected to many external excitations, such as engine vibration, exhaust vibration, suspension vibration, wind excitation, etc. Different from other excitation sources, wind excitation includes more middle- and high-frequency components. A good body panel should have minimum sound radiation into the interior under excitations by all sources. Therefore, the body panel design is critical for the structural borne path control of the wind noise. Readers can refer to Chapter 3 on the plate vibration and sound radiation.

6.3.2.3 Body Design Strategy for Wind Noise Control

The source and transfer path controls of the wind noise, and the body design should be considered simultaneously, so the wind noise can be controlled by following aspects:

1) First is the vehicle overall styling, including the body overall streamlined styling, interface design between front grille and engine hood, between engine hood and front windshield, between rear windshield and trunk lid, local design of A pillar region, roof, underbody, etc.

2) Second is the local structure design, including design of mirror, antenna, luggage rack, sunroof, etc.
3) Third is the dynamic sealing control, including sealing between the door and the body, sealing of the side mirror sail, sealing of the rear quarter window, etc.

6.4   Body Overall Styling and Wind Noise Control

The body overall styling has significant impact on the wind noise. During the body design, the influence of the overall styling and the local design on the wind noise should be fully considered, i.e. the styling from a vehicle front end to back, styling from left to right, and styling from top to bottom.

6.4.1   Ideal Body Overall Styling

The major factors considered in the vehicle styling include fashion, wind resistance, and wind noise. The fashion is to meet the customers' needs on aesthetic appreciation of the vehicle styling. In modern society, the vehicle styling is similar to fashionable dress and changes fast. The customers' aesthetic requirements change with the development of the times, and some of the aesthetic appreciations continuously change and may come back with times, and even become a cycle. The wind resistance is an important factor for fuel economy. Therefore, the body design should meet requirements of low wind resistance and low fuel consumption. The body design for low wind noise reflects the customers' pursuit of vehicle quality. The vehicles with low wind noise provide them comfortable driving environments and meet their pursuit on high quality of life.

The body consists of many connected surfaces. The transitions among the surfaces have significant influence on the wind noise. A longitudinal section of the body, as shown in Figure 6.21a, has the following transitional areas:

1) Transition from front grill to engine hood
2) Transition from engine hood to front windshield
3) Transition from front windshield to roof
4) Transition from roof to rear windshield
5) Transition from rear windshield to trunk lid
6) Transition from trunk lid to the tail.

A lateral section of the body, as shown in Figure 6.21b, has the following transitional areas:

7) Transition from front grill to front fender
8) Transition from front windshield to A pillar, then to side window
9) Transition from rear side window to C pillar, then to rear windshield
10) Transition from rear side panel to the tail.

In order to achieve low wind resistance and the low wind noise, the body surface is preferably designed to be a stretched streamline. Figure 6.5 shows an ideal streamlined contour of a vehicle body where the transition curvature at the corner is large enough and protrusions are avoided. The purpose of this design is to force the airflow to attach the body surface, to avoid the flow separation and reattachment, to reduce the pressure

(a)

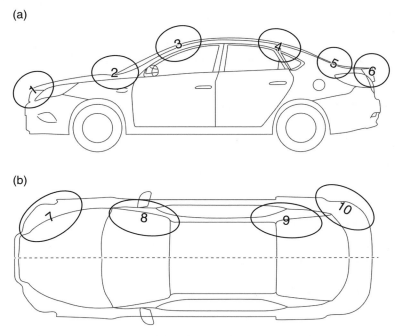

(b)

Figure 6.21 Transitions of body interfaced surfaces: (a) longitudinal section; (b) lateral section.

(a)

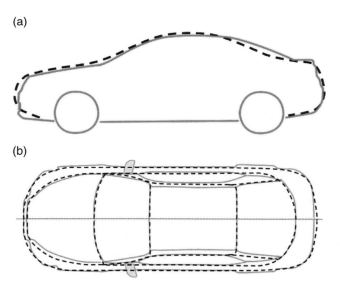

(b)

Figure 6.22 Ideal streamlined body styling: (a) longitudinal section; (b) lateral section.

fluctuation, and to achieve the low wind noise and the low wind resistance. The solid line in Figure 6.22 is the contour of the actual body, and there are protrusions on the line. If the body design is virtually optimized, the contour becomes the dot line. The dotted line is much smoother than the solid line and there are no protrusions, instead, it is the ideal streamline. If a vehicle body could be designed as the dotted lines, the wind noise will be greatly reduced.

The main function of a vehicle is a transportation tool. The passengers want to have large interior space; however, many systems such as engine, driveline, suspension, are installed in the vehicle, which limits the body to achieve a streamlined styling shown in Figure 6.22. Similarly, large spaces should be provided for the engine compartment, the interior, the trunk, etc. which pushes the body shape to be designed as close to a rectangle as possible. In addition, the fashion requirement makes the body styling to change continuously. In a word, it is impractical to design a vehicle body like the ideal streamlined shape.

Main function for a sporty car is to pursue the driving fun at high speed cruising. Because of its fast speed and requirements on the low wind resistance and the low noise, the transitions at the body surfaces should be much smoother than the passenger car. The body shape of the sporty car could be designed as close to the streamline as possible.

During a vehicle development, the body styling and the engineering application should be simultaneously considered, which means that functions including styling, wind resistance, wind noise, even marketing, etc. should be balanced. Some functions can be achieved simultaneously for one design. For example, the streamlined design is benefit for both the low wind resistance and the low wind noise. However, some pursuits for different functions are contradicted. For example, the antenna has little impact on the wind resistance, but has significant influence on the wind noise.

6.4.2   Design of Transition Region between Front Grill and Engine Hook

The airflow blows to the region of the front grille and the engine hood. Three locations could induce wind noise: the transition arc and the cut line between the grille and the hood, and the grille shape.

First, influence of the transition arc between the grille surface and the hood surface on the wind noise. The larger and more smooth the arc, the lower the wind noise. Figure 6.23 shows the local structures of the grilles and the hoods for two vehicles where

(a)   (b)

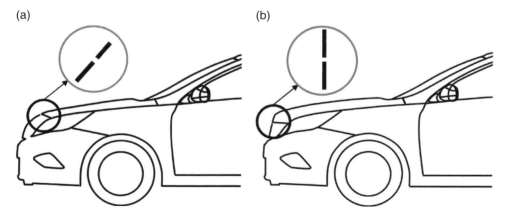

Figure 6.23 Local structures of grilles and hoods for two vehicles.

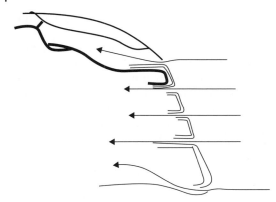

Figure 6.24 Airflow moving through grille gaps.

their transition arcs are different. Figure 6.23a shows that the transition area is smooth, so the noise will be lower. Figure 6.23b shows that the two surfaces are almost perpendicular each other, which will result in high wind noise.

Second, influence of the cut line between the grille and the hood on the wind noise. Figure 6.23a indicates that the cut line is on a oblique plane. Strong turbulence will be induced around the cut line when blown by the airflow, resulting in relatively high noise. Figure 6.23b indicates that the cut line is on horizontal direction. The airflow will smoothly move into the cut line, so the local turbulence is small and the wind noise will be low.

Third, influence of the grille on wind noise. When the airflow moves through the grille gaps as shown in Figure 6.24, whistle could be induced due to the shedding of the vortexes from the grille bars. The gaps, approach angle, airflow speed, etc. influence the whistle. The whistle can be reduced or eliminated by increasing the gaps, smoothing approach angle, adding hood tip spoiler, etc.

6.4.3 Design in Area between Engine Hood and Front Windshield

6.4.3.1 Analysis of Noise Sources

Wind noise can be generated at the areas between the front windshield and the engine hood. In the first instance, the airflow is separated from the hood when it reaches to its rear end, and then reattaches the windshield, as shown in Figure 6.25. The "separation-reattachment" induces huge pressure fluctuation in the separation region, generating the wind noise.

In the second instance, the cause of wind noise is that the airflow blows to the exposed wiper (as shown in Figure 6.25) and the vortex is produced, resulting in a strong pulsating noise.

6.4.3.2 Control of Noise Sources

In order to reduce the wind noise in this area, a streamlined transition surface between the hood and the front windshield is needed and the wiper should be not exposed in the flow field, which can be achieved by three methods:

1) The hood rear end is designed as a tilted arc in order to guide the airflow to smoothly flow to the windshield, as shown in Figure 6.26. The wiper is placed beneath the hood in order to avoid the airflow.

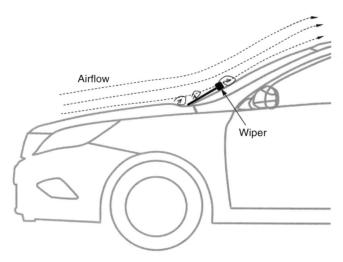

Figure 6.25 Airflow at area between front windshield and engine hood.

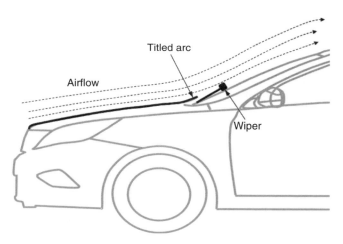

Figure 6.26 Engine hood rear end is designed as a tilted arc.

2) A deflector is installed in the front of the wiper, as shown in Figure 6.27. The deflector will guarantee that the air flows smoothly from the hood to the windshield, and simultaneously the wiper avoids the airflow.
3) A transition plate is installed between the hood and the windshield, as shown in Figure 6.28. In the local area, the hood, the transition plate, and the windshield could form a curved surface where the air moves smoothly. The wiper is also hidden and avoids the airflow, so wind noise will be significantly reduced.

6.4.4 Design of A-Pillar Area

6.4.4.1 Analysis of Noise Sources at A-Pillar Area

The wind noise around the A-pillar region is usually very high because the region where the front windshield, door, roof, side mirror, and front body intersect is very complicated.

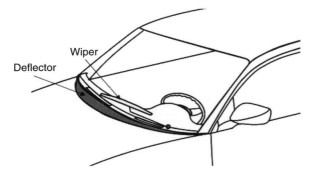

Figure 6.27 A deflector installed in the front of wiper.

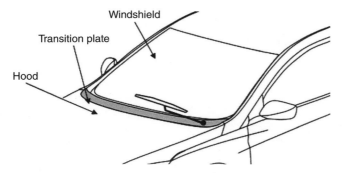

Figure 6.28 A transition plate between engine hood and windshield.

The side mirror is also complicated, and its shape, mounting with the body and gaps, have significant impact on the wind noise. The design principle of the mirror will be described in later material in this chapter.

The noise in this region is also complex, including pulsating noise, aspiration noise, and cavity noise.

The first type of noise is the pulsating noise. The airflow blows to the windshield, then separates and moves toward to the roof and to the side windows. In the area between the A pillar and the side window, a separation region is formed. After the airflow leaves the area, it reattaches the side window, as shown in Figure 6.29a. A section is cut from the A pillar, as shown in Figure 6.29b. It is clearly to see the separation region and the reattachment region. At the edge of the A pillar, the airflow is standstill. In the separation region, the vortex is strong.

Similarly, a separation region is formed after the airflow blows to the side mirror. The airflow reattaches the door and the side window to generate a reattachment region, as shown in Figure 6.30. In the separation region behind the mirror, a strong wake is formed.

The front windshield and A pillar are not on the same surface; instead, there is a height difference between them. When the airflow blows to the rain gutter, the airflow will be separated and reattached.

(a)

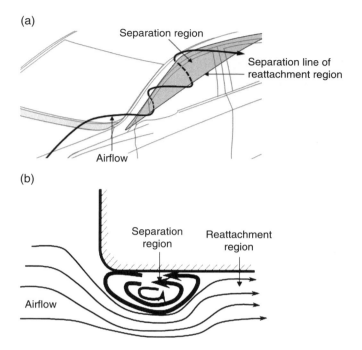

(b)

Figure 6.29 Airflow around A pillar: (a) side view, (b) a section view.

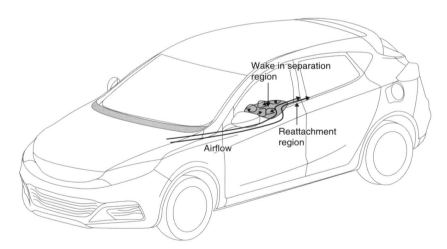

Figure 6.30 Airflow in side mirror area.

Three airflow separation regions and three reattachment regions coexist, so it is extremely important to control the wind noise around the A pillar.

The second type of wind noise is the aspiration noise. There are many seals in the A-pillar area, such as door seal and body seal, cutline seal, glass run and beltline seal in moving windows, and fixed triangle windows. Poor sealing results in apertures or gaps that induces the monopole tonal aspiration noise.

(a)

(b)

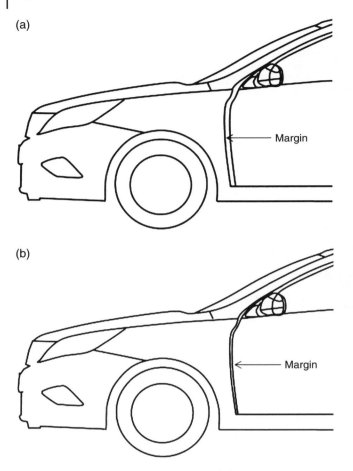

Figure 6.31 Outside cavity in A pillar area, (a) large margin; (b) small margin.

The third type of the wind noise is the cavity noise. In A-pillar area, there are many small gaps outside the body, such as margins between the door and the A pillar, as shown in Figure 6.31. The large margin and the space behind, as shown in Figure 6.31a, could form a cavity, resulting in the cavity noise. The margins should be as small as possible, as shown in Figure 6.31b.

The airflow speed at the A-pillar area is relatively large, which increases the wind noise. If the edges of the margins are sharp, the wind noise will be more obvious. The driver and the front passenger are close to the A pillar, so they are very sensitive to the outside wind noise. Thus, the design of A pillar area becomes critically important for wind noise control.

6.4.4.2 Design of A-Pillar Area to Achieve Low Noise

Many components in A pillar area could induce wind noise. Design for each component and their combinations must follow the principles of low noise design.

First, in A-pillar area, intersection arc among the component surfaces should be large enough and their transition should be smooth, as shown in Figure 6.32. There are four

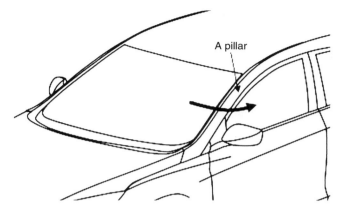

Figure 6.32 Transitions between surface and surface in A pillar area.

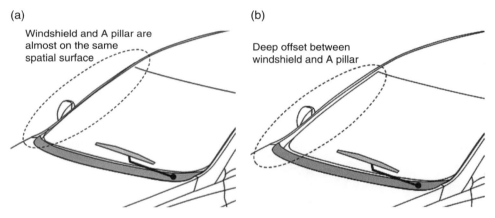

Figure 6.33 Transition between front windshield and A pillar: (a) on the same spatial surface, (b) deep offset.

surfaces in A-pillar area: the hood surface, the windshield surface, the roof surface, and the side body surface. There are four intersections for the surfaces: intersections between the hood surface and the windshield surface, between the windshield surface and the body side surface, between the windshield surface and the roof surface, between the roof surface and the body side surface. When the airflow blows the intersections, the separation region and the vortex should be as small as possible to achieve the lower wind noise.

Second, it is ideal to design these intersected surfaces and the A pillar on one spatial surface. Figure 6.33a indicates that the windshield and the A pillar are almost on the same spatial surface. The airflow lines will be smooth when it blows to the area. In Figure 6.33b, the windshield and the A pillar are not on a curved surface; instead, there exists a deep offset. When the airflow blows the area, a strong separation region and disordered vortex will be generated.

Third, the gaps around the A pillar should be as small as possible. A cavity is easily formed for large gaps, resulting in the cavity noise. The small gaps between the door and the body, and between body panels will be beneficial to low wind noise.

Fourth, the A pillar area must have good sealing. There are many seals in the A-pillar area, such as sealing between the door and the body, sealing between the moving glass window and the door, sealing between the side mirror and the body, sealing between interior cable inside mirror and the body, and sealing of the holes on the mirror. Poor sealing easily induces the aspiration noise.

6.4.5 Design of Transition Area of Roof, Rear Windshield, and Trunk Lid

In the areas of the roof, the rear windshield, and the trunk lid, several branches of the airflows intersect. The airflow attached on the roof separates with the rear windshield, then reattaches somewhere on the windshield or on the trunk lid. Two groups of the airflows attach the body sides to move, and partial airflows move to the windshield and the trunk lid. These airflows intersected, which makes the flow field in this area very complicated, as shown in Figure 6.34. A distinct separation region is formed in the intersected area between the rear windshield and the trunk lid.

During the body design phase, a smooth transition for intersected surfaces must be considered. In addition, the arcs of the intersected surfaces must be big enough to guide the airflow to move and to reduce the separation region and the vortex strength.

Sealing is also a problem in this area. The most important sealing is the seal on back of the trunk lid, and its purpose is to prevent the airflow to move into the trunk, as shown in Figure 6.35. If there is a gap in this area, the airflow will easily flow into the trunk and the aspiration noise is induced, which will be transmitted into the passenger compartment.

Figure 6.34 Airflow in area of roof, rear windshield and trunk lid.

Decklid seal

Figure 6.35 Seal on back of trunk lid.

6.4.6 Underbody Design

Many appendages are installed on an underbody, including power plant, exhaust pipe, fuel tank, oil pipe, and so on, as shown in Figure 6.36. When the airflow impacts these appendages, separation regions will be formed that induce the strong vortex, generating the dipole noise sources.

To reduce the disturbance of the underbody airflow, two methods are usually adopted. The first method is to add an air spoiler (or called air dam) at the front of the body bottom, as shown in Figure 6.37. The spoiler is composed of an inclined plate and a horizontal plate. The spoiler guides the airflow's direction in order to let the airflow move away from the appendages. The spoiler should be installed as close as possible to the front, and its height and width should be big enough.

The second method is to install a belly pan cover so that all appendages are hidden, as shown in Figure 1.36. The airflow does not directly blow the appendages; instead, it moves along the cover, so many potential small separation regions are avoided. Figure 6.38 shows the interior noise comparison with and without a belly pan cover, where a vehicle was tested in a wind tunnel with the wind speed of $120\,\mathrm{km\,h^{-1}}$. It is clear to see that the wind noise is significantly reduced after the belly pan cover is installed, especially at the low- and

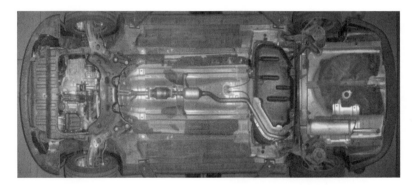

Figure 6.36 Appendages on underbody.

Figure 6.37 A underbody spoiler (air dam).

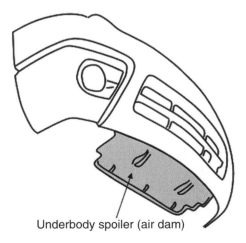

Underbody spoiler (air dam)

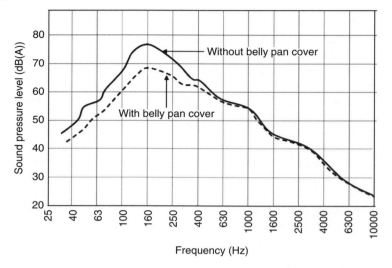

Figure 6.38 Interior noise comparison between with and without a belly pan cover.

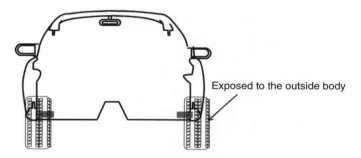

Figure 6.39 Relative positions between tires and body.

middle-frequency ranges. The tested results also explain that the major wind noise components caused by the underbody appendages are of low and middle frequencies.

6.4.7 Design in an Area of Wheelhouse and Body Side Panel

If a tire is exposed to the outside body (the right tire in Figure 6.39) and subjected to the airflow, a separation region appears and the vortex is intense, which induces huge local noise. Therefore, the tires must be covered by the wheelhouses and avoided directly to expose to the airflow, like the left side tire as shown in Figure 6.39.

6.5 Body Local Design and Wind Noise Control

6.5.1 Principles for Body Local Structure Design

In addition to the overall styling, design of local structures and accessories has huge impact on the wind noise. The principles for the overall styling include smooth airflow around the body, large enough arcs for the transition surfaces, flat underbody, no protrusions on the body. This ideal styling can reduce the airflow separation and reattachment, and decrease the vortex and the pulsating noise.

The body local structures could induce complicated wind noise. The protrusions induce the dipole pulsating noise. Poor sealing lets the airflow move into the interior, which generates the monopole aspiration noise. The outside gaps and cavities produce the cavity noise. The improperly designed sunroof induces the buffeting. Therefore, the design of the local structures and accessories is very important to reduce wind noise.

There are many local structures. The book focuses on major body appendages such as side mirror, antenna, roof luggage rack, and door handle, and their influences on the wind noise will be introduced.

Design of the body local structures and appendages must follow four principles:

1) *The local structures should be as far as possible to avoid the airflow.* For example, the door handle should be designed on the same plane as the door surface.
2) *The local structures and appendages should have sufficiently large and rounded arcs to reduce the separation regions and the vortex intensity.* For example, the windward side of the side mirror should have smooth surface, and the roof rack rod should be designed as an oval section.
3) *The airflow should move away from the sensitive areas.* For example, after a deflector is added at the front edge of a sunroof, the airflow moving direction is changed so that it does not enter the passenger compartment.
4) *The tonal noise induced by the turbulence must be avoided.* For example, exposed in the airflow, a round antenna generates the tonal sound, but the annoying tone can be eliminated by using a spiral antenna because the turbulence around the antenna is broken.

6.5.2   Design of Side Mirror and Its Connection with Body

The airflow around the A pillar is very complicated and its relative speed is very high. Unfortunately, the side mirror is in this area. The mirror is one of the most important parts on the vehicle body for inducing wind noise that contributes significantly to the interior noise. In a wind tunnel, a vehicle with and without the mirrors was tested and their interior sounds were compared, as shown Figure 6.40. The main difference between

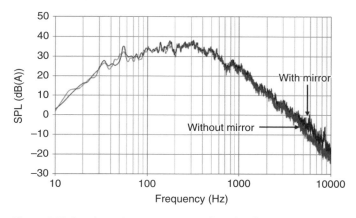

Figure 6.40 Interior noise comparison with and without mirror.

the two cases is in the middle- and high-frequency range. After the mirror is removed, the middle- and high-frequency noise is reduced and the sound quality is improved.

Influence of the side mirror on wind noise includes three aspects: the mirror's shape and geometry structure, mounting of the mirror on the body, and sealing of the mirror. Influence factors and control methods of the wind noise will be described in following materials.

6.5.2.1 Design of Side Mirror

A side mirror consists of mirror head, mirror glass, base, motor, etc. as shown in Figure 6.41. The base is usually composed of a triangular sail plate and a pedestal. The base is connected to the body or the door. The head usually can be folded and is mounted on the base. The motor mounted inside the head drives the mirror to rotate.

The factors related with the mirror structure influencing on the wind noise include: head shape, drain groove on the head, and folding margin between the head and the base.

The head is a windward part. The airflow blows to the head, and then separates, forming a wake behind the mirror. The head design should allow the airflow to move smoothly around its surface in order to obtain uniform flow in the separation region and low wake disturbance. Referring to Figure 6.42, the design principles of the side mirror are as follows:

1) The mirror head should be as small as possible while still meeting the vision requirements. The smaller the head, the less the airflow blows to this area, the lower the wind noise.

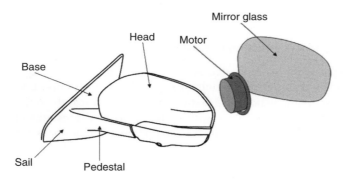

Figure 6.41 A side mirror structure.

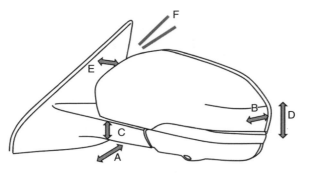

Figure 6.42 Outside structure and dimension of a side mirror.

2) The head shape should be streamlined. Good head shapes have smooth circular or curved windward surface to guide the airflow to flow smoothly and uniformly on its surface, which will reduce the pulsating noise.

3) Sharp edges should be avoided. The sharp edges increase the airflow turbulence and vortex. The edges must be in curved shapes.

4) Both the pedestal depth (A) and mirror head depth (B) should be designed as long as possible to let the airflow have a longer stable flow area where the airflow will be uniform and the turbulence is reduced.

5) The heights of both the pedestal (C) and the mirror head (D) should be as lower as possible to reduce the airflow separation region.

6) The head should be kept as far as possible from the door (E), so that the airflow speed in the area between the head and the door is reduced. The angle between the door and the head inner surface is smaller at the front and larger at the end, as shown in Figure F.

7) The folding margin should be as small as possible; otherwise, the whistle could be induced because the airflow blowing the margin is similar to whistling. Generally, the margin should be less than 0.5 mm.

8) The groove on the mirror head should not affect the airflow. It is better to avoid the groove on the shell because a streamlined shell easily guides the water to drop. For the head with drain groove, as shown in Figure 6.43, the groove should avoid the airflow.

9) Holes on the head should be avoided. Figure 6.43 shows that a folding head and the base are connected by bolts. After the airflow blows to the exposed holes, strong vortex will be induced, which generates huge noise. Existing holes must be covered and sealed.

6.5.2.2 Mounting of Mirror on Body

Because the airflow speed around A pillar is very high, and the flow field is very complex, so the ideal installation location of the mirror should be kept away from this area, i.e. the mirror should be installed as far behind the area as possible. However, the function of the mirror is to provide a good vision for a driver, so it is impossible to install the mirror beyond the A-pillar area, otherwise the installed location cannot meet the needs of the driver's visual angle. The flow field between the mirror and the

Figure 6.43 Local design of mirror surface.

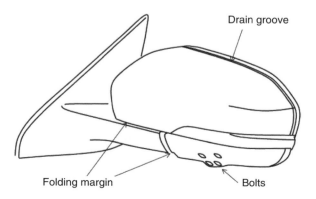

Drain groove

Folding margin

Bolts

body must be considered for installing the mirror. The principle is that the separation region and the wake behind the mirror should be as small as possible.

First, the mirror is preferably mounted on the metal plate of the door, instead of on the fixed glass of the body. High rigidity of the metal plate increases the modal frequencies of the mirror system, and the possibility to generate a resonance can be avoided. Installed on the glass, the mirror system's natural frequency is low, which easily induces resonance and even makes the vision on the mirror blurred.

Second, the inner surface of the head should be perpendicular to the ground and form an angle from the side body surface in order to guide the airflow to smoothly flow in the area and reduce turbulence.

Third, the distance between the mirror and the side window should be as large as possible to slow down the local airflow speed and reduce the local vortex.

Fourth, the sail and the door should be on the same surface. Large offset between the sail and the body easily induces the turbulence.

6.5.2.3 Sealing of Mirror

Sealing between the mirror and the body includes the outside sealing between the sail and the body and the sealing inside the mirror.

The sail and the triangle window metal sheet must be tightly fitted and sealed. Otherwise, the airflow will easily permeate the body through the unsealed apertures, generating the monopole noise.

Figure 6.44 shows a sail and the triangular body metal sheet. First, the sail and the metal sheet must be smoothly fitted; second, the holes on the metal sheet must be avoided; finally the two parts should be well sealed. To achieve the perfect sealing between the sail and the body, three layers of sealing can be used. The first layer sealing is to seal the sail edges using sponges or seals. The second layer sealing is to seal edges of inner protrusions using adhesive sealants. The third layer sealing is to seal the surroundings of the bolts.

The glass rotation is driven by a motor. The cables inside the body pass through the triangle metal sheet and the sail, and then connect with the motor installed inside the head. The locations where the cables pass through could bring poor sealing problems, so sealing these areas is very important. The best way is to install a socket on the triangle metal sheet and a plug on the sail. This way, where the plug is inserted into the socket, the opening on the metal sheet can be avoided, as shown in Figure 6.45.

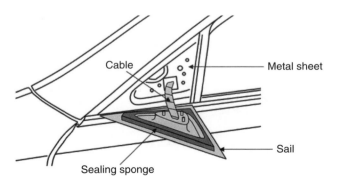

Figure 6.44 A sail and triangular body metal sheet.

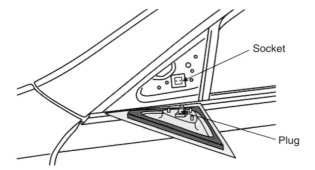

Figure 6.45 A socket on the triangle metal sheet and a plug on the sail.

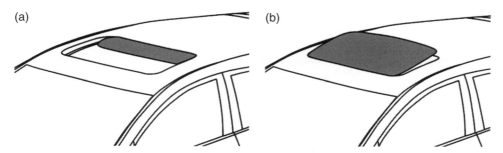

Figure 6.46 Two working modes for a sunroof: (a) retracted mode and (b) vented mode.

6.5.3   Sunroof Design and Wind Noise Control

The sunroof has two working modes: retracted mode and vented mode, as shown in Figure 6.46. The retracted mode (Figure 6.46a) refers to back-forward movement of the sunroof, and its opening size is controlled by moving the sunroof glass. When the glass is moved to the end, the sunroof is fully open. In vented mode (Figure 6.46b), the glass cannot be moved; instead, it can be only rotated.

Buffeting could be generated after the sunroof is opened. The airflow blows to the edges of the sunroof and repeats to move forward and back between the front and rear edges. When the movement frequency is consistent with the interior cavity mode frequency, the buffeting happens. So, the most important method to control the buffeting is to break the repeated forward-back movement of the airflow, or to avoid the airflow blowing into the sunroof opening area. The most common method to control the buffeting is to add a deflector at front edge of the sunroof. For the vented mode, the control method is to add side shields.

The deflector installed at the front edge of the sunroof can be rotated at the front end and the back end in free condition. When the sunroof is open, the deflector rotates to form an angle with the roof surface, as shown in Figure 6.47. Four principles must be followed for the deflector design:

1) The gap between the sunroof front edge and the deflector should be as small as possible – generally less than 3 mm.
2) The transition from the roof to the deflector should be smooth, and the cavity between the sunroof front edge and the deflector should be avoided.

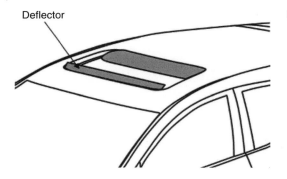

Deflector

Figure 6.47 A sunroof deflector.

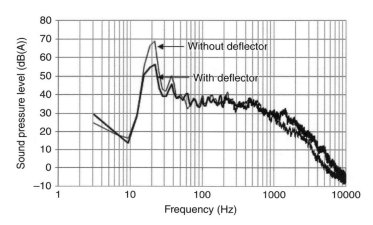

Figure 6.48 Interior noise comparison for a vehicle cursing at 80 km h⁻¹ with and without a deflector.

3) The edges of the deflector should be smooth, and the sharp edge should be avoided. The curvatures of the deflector's edges should be larger than 1 mm.
4) The deflector should be stiff enough.

The geometrical shape, height, and frontal angle of the deflector influence effectiveness of the buffeting noise reduction. Figure 6.48 shows an interior noise comparison for a vehicle cursing at 80 km h^{-1} with and without a deflector.

The buffeting noise at 21 Hz is reduced 12 dB with the deflector. The deflector solves the buffeting problem, but it induces high-frequency pulsating noise because it is exposed in the airflow. Figure 6.48 shows that the buffeting noise is reduced but the middle- and high-frequency noise increases when the deflector is used.

The pulsating noise can be reduced by a castellated sunroof deflector, as shown in Figure 6.49. The buffeting noise and the pulsating noise can be balanced by the modified deflector. The geometrical shape, frontal angle, grooves, and holes (as shown in Figure 6.50) influence the wind noise as well. According to characteristics of different buffeting frequency, these parameters can be optimally designed.

When the sunroof is in the vented mode, the airflow could enter into the body interior from its side opening area, as shown in Figure 6.51a, which could induce the buffeting noise or increase the interior sound pressure level. After side shields are added

Figure 6.49 Castellated sunroof deflector.

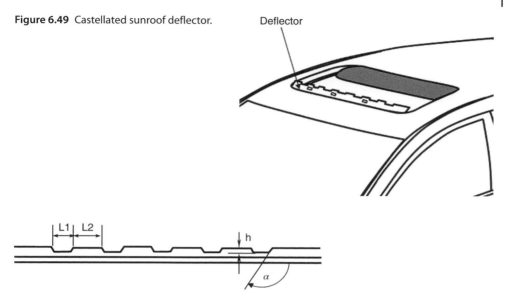

Figure 6.50 Parameters on the castellated sunroof deflector.

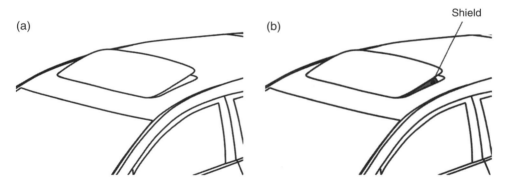

Figure 6.51 Sunroof in vented mode: (a) without side shields; (b) with side shields.

on both sides of the sunroof, as shown in Figure 6.51b, the side airflow is stopped by the shields and there is no airflow moving into the interior, so the buffeting is avoided.

The interior noise for a vehicle without the side shields is much higher than that with the side shields, and the sound level difference for the two cases increases with vehicle speed increase. Thus, the side shields are needed for the sunroof design. The gaps between the shields and the roof metal sheets should be as small as possible.

6.5.4 Antenna Design and Wind Noise Control

An antenna can be regarded as a slim cylinder, as shown in Figure 6.52a. When the airflow blows to the antenna, a strong conical vortex flow is formed. The vortex flow moves upward along the axis of the cylinder, forming an upward fluctuating lift force.

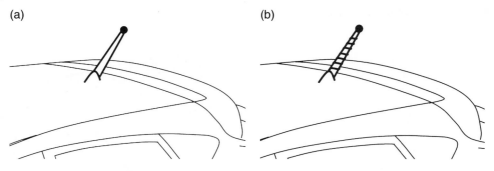

Figure 6.52 Antenna: (a) cylinder shape; (b) spiral shape.

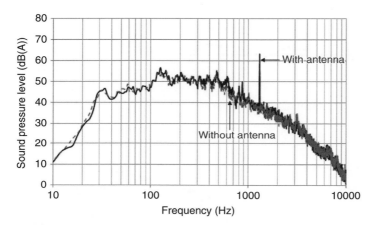

Figure 6.53 Interior noise comparison for a vehicle with and without an antenna.

The fluctuating lift force generates a dipole noise source, producing tonal noise. In daily life, it is easy to simulate the tonal sound. We hold a long and thin bar, and then wave it in the air. The bar generates a sound that is usually a high-frequency tone.

The antenna frequency excited by the airflow can be calculated by a Strouhal number, i.e.

$$S = \frac{fD}{u} \tag{6.11}$$

where, f is frequency; D is the antenna diameter; u is the airflow speed; S is Strouhal number, for an antenna fixed at one end, S is about 0.21.

The antenna frequency is usually between 1000 and 2000 Hz that is within the most sensitive frequency range of human hearing. Even the sound pressure level of the tone is very low, it can be easily perceived by the human ears. Figure 6.53 shows an interior noise comparison for a vehicle with and without an antenna. With the antenna, there was a peak at 1300 Hz, and it makes a "hissing" sound. After the antenna is removed, the peak disappears.

The starting point to control the antenna noise is to destroy the conical vortex flow, i.e. to prevent the unstable airflow to move upward along the axis of the cylinder, which will stop the generation of the fluctuating lift force. The easiest way to reach the purpose is to wrap a helical strake on the antenna surface. Because the turbulence is broken by the strake, the antenna cannot generate a tonal sound. Therefore, the antenna should be designed as a spiral shape, as shown in Figure 6.52b.

There are many cases in our daily life to generate the tonal sound like the antenna, such as waving a long and thin bar in the air. After the bar is wrapped by a layer of cloth, waving the bar in the air again, the tonal sound disappears.

In past 20 years, varieties of antennas have been developed, such as the hidden antenna, the shark fin antenna, etc. The hidden antenna is installed on the rear windshield surface. The small-sized shark fin antenna has a good streamlined shape, so it almost eliminates wind noise. At present, the protruding antenna shown in Figure 6.52 are becoming less and less common, but it is still often used in the economy vehicles.

The installed location and angle of the antenna on the vehicle body can also affect the noise. The body location to install the antenna should have high input point inertance.

6.5.5 Design of Roof Luggage Rack

Most SUVs and sporty vehicles are equipped with luggage racks on top of the roofs, as shown in Figure 6.54. There are two purposes for the racks. One is to increase luggage space and the other one is to increase the driving fun for sportiness.

Figure 6.55 shows a circular cross section of a rack bar. When the airflow blows the bar, turbulence is generated around it, a separation region is formed at the back, and vortex is induced, which generates the dipole noise.

If the cross section is designed to be elliptical, the separation region behind the rack bar is reduced, and the vortex flow is reduced as well. If a nub is added on the rear of the

Figure 6.54 Luggage rack.

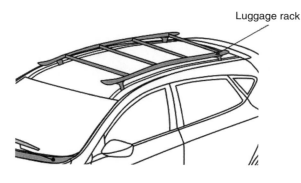

Luggage rack

Figure 6.55 Circular cross section of a rack bar.

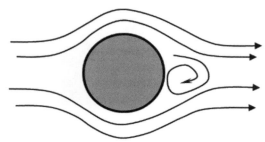

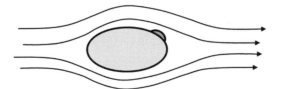

Figure 6.56 An elliptical cross section of rack bar with a nub.

elliptical bar, the vortex can be avoided and the wake is reduced as well, as shown in Figure 6.56. Therefore, the cross section of the rack bar should be designed to be an ellipse with a nub.

6.5.6 Control of Other Appendages and Outside Cavity

In addition to mirror, antenna, and luggage rack, there are other appendages on the vehicle body, such as door handle, wiper, etc. Two principles must be followed for design of the appendages. The first one is to prevent the airflow from hitting them directly. The second one is to reduce the vortex to the lowest level if they are blown; i.e. they must have good streamlined design.

Take a door handle as an example to illustrate these design principles. Figure 6.57 shows three designs of the door handles. The protruding handle in Figure 6.57a is exposed

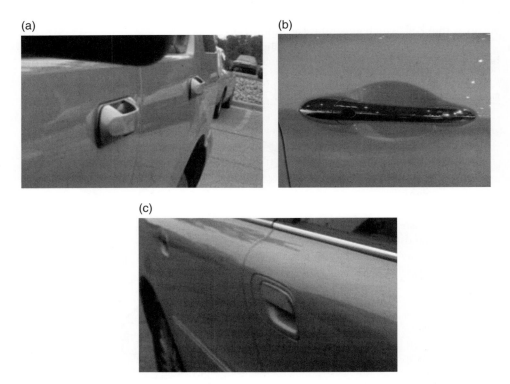

Figure 6.57 Three designs of door handles: (a) protruding handle; (b) handle with streamlined shape, (c) flush handle.

to the airflow, which easily induces noise. The handle in Figure 6.57b is also exposed to the airflow, but the transition between the handle and the door surface is smooth, i.e. the handle has good streamlined shape, so the noise induced in this design is lower than that in Figure 6.57a. The flush handle in Figure 6.57c is on the same plane as the door. It is hard for the airflow to blow it, so only very low levels of noise could be induced.

The cavity on the body surface, such as the cavity on top of B pillar shown in Figure 6.15, is the source of cavity noise. In the top area of the B pillar, the front door, rear door, B pillar, roof cross frame, and seals intersect, so the cavity could be formed. When the airflow blows to this area, the cavity noise will be generated. Even if the static sealing and the dynamic sealing in the area are very good, if the outside noise source is close to the occupants in the front row, then the cavity noise is easily heard.

Eliminating cavities is at the core of cavity noise control. Usually, there are two methods to eliminate cavities: fill them, as shown in Figure 6.17a, or cover them, as shown in Figure 6.17b.

6.6 Dynamic Sealing and Control

6.6.1 Dynamic Sealing and Its Importance

Static sealing was introduced in detail in Chapter 4. The static sealing refers to the sealing for a vehicle at standstill condition, which includes two parts. The first part is to seal the holes and openings on the fixed components. The second part is to seal relatively moving components such as the door and body. When a vehicle travels down the road, the sealing status on the fixed components changes little, but the sealing status between the relatively moving components could have big change.

The dynamic sealing refers to the sealing between the relatively moving components for a moving vehicle. A good static sealing cannot guarantee good dynamic sealing. Gaps or openings between relatively moving components could emerge for the driven vehicle even if the "static sealing" is very good. A good seal should be able to dynamically compensate the gap generated by the components' movement.

When a vehicle is driven on a road, the flexible components will deform because they are subjected to the excitations from road, engine, and wind. Take the body and the door as an example to illustrate their relative motion and sealing. Subjected to the outside excitations, the deformations and movements for the body and the door are different because their stiffness, modal parameters, etc. are different. Their original distances at standstill condition will change. When the relative deformation between the door and the body frame is larger than the seal compressed deflection, gaps will emerge.

The locations where the dynamic sealing problems could happen include:

- Body seals
- Door seals
- Cutline seals
- Glass run
- Beltline seals
- Rocker seal
- Trunk lid seal
- Liftgate seal

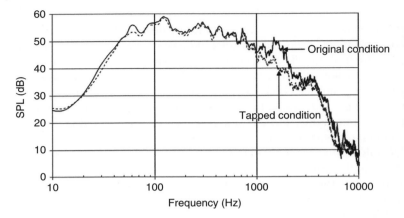

Figure 6.58 Interior noise comparison for a vehicle at original condition and body tapped condition.

The static sealing is the foundation of vehicle noise and vibration control, and the dynamic sealing is based on the static sealing. Poor dynamic sealing will significantly influence the vehicle noise control, especially the sound quality. If the dynamic sealing loses its function, i.e. large gaps or openings appear for a moving vehicle, the outside sound directly transmits into the interior, and the interior noise will increase significantly, especially at high speed. Even if there are no large openings on the body, instead, only tiny apertures appear, the interior sound level doesn't change too much, but the sound quality will deteriorate. The small apertures inducing the aspiration noise have little impact on the overall sound level but reduce the articulation index.

A vehicle was tested in a wind tunnel for two conditions. One is the original condition, and the other one has all gaps and apertures on body surface sealed by tapes. Figure 6.58 shows the interior noise comparison for the two conditions. At the middle- and high-frequency range, taping reduces the interior noise.

Wind noise accounts for 10–20% of the problems revealed in feedback from market survey and customer complaints, and poor dynamic sealing is the main reason. Body sealing must satisfy both static sealing and dynamic sealing. Therefore, only when the dynamic sealing meets the body design requirements can the sealing be considered completed and satisfy the customers' demands for sound quality.

6.6.2 Expression for Dynamic Sealing

Two parts, such as a door and a body frame, contact each other at standstill conditions and there is no gap between their contacted surfaces, as shown in Figure 6.59a. But after they move, gaps could appear between the contacted surfaces because their relative displacements are different, as shown in Figure 6.59b. The gaps induced by the relative movement of different parts must be sealed. The sealing relating with the motion is called dynamic sealing.

The body and the door are metal parts. If the door is pushed to impact the body frame directly, a huge metal impact sound will be generated. The gap between the body and the door is filled by the seals that have two functions. One is to make the door and the body contacted and sealed tightly, and the other one is to obtain good sound quality and relaxing sensation for door-closing action.

(a) (b)

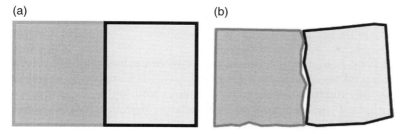

Figure 6.59 Two contacted parts: (a) at standstill condition; (b) after moving.

Figure 6.60 shows the geometric relationship of a body frame, a door, and a seal. Contact points between the body and the seal and between the seal and the door are used to analyze their displacement relationship. The static gap between the door and the body is d_s. The dotted ellipse represents a section of the seal at free status and its thickness is l. When the seal is placed between the body and the door, it is compressed and its cross section is changed and represented by the solid ellipse. The seal compressed deflection is

$$\delta_s = l - d_s \tag{6.12}$$

Because the door and the body are flexible structures, they are deformed when subjected to the outside excitations, and the static gap will change. Assume displacements of the door and the body are u_d and u_b, respectively, as shown in Figure 6.61. Under the vehicle running condition, the operational gap becomes

$$d_d = d_s + u_d - u_b \tag{6.13}$$

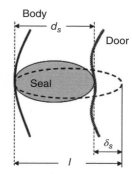

Figure 6.60 Geometry relationship of a body, a door and a seal at standstill condition.

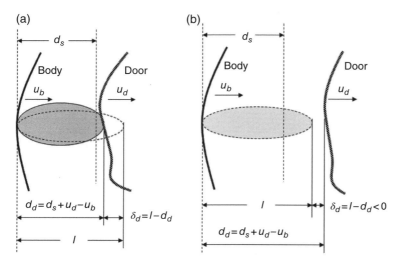

Figure 6.61 (a) Seal is compressed; (b) seal is not compressed.

The seal compressed deflection becomes

$$\delta_d = l - d_d \tag{6.14}$$

When $\delta_d = l - d_d \geq 0$, i.e. $l \geq d_d$, the operational gap between the body and the door is shorter than the seal's thickness, which means that the seal is still compressed, as shown in Figure 6.61a. The body and the door are in good sealing condition.

When $\delta_d = l - d_d < 0$, i.e. $l < d_d$, the operational gap is larger than the seal's thickness, which means that the seal is not compressed; instead, there is a gap between the seal and the door, as shown in Figure 6.61b. The seal loses its function, so a dynamic sealing problem is generated.

Thus, for a moving vehicle, the condition to guarantee that the door, the body, and the seal remain compressed is $\delta_d \geq 0$. There are many contact points on the body and the door. After every point is analyzed, the dynamic sealing status will be identified.

6.6.3 Dynamic Sealing between Door and Body

6.6.3.1 Types of Dynamic Sealing

The dynamic sealing involves many components. Sealing between the body and the door is taken as an example to illustrates types of dynamic sealing. The seals between the door and the body are divided into three categories: primary seal, secondary seal, and auxiliary seal.

The primary seal is the first continuous barrier for the body to isolate itself from the outside environment. Its main function is to prevent outside noise, water, and dust to enter into the interior. The primary seal is a continuous sealing loop around the door, as shown in Figure 6.62. In addition to the sealing function, the primary seal also should provide good sound quality of closing door and prevent cluck and squeak.

In the case of existence of the primary seal, another continuous bulb seal is placed on the body flanges, as shown in Figure 6.63, which is called the secondary seal. Its function is to further prevent the outside noise, water, and dust from entering into the interior, aid the door closing, and increase the body sound insulation. In the case of a function failure of the primary seal, the secondary seal plays the role of the primary seal.

Door seal (primary)

Figure 6.62 Primary seal installed on door.

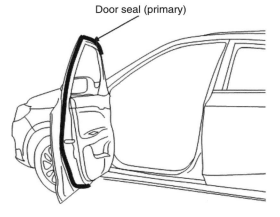

Figure 6.63 Secondary seal installed on body.

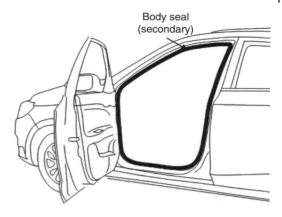

Body seal
(secondary)

Figure 6.64 Auxiliary seals on body.

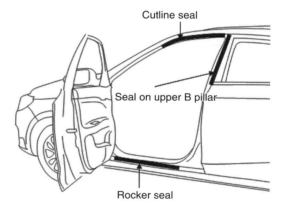

Cutline seal

Seal on upper B pillar

Rocker seal

Based on the primary seal and the secondary seal, additional seals called auxiliary seals are placed on specialized local areas to cover the margins at A pillar, B pillar, and C pillar, further preventing the outside noise, water, and dust from entering into the interior. The auxiliary seals include the seals on A, B, and C pillars, the rocker seal, and the cutline seal, as shown in Figure 6.64. For example, the seal on the B pillar shown in Figure 6.17a, and the cutline seal shown in Figure 6.17b are used to fill the cavity on the top of the B pillar area. The rocker seal is used to insulate the noise from the road and chassis.

6.6.3.2 Configuration of Dynamic Sealing

Combination of different seals can be used to form different sealing configurations. There are four commonly used sealing configurations.

The first configuration is that the seals are only installed on doors, as shown in Figure 6.65. Each door has a closed-loop seal around its surrounding and a cutline seal on the upper area of the door. This seal structure is simple and cheap, but the sealing can only achieve ordinary performance, which is usually used in the economy vehicles.

The second configuration is that there is a primary seal on the body and auxiliary seals on the door. Figure 6.66 shows a closed-loop seal installed on the body flange, and its function is as a primary seal. A B-pillar cutline seal is placed on the front edge of the

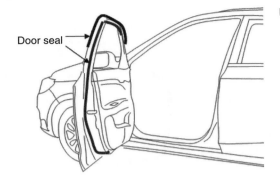

Figure 6.65 Seals only installed on door.

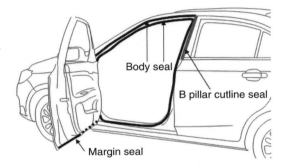

Figure 6.66 Primary seal installed on body and auxiliary seals on doors.

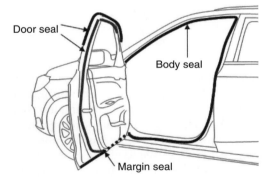

Figure 6.67 A primary seal on door, a secondary seal on body and auxiliary seals on door.

rear door. A margin seal is placed on the bottom of the front door. This structure is simple and cheap. Its sealing effectiveness is better than the first configuration, which is usually used in the economy and middle-class vehicles.

The third configuration is that there is a primary seal on the door, a secondary seal on the body, and auxiliary seals on the door. Figure 6.67 shows that the closed-loop seals are placed on both the door and the body. In addition, there are cutline seals on top and bottom of the door. This seal configuration is complex and expensive, but it has good sealing effectiveness, which is widely used in middle-class and luxury vehicles.

The fourth configuration is that there is a primary seal on the door, a secondary seal on the body, and auxiliary seals on both the door and the body. Figure 6.68 shows that the closed-loop seals are installed on both the door and the body. In addition, there are

Figure 6.68 A primary seal on door, a secondary seal on body and auxiliary seals on both door and body.

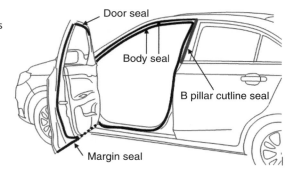

cutline seals on top of the body, and on the bottom and top of the door. This sealing structure is complex and expensive, but it has excellent sealing effectiveness, which is widely used in the middle class and luxury vehicles.

6.6.4 Control of Dynamic Sealing

Dynamic sealing relates to the static gap and relative motion between the body and the door, and seal compression, so the dynamic sealing can be controlled from three aspects.

First, the static gap between the body and the door is the basis for the seal design. The gap must be reasonable. It is impossible to seal a big gap. It is also very hard to seal a small gap where the seal is severely compressed, which influences sound quality and the force of door closing. Usually, the gap between the body and the door is around 14 mm.

Second, the relative motion between the body and the door must be controlled within a certain range. If the relative motion is too large, the dynamic gap may exceed the static gap, so good sealing effectiveness cannot be achieved. The relative motion may also generate rattle and squeak, so, stiffness, modes, and deformation of the body and the door should be controlled within certain ranges. Please refer to Chapters 2 and 3 about body structure analysis.

Third, the seal design should meet requirements of the dynamic sealing. The seal design includes selections of material, section shape, and load analysis, etc. The load analysis and section design will be introduced in the following materials.

The seal compression can be represented by compression load deflection (CLD). CLD, the most important characteristic for a seal design, refers to the seal resistance force per 100 mm seal length, as shown in Figure 6.69. The abscissa is the deflection of the cross section, and the ordinate is the CLD value. The CLD value is a very important index to measure the seal quality.

Figure 6.69 Compression Load Deflection (CLD) of a seal.

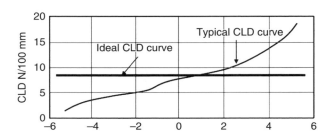

An ideal CLD curve is a flat line, as shown in Figure 6.69, which means that the load applying on the seal during the compression process is a constant and is independent of the compression deflection. If the CLD curve is not a flat line, instead, it changes with the compression deflection, problems relating to the sealing quality and the door closing sound quality could be induced.

The seal cross-section has a great influence on the CLD values. The sections include two typical structures: single bulb seal and double bulb seal. Figure 6.70a shows a section of a single bulb seal that has simple structure and is cheap. Its CLD curve is not flat line; instead, it increases with increase of the compression deflection, as shown in Figure 6.70b. The structure is easy to be folded in the corner, so the single bulb seal is suitable for small gap areas. In addition, it is not good for the door-closing sound quality.

Figure 6.71a shows a section of a double bulb seal. Compared with the single bulb seal, the structure is complex and the price is higher. The CLD curve is relatively flat, as shown in Figure 6.71b; that is, the CLD values are almost a constant as the compression deflection increase. This structure is good for both the dynamic sealing and the door-closing sound quality, which is usually used in the large gap areas.

The CLD value indicates the load required to compress a seal. A large load forces the seal to tightly compress to keep the door and the body in good compression state when subjected to external excitations. However, the large compression load requires large effort to open and close the door, which is often uncomfortable for those opening and closing the door. In addition, it could lead to deformations in the door.

Small CLD value indicates that a small load is required to compress the seal. The small load makes it easy to close the door, which customers could view as a sign of comfort and

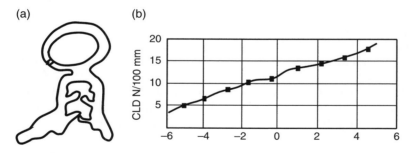

Figure 6.70 (a) Section of a single bulb seal; (b) CLD curve.

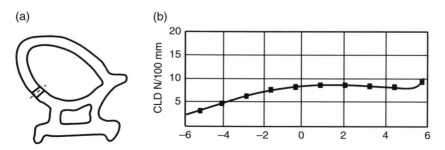

Figure 6.71 (a) Section of a double bulb seal; (b) CLD curve.

even luxury. However, excessive small load can easily induce a dynamic gap, which brings the dynamic sealing problems.

The CLD curve should not only be a flat line but also be limited within a certain range. The CLD values of the primary seal, secondary seal, and auxiliary seal are generally limited around 6 ± 2, 4 ± 2, and $2 \, \text{N}/100 \, \text{mm}$, respectively.

The CLD can be calculated by using nonlinear CAE method. Deflection of a seal and deformations of the door and the body should be calculated simultaneously. Only the deflection and deformations are considered simultaneously, can the dynamic sealing quality be evaluated for a moving vehicle.

6.7 Measurement and Evaluation of Wind Noise

The first task for the wind noise analysis and control is to measure it and understand its characteristics. The wind noise can be measured in a wind tunnel and/or on road. During a vehicle development, both wind tunnel testing and road testing are needed.

6.7.1 Wind Noise Testing in Wind Tunnel

6.7.1.1 Wind Tunnel

There are two kinds of wind tunnels: environmental wind tunnel and aerodynamic wind tunnel. In the environmental wind tunnel, due to installing sunshine simulation system, rain simulation system, etc. the ambient temperature can be tuned from minus 40 °C to plus 50 °C. The environmental wind tunnel is used to measure vehicle performance in various environmental conditions, such as power performance, fuel consumption, etc. The aerodynamic wind tunnel is used to measure drag coefficient and other aerodynamic performance.

The aerodynamic wind tunnel is divided into: aeroacoustic wind tunnel and non-aeroacoustic wind tunnel. In the non-aeroacoustic wind tunnel, because there is no special acoustic treatment inside the tunnel, the background noise is relative high, so the tunnel can be used to measure aerodynamic drag, lift force, lateral force, yaw, roll and pitching moments, and other aerodynamic performances. Inside the aeroacoustic wind tunnel, fans, turning vans, flow screens, inlet, walls, etc. are specially treated by sound absorptive materials and structures, and blowers and blades are specially designed to meet low noise requirements, so its background noise is low.

In the low noise wind tunnel, not only the aerodynamic performances but also wind noise can be measured. The wind tunnel mentioned in this book refers to the aeroacoustic wind tunnel. Figure 6.72 provides background noise levels of some wind tunnels at different wind speed. Figure 6.72a and b show the measured background noise for inflow and outflow area, respectively. The sound levels in the inflow are higher than those in the outflow area because the airflow induces the noise inside the inflow area.

Figure 6.73 shows a schematic diagram of a wind tunnel. The wind tunnel is composed of a tubular air passage where a blower, flow screens, and acoustic turning fans are installed inside. In the central testing area, there is a circular turntable that can be rotated. The turntable is the most sophisticated equipment in the wind tunnel. When a

(a)

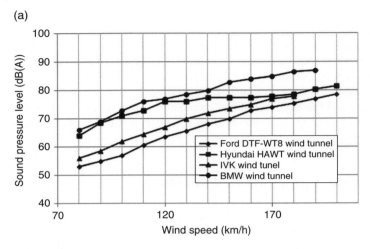

(b)

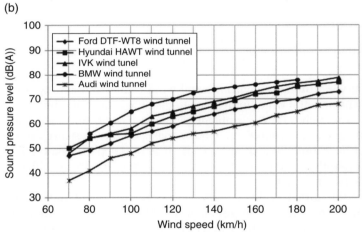

Figure 6.72 Background noise levels of some wind tunnels at different wind speed (a) Inflow (b) Outflow.

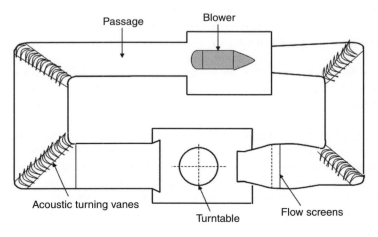

Figure 6.73 A schematic diagram of a wind tunnel (top view).

(a)   (b)

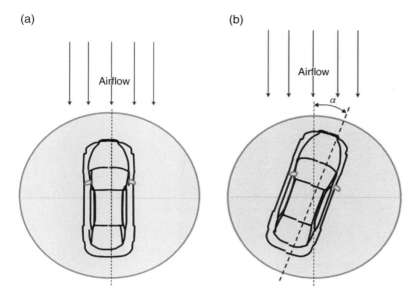

Figure 6.74 Angle between vehicle and airflow: (a) 0°; (b) angle α.

vehicle is placed on the turntable, the turntable should be tuned to achieve a balanced level by exquisitely adjustment. There are two belts on the table on which the wheels are placed, and they can simulate the relative motion between the tires and the road. For some sophisticated turntables, four or six belts are installed to achieve more accurate results. The longitudinal axis of a vehicle body is parallel to the center line of the turntable. When the center line is parallel to the central axis of the nozzle, the vehicle is in windward direction; that is, the angle between the vehicle and the airflow is 0°, as shown in Figure 6.74a. When the turntable is rotated to be an angle α, the center line and the center axis also have the angle α, i.e. the angle between the vehicle and the airflow is α, as shown in Figure 6.74b. The wind noise for windward and sideward can be measured by adjusting the turntable's angle.

Due to avoidance of engine noise, road noise, and other noise, testing the wind noise in a wind tunnel has the following advantages. Wind noise sources and its characteristics can be precisely identified. Both inside and outside noise can be measured. Also, both a real vehicle and a styling model can be measured. However, wind tunnel testing also has limits, such as availability of the wind tunnels and high cost.

6.7.1.2 Measuring Outside Wind Noise

Wind noise is generated by pressure fluctuation of the airflow on the body surface, so it is very important to measure the vehicle outside noise to identity wind noise sources. Three ways are used to measure the outside wind noise: beamforming measurement, acoustic mirror measurement, and surface sound pressure measurement. In addition, a laser vibrometer is used to measure vibration at the vehicle surface.

In beamforming measurement, a microphone array is placed in the outflow area, as shown in Figure 6.75. The mechanism of the beamforming technique has been introduced in Chapter 3. The beamforming is also called an acoustic camera because the

Figure 6.75 Beamforming is used to measure wind noise.

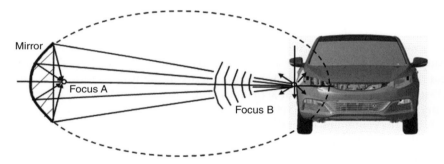

Figure 6.76 Diagram of body surface sound measurement by an parabolic mirror.

tested results can be presented in the form of pictures. This method is convenient and has been widely used to measure the external wind noise.

The acoustic mirror measurement is to measure sound by focusing it onto a parabolic surface. Figure 6.76 shows a mirror and a tested vehicle. The section of the mirror is part of an ellipse that has its two focuses, A and B. A microphone is placed at focus A and the sound source is placed at focus B. The sound source is transmitted to the mirror and then reflected and focused on point A, so the sound is measured by the microphone at point A.

The parabolic mirror is placed in the outflow field so that it is not affected by the airflow. The distance between the two focuses should be much longer than the wavelengths of the measured acoustic waves in order to obtain the desired frequency signals. In the test, the distance and angle of the mirror are constantly adjusted until the microphone and the measured body point are on the two focuses.

In the last decade, due to the widely application of the beamforming (acoustic cameras), the use of the acoustic focusing method is becoming less and less.

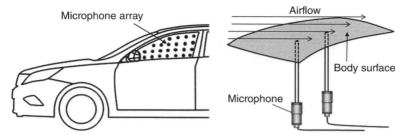

Figure 6.77 Traditional microphone array on a specially made body surface.

Surface sound pressure measurement measures near-field sound pressure distribution on the body surface. Two types of sensors are used. The first one is to place many traditional microphones on a specially made body part and make their heads and the body surface on the same surface, as shown in Figure 6.77. The second one is the surface microphone, as shown in Figure 6.78, that is directly attached to the body surface.

Figure 6.78 Surface microphone.

The laser vibration measurement measures vibration of the body surface, which was introduced in detail in Chapter 3. In a wind tunnel, a laser vibrometer is placed in outflow field and the body surface vibration is measured. The airflow causes the body panels to vibrate and then radiates noise into the interior, so the wind noise is transmitted through the structural borne path. By measuring body surface vibration, the contribution of each panel to the structural-borne sound can be identified.

6.7.1.3 Inside Wind Noise Measurement

There is no difference between interior wind noise measurement and the regular interior noise measurement, and the layouts of the microphones and testing systems are the same. The interior wind noise measurement has two purposes: to evaluate the wind noise levels and to identify the sources. The evaluation of the wind noise will be introduced in detail in later materials of this section. Identification of the wind noise contribution sources includes frequency spectrum analysis method and the exclusion method.

The exclusion method decomposes the main factors inducing the wind noise step by step until the contribution of each noise source can be identified. Figure 6.79 shows an interior noise comparison for different decomposed structures of a vehicle. First, all accessories attached on the outside body are removed and all seals are covered by adhesive tape so the tested noise is contributed by the body itself – this body is called "quietest body." Because there are no accessories and no apertures, the tested noise, called quietest wind noise or base body noise, is the lowest noise level that can be achieved for the vehicle. The quietest wind noise is only contributed by the surface pulsating.

Second, all accessories are installed back to the body. The body is called "sealed body." The tested noise is called "sealed body noise," which is higher than the quietest wind noise. The major increased band is at the middle- and high-frequency range, especially a peak at 1800 Hz contributed by the antenna.

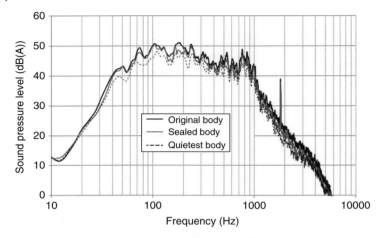

Figure 6.79 A vehicle interior noise comparison for quietest body, sealed body, and original body.

Finally, the adhesive tapes are removed and the vehicle body is restored to its original state, called "original body." The airflow could move into the interior through apertures and openings. The overall noise increases due to existence of the aspiration noise.

Figure 6.79 shows the interior noise comparison for quietest body, sealed body, and original body.

Through the step-by-step decomposed measurement, contributions of the appendages and the dynamic sealing to the wind noise are obtained. More detailed work can be implemented in the decomposition processing. For example, for the processing of removing the appendages, the mirror, the antenna, and other parts can be removed one by one in order to obtain each appendage's contribution to the wind noise. Similarly, in the process of removing the adhesive tapes, the tapes on the windows and on the doors can be removed one by one, so contribution of each sealing part to the wind noise can be tested.

Similarly, a reverse processing can be used to assess the influence of attached accessories and the dynamic sealing on the wind noise. First, the original body is tested and then the sealed body is tested; finally, the attached accessories are removed.

6.7.2 Wind Noise Testing on Road

The road test is to measure the interior noise when a vehicle is driven on road. The road test is divided into testing on proving ground and testing on public highway. Both objective testing and subjective evaluation can be implemented for the road testing. A series of microphones are placed inside a vehicle, and the interior noise is tested for different vehicle speeds, so the wind noise is obtained.

The benefit of the road test is simple and cheap. Because there is no limitation (such as there is using a wind tunnel), the road test can be easily implemented on a proving ground or a public highway. This method is widely used in identification and analysis of the wind noise, especially for comparison of different solution designs and subjective evaluation. However, there are three deflects for the road test. First, the test is influenced by environment factors. Because the tested data include wind noise, engine noise, and road noise, sometimes it is difficult to distinguish their contributions. For a vehicle installed with a quite engine and tested on a smooth road where the engine noise and the road noise are much lower than the wind noise, it is easy to identify the wind noise. However, for a vehicle with high engine noise or running on coarse road, it is almost impossible to identify the wind noise. Second, the outside wind noise, such as the wind noise at the mirror area, cannot be tested. Third, test consistency and repeatability is difficult to achieve.

6.7.3 Evaluation of Wind Noise

Wind noise evaluation includes subjective evaluation and objective evaluation.

The subjective evaluation includes subjective scoring and subjective description. Subjective scoring is to score a wind noise level based on the evaluators' subjective sensations. Usually, a 10-point scoring system is used, as shown in Table 6.1. The higher the score, the better the sound quality of the wind noise. Subjective description refers to the detailed descriptions of wind noise performances and characteristics by the evaluators, for example. Evaluators describe the wind noise using words like whistling, booming, hissing, buffeting, and so on.

The objective evaluation is to test and analyze the level, frequency, and sound quality characteristics of the wind noise. The objective evaluation indexes of the wind noise include sound pressure level, loudness, articulation index, etc. These objective indexes provide the characteristics of the wind noise that facilitates comparison for different wind noise and offers guidelines for body structure design.

The customers' perception about the wind noise is divided into two levels. The first level is loud wind noise and the second level is low-level wind noise but poor sound quality.

When a body structure and its accessories are not properly designed, huge pulsation noise and cavity noise will be generated. When the dynamic sealing is very bad, aspiration noise will be induced. In these cases, the customers complain about the huge wind noise level, and the indexes used to evaluate the wind noise are sound pressure level or loudness level. The evaluation using sound pressure level or loudness level includes not only overall sound level but also spectral components, which can help to identify the parts inducing the wind noise. Figure 6.79 shows the sound pressure level in different body states.

When a body, appendage structures, and sealing are properly design, the wind noise level will be low. However, if some local structures or accessories or dynamic sealing are not well designed or installed, some special frequency noise could be induced. Even the

Table 6.1 Subjective scoring of wind noise.

Rating Scales	1	2	3	4	5	6	7	8	9	10
	Unacceptable				Borderline acceptable		Acceptable			
Conditions Noted By	All customers	Average customers			Critical customers		Trained observers		Not perceptible	

noise level is low, but it could annoy the customers. In this case, the evaluation index is an articulation index (AI), and sound pressure level is used as an auxiliary index. Even if the annoying noise is eliminated, the overall sound pressure level or loudness level will not change, but the articulation index is obviously changed.

For example, if a car is cruising at $120\,\mathrm{km\,h^{-1}}$ and the interior noise level is 67 dB, this represents a relatively quiet vehicle. In a high-frequency range, there are several peaks where the amplitudes are only 30–40 dB(A), and their contribution to the overall sound pressure level can generally be neglected. However, let's say that the customers can clearly hear the tonal "hissing" sound and perceive the poor sound quality. After diagnosis, a small aperture between the mirror and the body that generates the aspiration noise is found. After the aperture is covered by the adhesive tape, the uncomfortable hissing noise disappears and sound quality is improved, but the measured sound pressure level doesn't change.

6.8 Analysis of Wind Noise

6.8.1 Relationship Between Aerodynamic Acoustics and Classical Acoustics

6.8.1.1 Difference Between Aerodynamic Acoustics and Classical Acoustics

Significant differences exist between the aerodynamic acoustics and the classical acoustics, and the major difference is at their sound generation mechanism.

In classical acoustics, acoustic wave is generated by surface vibration of an object. The vibration on the object surface causes the fluid on its boundary to compress and expand continuously, which induces density and pressure of the fluid change constantly, so a sound is generated. The vibrated object doesn't move and the sound source is on its surface. Intensity of the sound radiation depends on magnitude of the vibration.

In aerodynamic acoustics, the sound is generated by action of a moving object in a fluid or interaction between fluids. The action and interaction can be divided into three categories. The first category is that an object is inserted into and removed from a fluid continuously to force the local fluid to be continuously compressed and expanded, which generates sound. The second category is that a moving object forces the fluctuating lift force on its surface to change that generates a pulsating thrust acting on the fluid at its surface. The third category is that turbulent motion of a fluid induces interaction between the fluids, which generates sound. For example, for an engine running at high speed, when the airflow inside an exhaust system reaches the tailpipe, the fluids interact one other, so a huge jet noise is generated. Similarly, inside the separation regions on a body surface, the pulsating noise is generated by the interaction between the fluids.

6.8.1.2 Analogy Between Aerodynamic Acoustics and Classical Acoustics

The analysis and solution of the aerodynamic acoustics is very difficult. Scientists have found a way to use the theories of the classical acoustics to analyze the aerodynamic acoustics.

The first method is to analogize the aerodynamic sound source to the spherical pulsating source in the classical acoustics. Fluid motion can be regarded as continuously adding fluid mass to an object surface to form pulsation mass sources. Many small pulsation balls on the object surface that are similar to monopole sound sources constitute the surface aerodynamic sound sources, so the pulsating mass source belongs to the monopole sound source.

The second method is to analogize the aerodynamic sound source to vibrational ball sources in the classical acoustics. The object continuously pushes the fluid to force the fluid particle velocity to continuously change, generating sound. The separated flow and the vortex flow caused by the airflow on the object surface also induce fluctuated lift force on its surface, generating sound. This aerodynamic sound source is a pulsating force source, which belongs to the dipole sound source.

The third method is that the aerodynamic sound source is regarded as internal stress change due to the interaction among fluids. A dipole sound source is formed by applying a pulsating thrust to the fluid element. The interaction among the fluids can be regarded as the interaction between two dipole sound sources that are equal in magnitudes and opposite in phases. The interaction of the two dipole sound sources generates a stress, so this aerodynamic sound source is the flow turbulent stress source, which belongs to the quadrupole source.

6.8.2 Lighthill Acoustic Analogy Theory

The compressible Navier-Stokes equation (simply noted as N-S equation) can be used to calculate aerodynamic sound directly. N-S equation is the equation of conservation of momentum. That is, the ratio of the momentum of a micro-fluid element to time is equal to the total external forces acting on the element. The N-S method requires a large amount of volume elements to calculate near-field sound sources and far-field sound radiation response, which requires huge computation capacity and time, so this method can be used only for the small- and low-frequency acoustic problems.

In 1952, to overcome the limitation of the aerodynamic acoustic calculation, the British scientist James Lighthill proposed an analogy theory of aerodynamic acoustics. During his research work on the jet emission acoustics, Lighthill analyzed and modified the N-S equation and obtained the Lighthill equation, as follows:

$$\frac{1}{c^2}\frac{\partial^2 \rho}{\partial t^2} - \nabla^2 \rho = \nabla^2 T_{ij} \tag{6.15}$$

where, T_{ij} is the Lighthill turbulence stress tensor, and expressed as follows:

$$T_{ij} = \rho u_i u_j - \tau_{ij} + \delta_{ij}\left[(p-p_0) - c^2(\rho-\rho_0)\right] \tag{6.16}$$

where $\rho u_i u_j$ is the Reynolds stress generated by the velocity change; τ_{ij} is the stress generated by the fluid viscosity; $\delta_{ij}[(p-p_0) - c^2(\rho-\rho_0)]$ is the stress generated by the heat conduction.

From Eq. (6.15), the left side of the equation is in the form of the classical acoustic equation, and the right side represents the external force caused by the fluid interaction. It can be regarded as the sound source. This way, the relationship between the aerodynamic acoustics and the classical acoustics is established. Eq. (6.15) is called the Lighthill analogy equation. The variables of the acoustic field on the left side of the equation do not affect the fluid parameters on the right side of the equation, which means that the sound wave and the fluid motion don't impact each other.

In the process of injection, the turbulence stress tensor is generated by the interaction among the fluids, so the turbulence stress is a quadrupole sound source.

The left side of Eq. (6.15) is an expression in the form of the classical acoustics, and the variable is the density (or pressure or velocity) relating to the acoustic field. However, the right side also contains the sound field parameters, so, the equation cannot be solved. However, Lighthill solved this problem by using the experimental method. Through testing, he obtained the sound sources, that is, the input terms on the right side of the equation, so the equation could be solved and sound field could be obtained. Lighthill proposed the acoustic analogy theory by combining theoretical analysis and experiment method.

Although Lighthill pioneered the aerodynamic acoustics, a new branch of acoustics, two problems were left. The first problem is that he did not consider the sound field caused by the interaction between fluid and solid. The second problem is that if the sound source is not available, the Lighthill equation cannot be solved.

6.8.3 Lighthill-Curl Acoustic Analogy Theory

In order to overcome the first problem left by Lighthill, in 1955, Curle studied the interaction between fluid and solid. He used the Kirchhoff integral to extend the Lighthill equation to the interaction between fluid and solid at their boundary. He included the compression and expansion motion of the fluid on the solid surface, as well as the lift force and the impulse force generated by the solid on the surface of the fluid. Curle further developed Lighthill's theory and formed the Lighthill-Curle analogy theory. The sound source terms include a monopole, dipole, and quadrupole.

After solving the Lighthill-Curle equation, the sound pressure can be written as

$$
p(\vec{x},t) = \frac{\partial^2}{\partial x_i \partial x_j} \int_V \frac{T_{ij}\left(\vec{y}, t - \frac{|\vec{x}-\vec{y}|}{c}\right)}{4\pi|\vec{x}-\vec{y}|} d^3\vec{y} - \frac{\partial}{\partial x_i} \int_S \frac{P\left(\vec{y}, t - \frac{|\vec{x}-\vec{y}|}{c}\right)}{4\pi|\vec{x}-\vec{y}|} d^2\vec{y}
$$

$$
+ \frac{\partial}{\partial t} \int_S \frac{\rho\bar{u}\left(\vec{y}, t - \frac{|\vec{x}-\vec{y}|}{c}\right)\vec{n}}{4\pi|\vec{x}-\vec{y}|} d^2\vec{y}
$$

(6.17)

where S and V are the area and volume of the sound sources; $\vec{y}$ is the position vector of the sound source, $\vec{x}$ is the position vector of the measured point; and $t - \frac{|\vec{x}-\vec{y}|}{c}$ is the time delay.

Three terms on right side of Eq. (6.17) represent the sound pressures generated by the quadrupole source, dipole source, and monopole source, respectively. The quadrupole source is a volume integral of the sound source region, and the dipole and monopole sources are surface integrals of the sound source region. The relative speed between the airflow and a vehicle is much lower than sound speed, so the wind noise problem belongs to small Mach number issue. For small Mach numbers, the fluid radiation capability of the quadrupole source is very low, so its contribution can be neglected. For the movement between the airflow and the vehicle body, the influence of the monopole sound source is small, so it also can be ignored. Therefore, the main contribution of the wind noise source is from the dipole sound source.

6.8.4   Solution of Aerodynamic Equations

The second problem left by the Lighthill equation is how to get the sound source. He obtained the sound source from experiment, and then solved the equation. Another way to obtain the sound source is from CFD calculation. So, solving the aerodynamic acoustic equation can be cascaded into two steps. The first step is to solve the sound source in the near field and the second step is to solve the sound radiation in the far field.

First step: calculation for near field sound source. Although the aerodynamic sound source and the radiation sound can be obtained directly by solving the compressible N-S equation, because huge grids and computational capacity are need, in real work, people don't do it this way. The automobile wind noise analysis belongs to a small Mach number problem, and the noise source is within only a very thin layer. In this layer, the flow field is not affected by the sound field, and the fluid is assumed to be incompressible fluid. Under such conditions, only the flow field in the thin layer is calculated; the sound source can be calculated by using N-S equation. The method using N-S equation to solve near-field sound source is called direct numerical simulation (DNS).

There are many other methods to calculate sound sources, such as the Large Eddy Simulation method (LES) and Reynolds Averaged Navier Stockton method (RANS).

Second step: solution for pulsating sound pressure and sound radiation. The aerodynamic noise is calculated using the sound sources calculated from the first step. The solution methods include Lighthill analogy method, Lighthill-Curle analogy method, Ffowcs Williams-Hawkings (FWH) analogy method, and the Kirchoff method.

Lighthill analogy method is suitable for the calculation of the quadrupole noise generated by relative motion between fluids. The Lighthill-Curle method is suitable for calculating the motion between fluids and between fluid and solid, including monopole, dipole, and quadrupole.

6.8.5   Simulation of Wind Noise

Commercial software to calculate the wind noise is gradually mature. The methods used in the software include traditional methods and new methods. The traditional methods are based on the N-S equation, using Lighthill-Curle analogy method and FWH, for example, FWH method is used in FLUENT. The new method is to use the Lattice Boltzmann Method (LBM), and PowerFlow uses this method.

Using the available commercial software, the overall wind noise and local wind noise of a vehicle body can be calculated. For example, Figure 6.80 shows a calculated wind noise distribution in the mirror area of a vehicle.

If this software is combined with SEA analysis, the transfer of wind noise into the interior can be analyzed. The analysis method is the same as the one to analyze the sound package. See Chapter 4 for detailed information.

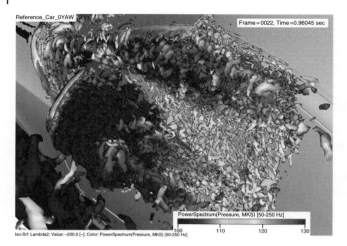

Figure 6.80 Calculated wind noise distribution in the mirror area.

Bibliography

Alam, F., Watkins, S., Zimmer, G. et al. (2003). Mean and time-varying flow measurements on the surface of a family of idealized road vehicles. *Experimental Thermal and Fluid Science* 27 (2003): 639–654.

Alam, F., Watkins, S., Zimmer, G. (2001). Effects of Vehicle A-pillar Shape on Local Mean and Varying Flow Properties. SAE Paper; 2001-01-1086.

Allen, M.J. and Vlahopoulos, N. (2001). Computation of Wind Noise Radiated from a Flexible and Elastically Supported Panel. SAE Paper; 2001-01-1495.

An, C.F., Alaie, S.M., Sovani, S.D. et al. (2004). Side Window Buffeting Characteristics of an SUV. SAE Paper; 2004-01-0230.

Balasubramanian, G., Mutnuri, L.A.R., Sugiyama, Z. et al. (2013). A Computational Process for Early Stage Assessment of Automotive Buffeting and Wind Noise. SAE Paper; 2013-01-1929.

Beranek, L.L. (1996). *Acoustics.* Acoustical Society of America.

Beigmoradi, S., Jahani, K., Keshavarz, A. et al. (2013). Aerodynamic Noise Source Identification for a Coupe Passenger Car by Numerical Method Focusing on the Effect of the Rear Spoiler. SAE Paper; 2013-01-1013.

Blanchet, D. and Golota, A. (2014). Wind Noise Source Characterization and How It Can Be Used To Predict Vehicle Interior Noise. SAE Paper; 2014-01-2052.

Blanchet, D. and Golota, A. (2015). Combining Modeling Methods to Accurately Predict Wind Noise Contribution. SAE Paper; 2015-01-2326.

Blommer, M., Amman, S., Abhyankar, S. et al. (2003). Sound Quality Metric Development for Wind Buffeting and Gusting Noise. SAE Paper; 2003-01-1509.

Bremner, P.G. and Zhu, M. (2003). Recent Progress using SEA and CFD to Predict Interior Wind Noise. SAE Paper; 2003-01-1705.

Cogotti, A., Cardano, D., Carlino, G. et al. (2005). Aerodynamics and Aeroacoustics of Passenger Cars in a Controlled High Turbulence Flow: Some New Results. SAE Paper 2005-01-1455.

Crocker, M.J. (2007). *Handbook of Noise and Vibration Control.* Wiley.

Crouse, B. and Senthooran, S. (2006). Experimental and Numerical Investigation of a Flow-Induced Cavity Resonance with Application to Automobile Buffeting. 12th AIAA/CEAS Aeroacoustics Conference (27th AIAA Aeroacoustics Conference); 8-10 May 2006, Cambridge, Massachusetts.

Crouse, B., Freed, D., Senthooran, S. et al. (2007a). Analysis of Underbody Windnoise Sources on a Production Vehicle using a Lattice Boltzmann Scheme. SAE Paper; 2007-01-2400.

Crouse, B., Senthooran, S., Balasubramanian, G. et al. (2007b). Computational Aeroacoustics Investigation of Automobile Sunroof Buffeting. SAE Paper; 2007-01-2403.

Deaton, L., Rao, M., and Shih, W.Z. (2007). Root Cause Identification and Methods of Reducing Rear Window Buffeting Noise. SAE Paper; 2007-01-2402.

DeJong, R.G., Bharj, T.S., and Lee, J.J. (2001). Vehicle Wind Noise Analysis Using a SEA Model with Measured Source Levels. SAE Paper; 2001-01-1629.

DeJong, R.G., Bharj, T.S., and Booz, G.G. (2003). Validation of SEA Wind Noise Model for a Design Change. SAE Paper; 2003-01-1552.

Duell, E.G., Walter, J., Yen, J. et al. (2013). Progress in Aeroacoustic and Climatic Wind Tunnels for Automotive Wind Noise and Acoustic Testing. SAE Paper; 2013-01-1352.

Duncan, B.D., Senthooran, S., Hendriana, D. et al. (2007). Multi-Disciplinary Aerodynamics Analysis for Vehicles: Application of External Flow Simulations to Aerodynamics, Aeroacoustics and Thermal Management of a Pickup Truck. SAE Paper; 2007-01-0100.

Everest, A.F. (2001). *The Master Handbook of Acoustics.* McGraw-Hill.

Francescantonio, P.D., Hirsch, C., Ferrante, P. et al. (2015). Side Mirror Noise with Adaptive Spectral Reconstruction. SAE Paper; 2015-01-2329.

Ganty, B., Jacqmot, J., Zhou, Z. et al. (2015). Numerical Simulation of Noise Transmission from A-pillar Induced Turbulence into a Simplified Car Cabin. SAE Paper; 2015-01-2322.

Gaylard, A.P. (2006). CFD Simulation of Side Glass Surface Noise Spectra for a Bluff SUV. SAE Paper; 2006-01-0137.

Graf, A., Lepley, D., and Senthooran, S.A (2011). Computational Approach to Evaluate the Vehicle Interior Noise from Greenhouse Wind Noise Sources – Part II. SAE Paper; 2011-01-1620.

Hamamoto, N., Okutsu, Y., and Yanagimoto, K. (2013). Investigation for the Effect of the External Noise Sources onto the Interior Aerodynamic Noise. SAE Paper 2013-01-1257.

He, Y.Z. and Yang, Z.G. (2012). An experimental investigation of sunroof buffeting characteristics of a sedan. *Applied Mechanics and Materials* 226–228: 247–251.

He, Y.Z., Yang, Z.G., and Wang, Y.G. (2012). An experimental investigation of automobile interior wind noise using a production vehicle. *Applied Mechanics and Materials* 105–107: 1860–1866.

Herpe, F.V., D'Udekem, D., Jacqmot, J. et al. (2012). Vibro-Acoustic Simulation of Side Windows and Windshield Excited by Realistic CFD Turbulent Flows Including Car Cavity. SAE Paper; 2012-01-1521.

Hucho, W.H. (1998). Aerodynamics of Road Vehicle: From Fluid Mechanics to Vehicle Engineering, SAE International.

Hou, H.S. (2015). Automobile Wind Noise Speed Scaling Characteristics. SAE Paper; 2015-01-1531.

Hou, H.S. and Yue, G. (2015). Impact of Sunroof Deflector on Interior Sound Quality. SAE Paper; 2015-01-2324.

Iacovoni, D.P., Zeuty, E.J., and Morello, D.A. (2003). Wind Noise and Drag Optimization Test Method for Sail-Mounted Exterior Mirrors. SAE Paper; 2003-01-1702.

Karbon, K.J. and Dietschi, U.D. (2005). Computational Analysis and Design to Minimize Vehicle Roof Rack Wind Noise. SAE Paper; 2005-01-0602.

Kato, Y. (2012). Numerical Simulations of Aeroacoustic Fields around Automobile Rear-View Mirrors. SAE Paper; 2012-01-0586.

Kim, G.S., Park, H.K., Jung, S.G. et al. (2001). Development of Acoustic Holography and Its Application in Hyundai Aeroacoustic Wind Tunnel. SAE Paper; 2001-01-1497.

Kounenis, C., Sims-Williams, D., Dominy, R. et al. (2015). The Effects of Unsteady Flow Conditions on Vehicle in Cabin and External Noise Generation. SAE Paper; 2015-01-1555.

Kumarasamy, S. and Karbon, K. (1999). Aeroacoustics of an Automobile A-Pillar Rain Gutter: Computational and Experimental Study. SAE Paper; 1999-01-1128.

Kunstner, R., Potthoff, J., and Essers, U. (1995). The Aero-Acoustic Wind Tunnel of Stuttgart University. SAE Paper; 950625.

Li, Y., Kasaki, N., Tsunoda, H. et al. (2006). Evaluation of Wind Noise Sources Using Experimental and Computational Methods. SAE Paper; 2006-01-0343.

Lokhande, B., Sovani, S., and Xu, J. (2003). Computational Aeroacoustic Analysis of a Generic Side View Mirror. SAE Paper; 2003-01-1698.

Manning, P., Manning, J., Musser, C., et al. (2013). Evaluation of Ground Vehicle Wind Noise Transmission through Glasses Using Statistical Energy Analysis. SAE Paper; 2013-01-1930.

Matsushima, Y. and Kohri, I. (2013). Experimental Study for Applicable Limit of Acoustic Analogy to Predict Aero-Acoustic Noise of Commercial Vehicles. SAE Paper; 2005-01-0606.

Mendonca, F.G. CFD/CAE Combinations in Open Cavity Noise Predictions for Real Vehicle Sunroof Buffeting. SAE Paper; 2013-01-1012.

Mendonca, F.G., Shaw, T., and Mueller, A. (2013). CFD-Based Wave-Number Analysis of Side-View Mirror Aeroacoustics towards Aero-Vibroacoustic Interior Noise Transmission. SAE Paper; 2013-01-0640.

Moron, P., Hazir, A., Crouse, B. et al. (2011). Hybrid Technique for Underbody Noise Transmission of Wind Noise. SAE Paper; 2011-01-1700.

Müller, G., Jany, J., and Neuhierl, B. (2012). Reducing a Sports Activity Vehicle's Aeroacoustic Noise using a Validated CAA Process. SAE Paper; 2012-01-1552.

Müller, G., Grabner, G., Wiesenegger, M. et al. (2014). Assessment of the Vehicle's Interior Wind Noise Due to Measurement of Exterior Flow Quantities. SAE Paper; 2014-01-2050.

Musser, C.T., Manning, J.E., and Min, S. (2012). Road Test Measurement and SEA Model Correlation of Dominant Vehicle Wind Noise Transfer Paths. SAE Paper; 2012-36-0624.

Mutnuri, L.A.R., Senthooran, S., Powell, R. et al. (2014). Computational Process for Wind Noise Evaluation of Rear-View Mirror Design in Cars. SAE Paper; 2014-01-0619.

Oettle, N., Mankowski, O., Sims-Williams, D. et al. (2013). Evaluation of the Aerodynamic and Aeroacoustic Response of a Vehicle to Transient Flow Conditions. SAE Paper; 2013-01-1250.

Ono, K., Himeno, R., and Fukushima, T. (1999). Prediction of wind noise radiated from passenger cars and its evaluation based on auralization. *Journal of Wind Engineering and Industrial Aerodynamics* 81: 403–419.

Patel, P. and Vijayakumar, S. (2001). External Flow Analysis Over a Car to Study The Influence of Different Body Profiles Using CFD. SA E Paper; 2001-01-3085.

Peng, G.C. (2007). SEA Modeling of Vehicle Wind Noise and Load Case Representation. SAE Paper; 2007-01-2304.

Peng, G.C. (2011). SEA Wind Noise Load Case for Ranking Vehicle Form Changes. SAE Paper; 2011-01-1707.

Powell, R., Moron, P., Balasubramanian, G. et al. (2013). Simulation of Underbody Contribution of Wind Noise in a Passenger Automobile. SAE Paper; 2013-01-1932.

Schell, A. and Cotoni, V. (2015). Prediction of Interior Noise in a Sedan Due to Exterior Flow. SAE Paper; 2015-01-2331.

Senthooran, S., Crouse, B., Noelting, S. et al. (2006a). Prediction of Wall Pressure Fluctuations on an Automobile Side-glass using a Lattice-Boltzmann Method. 12th AIAA/CEAS Aeroacoustics Conference (27th AIAA Aeroacoustics Conference); 8-10 May 2006, Cambridge, Massachusetts.

Senthooran, S., Crouse, B., and Noelting, S. (2006b). Prediction of Wall Pressure Fluctuations on an Automobile Side-glass using a Lattice-Boltzmann Method. 12th AIAA/CEAS Aeroacoustics Conference (27th AIAA Aeroacoustics Conference); 8-10 May 2006, Cambridge, Massachusetts.

Senthooran, S., Duncan, B.D., Freed, D. et al. (2007). Design of Roof-Rack Crossbars for Production Automobiles to Reduce Howl Noise using a Lattice Boltzmann Scheme. SAE Paper; 2007-01-2398.

Senthooran, S., Mutnuri, L.A.R., Amodeo, J. et al. (2013a). A Computational Approach to Evaluate the Automotive Windscreen Wiper Placement Options Early in the Design Process. SAE Paper; 2013-01-1933.

Senthooran, S., Mutnuri, L.A.R., Amodeo, J. et al. (2013b). A Computational Approach to Evaluate the Automotive Windscreen Wiper Placement Options Early in the Design Process. SAE Paper; 2013-01-1933.

Shorter, P., Blanchet, D., and Cotoni, V. (2012). Modeling Interior Noise due to Fluctuating Surface Pressures from Exterior flows. SAE Paper; 2012-01-1551.

Sovani, S.D. and Chen, K.H. (2005). Aeroacoustics of an Automotive A-Pillar Raingutter: A Numerical Study with the Ffowcs-Williams Hawkings Method. SAE Paper; 2005-01-2492.

Thompson, M. and Watkins, S. (2013). Wind-Tunnel and On-Road Wind Noise: Comparison and Replication. SAE Paper; 2013-01-1255.

Ver, I.L. and Beranek, L.L. (2006). *Noise and Vibration Control Engineering: Principles and Applications*. Wiley.

Watkins, S., Thompson, M.A., and Alam, F. (2013). Transient Wind Noise. SAE Paper; 2013-01-0096.

Walter, J., Duell, E., Martindale, B. et al. (2001). The Driveability Test Facility Wind Tunnel No. 8. SAE Paper; 2001-01-0252.

Walker, R. and Wei, W. (2007). Optimization of Mirror Angle for Front Window Buffeting and Wind Noise Using Experimental Methods. SAE Paper; 2007-01-2401.

Zou, T., Mahadevana, S., and Mourelatos, Z.P. (2003). Reliability-based evaluation of automotive wind noise quality. *Reliability Engineering and System Safety* 82: 217–224.

Zou, T., Mourelatos, Z.P., and Mahadevan, S. (2004). Simulation-Based Reliability Analysis of Automotive Wind Noise Quality. SAE Paper; 2004-01-0238.

7

Door Closing Sound Quality

7.1 Vehicle Sound Quality

7.1.1 Concept of Sound Quality

We live in a world filled with all kinds of sounds. People enjoy sweet sounds, such as a warble, and hate noise, such as airplane booming or noise at construction sites. People usually use special words to describe the sounds: love or hate, enjoy or dislike, prefer or ignore, etc. Their sensations on the sounds have no direct interrelation to the sound loudness; instead, they are strongly related with their subjective perception, which introduces a "quality" concept of the sounds.

Quality is a feature of an object that is different from other objects. Sound is the reception of audible mechanical waves and their perception by the brain, i.e. the human auditory impression. "Sound quality" is the combination of "sound" and "quality," that is, the unique sensations of hearings. Sound quality is typically an assessment of the accuracy, enjoyability, or intelligibility of the sound.

According to its definition, *sound quality* reflects human subjective sensations. Even for the same sound, each person's feelings might differ. Figure 7.1 shows the cartoon pictures of a violin and a tractor. Most people love melodious music from the violin but dislike the irritable and booming noise from the tractor. Do all the people dislike the noise of the tractor? No. During planting and harvesting seasons, farmers drive their tractors in the fields, and the roar of the tractors brings joy and excitement to many because it represents hope and wealth.

Even for music, different people have different preferences. Young people usually like pop music, but older people might not like it. Some people like the lyric piano sound, while others revel in the joy brought by the dynamic percussion.

Although the sound quality reflects subjective sensations and different people have different preferences, most people still have some common understandings. For example, most people like pure tonal, vigorous, romantic, vibrate, lyric sounds, or music. Therefore, the sound quality contains some kinds of aesthetic connotation, which can be called "hearing aesthetic taste."

Noise and Vibration Control in Automotive Bodies, First Edition. Jian Pang.
© 2019 China Machine Press. All rights reserved. Published 2019 by John Wiley & Sons Ltd.

Figure 7.1 Cartoon pictures of a violin and a tractor.

7.1.2 Automotive Sound Quality

Sound is composed of many single tones. Combination of different frequency tones provides different hearing effects; for example, the female voice has more high-frequency components than that of most males.

Automotive sound has unique characteristics as well, which are closely related to the engine's speed and order. For example, two sounds with the same sound pressure level (SPL) and the same frequency contents, but with different order components, provide totally different hearing effects. Luxury cars, comfort cars, powerful cars, and sporty cars have their own order and frequency characteristics. Because of the influence of the vehicle's order and speed, automotive sound quality becomes a special sound issue.

Automotive sound quality refers to the customer subjective sensations on the automobile sounds, such as good and bad, quiet or loud, sporty or conservative, pleasant or annoying, luxurious or bland, exciting or boring, or refined or harsh. The research contents of the sound quality include subjective evaluation, objective evaluation index, realization of specific sound quality, etc.

Automotive sound quality can be divided into three categories: power train sound quality, electrical sound quality, and body closure sound quality.

The sound generated by engine and associated system (such intake system, exhaust system, driveline system, mounting system, etc.) is attributed to the power train sound quality category, which is directly related to the engine's speed and order. Different quality sounds can be achieved by adjusting order combination relating to engine speed and sound frequency. For example, the sound containing more firing orders is associated with melody, comfort, and peace, while the sound containing both the firing orders and half orders sounds sporty.

The sound generated by the electrical systems or components is attributed to the electrical sound quality. There are many electrical devices on an automobile, such as generator, fuel pump, wiper motor, HVAC motor, and mirror motor. The sounds of these electrical apparatus have their own characteristics. For example, the rotating components in the electric devices generate high-frequency order noise, which sounds like a scream. In addition, sound generated by the turn signal, the horn sound, etc. are also included in the scope of the sound quality of the electrical apparatus.

The body closure sound quality refers to the sound quality of opening and closing the closure parts. Door, trunk lid, liftgate, and engine hook are connected with the body by hinges. When they are opened and closed, they impact the body, generating sound. Among the closure parts, the sound of opening and closing the doors is the most important, because the actions to open and close the doors are much more frequent than others, and the sounds of opening and closing doors are more noticeable and attracted by the customers. The sound of closing doors usually provides them deep impressions on the automobiles' quality and reliability, and even from the sound, they could judge the luxury or cheapness, etc. of the vehicle.

This book only describes the body closure sound quality and focuses on the sound quality of closing door.

7.1.3   Importance of Automotive Sound Quality

In modern society, people have to deal with a variety of transport tools every day. Driving a vehicle to work, flying an airplane from Shanghai to New York, riding a boat across the Mississippi River, all the tools generate sounds. The airplane engine generates a huge roar. Even inside the cabin, the passengers still hear the large booming noise that makes them uncomfortable – it might even be painful to the ear. Sometimes, people have to swallow saliva or yawn to release the pressure to get temporary relief. The sound generated by a ship's engine is also huge, making people uncomfortable. However, few people mention the "airplane sound quality" and "ship sound quality." Why do people care much more about automotive sound? Why is the sound quality regarded as one of the most important vehicle core technologies?

Here are several important reasons: first, automobiles and people's daily lives are closely linked; second, the automobile interior space is too small; third, almost everyone in the developed world owns an automobile.

There are no transportation tools like the automobiles that are so closely related to people's daily life. It takes only one hour to fly from Washington, DC, to New York. Even for long-distance travel, for example, from Beijing to New York, the flight takes just 16 hours. Passengers spend 10–20 minutes to cross the Mississippi River by boat. Even for a long sea journey, for example, people might spend a week on a luxurious cruise from Alaska to Vancouver to enjoy picturesque scenery. However, after they arrive at their destinations, they leave behind the airplanes and the boats and have nothing to do with them. These transportation tools are part of people's lives for only a short time.

However, automobiles are totally different. In modern society, people are driving or riding automobiles every day. Driving to work, after arriving at the destinations, they leave the automobiles. But, after work, they have to ride in an automobile again. During holidays, they might drive to visit family or see another city. Thus, automobiles are much more interrelated with our lives than the airplane, boats, and other forms of transportation, and because of that, people are very concerned about the automobile sound that is closely related to their life.

The automobile interior space is much smaller than that of a plane or a ship. The airplane noise is huge, but people generally can tolerate a few hours "temporary" loudness, and some people can even sleep and do not hear the noise. The ship's space is very big, and the passenger cabins are usually far away from the engine cabin. However, inside the small interior space, people can only sit and drive an automobile, so they have to tolerate the noise all the time.

Ordinary people cannot afford to buy airplanes and ships, but they can own cars. The automobiles are private properties, and sometimes, are regarded as tools to improve the quality of life. Therefore, people pay more attention to the automotive sound quality than to the airplane and ship.

In modern society, people not only satisfy their desire to own an automobile but also pursue personalized vehicles, and sound quality is one of the personalized indexes. One of the goals in pursuing a personalized vehicle is to make its sound different from others. So, the sound is an important factor to characterize the vehicle's personality – it becomes an important part of the automobile's DNA. Different customers demand different sounds. For example, the passenger car owners, especially the luxury car customers, expect the harmonious and low level sound, while sporty vehicle customers pursue dynamic, stimulating sound, even roaring sound.

World-famous brand automobile manufacturers attribute the vehicle sound to an important part of their brand DNA. They expect that the customers can distinguish their brands only by listening to the vehicles' sounds. When driving vehicles, the customers desire not only to hear the luxury sound or sporty sound but also to experience their difference from other brand vehicles.

7.1.4   Scope of Sound Quality

NVH engineers and scientists' work can be categorized in three levels: reducing vibration and noise, controlling sound quality, and designing branded sound.

At the first level, the NVH engineer's task is to find and solve the noise and vibration problems, that is, to identify noise and vibration sources, to reduce vibration and noise levels, and to eliminate annoy and uncomfortable noise. In the past long history, the engineers did work at this level.

In the second level, the NVH engineers' major task is to control the sound quality. In the last decades, with the technology development of engine, power train systems, and electronic devices, many new problems of noise and vibration have been generated. So the engineers' work is far from just reducing noise decibels and vibration levels; instead, their work focuses on solving the sound quality problems induced by the new technologies and making good "sound quality" vehicles.

In the third level, the major task for NVH engineers and scientists is to design sound quality according to the vehicle features – to design a specific sound and make it become an important part of the vehicle's DNA. At this level, the engineers should not only realize good sound quality but also design a number of specific acoustic components to achieve a unique sound.

The concrete work at second and third levels include subjective evaluation of sound quality, objective testing and analysis index, consistent analysis between the subjective evaluation and the objective testing, sound quality control of each system (such as power train, electric devices, body), and design of the specific sound quality control components, etc.

The subjective evaluation of sound quality is to invite a specific group of people to drive the vehicles or to listen to sound recordings, and then to provide feedback and to score the sound quality levels. After processing the statistics of the assessments and scores, the subjective evaluation results of the vehicles are obtained.

The objective testing and analysis index is to find the appropriate indexes to characterize the sound quality based on objective testing data. For example, loudness and sharpness are indexes to represent the sound quality of a door closing. Sound pressure level and order are indexes to characterize sound quality of a power train. Loudness, pure tone, fluctuation, and roughness are indexes to describe sound quality of electric devices.

The goal of the objective evaluation is to quantitatively describe the sound quality, and to verify the results of the subjective evaluation. So, only if the subjective evaluation is consistent with the objective evaluation can the subjective evaluation be verified. After the verification, even without the objective test data, the subjective evaluation can be used to evaluate the automotive sound quality.

The sound quality control is to control the noise and vibration levels for each system under the corresponding sound quality targets, rather than simply reducing the levels of the vibration and noise. For example, compared with the overall interior sound level, the SPL of the transmission rattle is so low that it can be neglected, but the rattle is easily heard. For example, the SPL of high-frequency tones of a motor or an oil pump noise is very low, but the harsh sounds can be easily perceived in a quiet environment. Therefore, reduction of the transmission rattle and the high frequency tones is not simply to reduce the noise levels but to improve the sound quality.

The sound quality design refers to the design to achieve the unique sound according to characteristics of a vehicle's DNA. For example, for sporty vehicles, the engine half order sound should be highlighted, and even the resonance sound could be added. For luxury cars, the engine half order sound should be eliminated as low as possible. Design of the power train sound quality involves design of engine combustion performance, design of intake and exhaust manifolds, design of exhaust system, etc.

7.2 Evaluation Indexes of Sound Quality

7.2.1 Description of Psychoacoustics

7.2.1.1 Scope of Psychoacoustics

Psychoacoustics, an interdisciplinary of acoustics, physiology and psychology, is a scientific discipline to study the relationship between the physical quantity of sound and the psychological reaction. It is the scientific study of the acoustic characteristics under the premise of psychological sensations. The research scope of psychoacoustics includes directivity of sound transmission to the ears, sound propagation processing inside the internal ear and ear's structural vibration, stimulated effort of sound and vibration on the human neural structure and psychological reaction, and subjective evaluation indexes.

7.2.1.2 Relationship between Ear Structure and Sound Transmission

The human ear is the organ of hearing, which is composed of the outer ear, middle ear and inner ear, as shown in Figure 7.2. The ear is a nonlinear structure.

The outer ear is the most external portion of the ear, which is composed of the pinna (also called auricle) and the external auditory canal. The function of the pinna is to

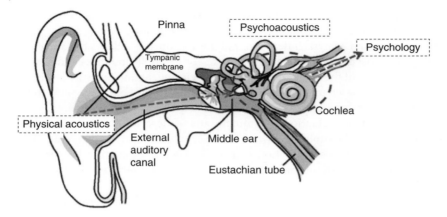

Figure 7.2 Ear structure.

collect sound energy and detect its location. The sound energy received by the pinna in different directions is different, which indicates that the pinna is a sound directional organ. The outer auditory canal is similar to a varying-section pipe, so it works like a filter, amplifying the middle frequency sound levels up to +15 dB and attenuating low and high frequency sound levels up to −30 dB. The outer ear is just a physical structure, independent of the human psychology.

The middle ear is composed of tympanic membrane, ossicular chain, and auditory tube. In the middle ear, the acoustic impedance is matched with the impedance of the medium consisting of the air and the liquid, and the mechanical vibration is converted to the liquid movement.

The inner ear comprises the oval window, semicircular canal, and cochlea. In the inner ear, the liquid movement is converted into electrical signals. The electrical signals are transmitted to the brain through the auditory nerve, so people hear the sound.

The listening process can be summarized as follows. The outer ear possessing directivity and filtering function is a passageway of sound transmission that belongs to the physical acoustics. In the middle ear, the mechanical movement drives the liquid to move, so the acoustic energy is transferred into the mechanical energy. In inner ear, the liquid movement is transmitted into electrical signals, and then to brain nerves. In the middle ear and inner ear, the transmission of sound waves and the human psychology are connected, so the study of the relationship between acoustics and human perception induces the birth of psychoacoustics.

The human ear hearing system is very complex. Scientists have not fully understood the problems relating to the hearing, such as hearing transmission, hearing mechanism, and psychological reaction. In the field of acoustics and vibration, it is still a hot research topic to explore the mechanism of hearing. For example, some scientists are working on the structural modes of the inner ear and trying to use laser technology to measure the modal characteristics of the inner ear.

7.2.1.3 Auditory Field

Auditory field refers to the scope of the audible sound, that is, the range from the threshold of quiet to threshold of pain, as shown in Figure 7.3. Three curves are plotted

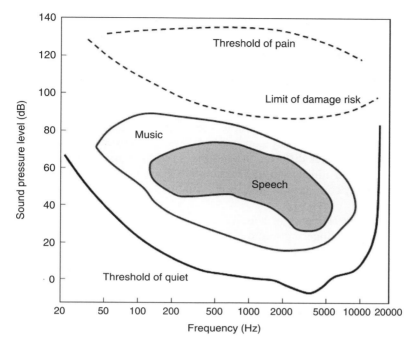

Figure 7.3 Hearing threshold curve.

in the figure. The bottom curve is the threshold in quiet, which indicates the SPL of a pure tone that is just audible. The middle curve is the limit of damage risk, which specifies that hearing damage could happen. The damage risk depends on sound levels and exposure time. The upper curve is threshold of pain, which means that people feel pain in their ears at the sound level and corresponding frequency.

The hearing threshold is a function of frequency. The most sensitive frequency of human ear is from 500 to 4000 Hz, so the hearing threshold reaches the lowest level in this frequency range. The most sensitive frequency at the eardrum is 1000 Hz, while the most sensitive frequency at the outer ear is 4000 Hz. The hearing threshold at low and high frequencies is higher than that at the middle frequency range. For example, the hearing threshold for pure tone at 20 Hz is 70 dB, which is higher than that at 1000 Hz.

There exist differences for individual hearing thresholds, and the difference increases as people get older. Figure 7.3 shows the average statistical values of the hearing threshold range of the normal young people.

7.2.2 Evaluation Indexes of Psychoacoustics

7.2.2.1 Subjective Evaluation Indexes

Usually, people subjectively evaluate their heard sound from three aspects: loudness, pitch, and timbre, which are also known as the three elements of sound.

Loudness is the characteristic of a sound that is primarily a psychological correlate of physical strength, which depends on strength and frequency of the sound. Since the human ear is a nonlinear structure, for the sounds with the same strength but different frequency, the loudness is different. The human ear is most sensitive to the middle frequency, so for the same strength pure tones, people perceive middle frequency tones as louder than the low-frequency tones and high-frequency tones.

The pitch is perceived as how "low" or "high" of a sound. The pitch is mainly determined by the frequency, and also relates to the sound strength. The high-frequency pitch increases with increase of sound strength, while the low-frequency pitch decreases with increase of sound strength.

The timbre refers to sound quality of people's sensations on hearing the sound. Sound is made up of many frequency components, and the sound with different frequency components provides different sensations. Even for musical instruments generating the same tone, in addition to its fundamental frequency, due to existing of many harmonic frequency components, the sound qualities generated by the instruments are different.

In addition to the three elements of sound, the sound subjective evaluation also includes sound masking and lingering sound.

In addition, the subjective evaluation can be implemented by subjective description and scoring of the sound, which will be described in detail in later sections of this chapter.

7.2.2.2 Objective Evaluation Indexes

The objective evaluation is to describe the sound by measured and analyzed physical parameters. The physical parameters corresponding to loudness, pitch, and timbre, the subjectively descried parameters, are the sound amplitude, frequency, composition, and phase of the frequency, respectively. In order to describe the sound features more accurately, some objective evaluation indexes of psychoacoustics have been developed, such as:

- Loudness
- Sharpness
- Modulation
  - Fluctuation
  - Roughness
- Tonality
- Articulation index
- Masking

By using these objective indexes, scientists and engineers can more accurately and quantitatively identify the influence of the sound on the human psychology.

7.2.2.3 Relationship between Subjective Evaluation and Objective Evaluation

The goal of the relationship study is to pursue consistency between the subjective evaluation and objective evaluation. Generally, the data from the objective testing and analysis can verify the subjective evaluation and description. However, sometimes, they are inconsistent – for example, four persons leaning against a wall, as shown in Figure 7.4. The subjective evaluation is to describe and rank their heights, while the objective evaluation is to measure the heights. At a glance, many people's first impressions are that the tallest person is one standing at the most left and the shortest one is at the most right, which reflects their subjective judgment. However, a ruler is used to

Figure 7.4 Four persons leaning against a wall.

measure their heights and the result is that all four persons have the same heights. Why is the difference between the subjective and objective evaluations?

The figure provides the observers a three-dimensional picture. The lines on the left side are denser, while the right lines look loose. There are many lines on the wall for the leftmost person, while only four lines are on the back of the most right person. So, the persons staggered with lines provide the observers the visual illusion where their impressions are that the left-side person is obviously taller than the right person. The example illustrates the inconsistency between the subjective and objective evaluations.

Similar phenomena exist for the evaluation of automobile noise and vibration. For example, the customers trying out two cars feel that the interior noises have a significant difference, but the measured sound pressure levels are the same. The subjective and objective evaluations are inconsistent, which demonstrates the SPL is not an exact objective index for this case.

NVH engineers' task is to measure the objective data and then make subjective judgment, but their more important task is to find consistent indexes for subjective and objective evaluations. Thus, it becomes a research direction for the engineers and scientists to find the objective indexes that can reflect the subjective perception.

7.2.3 Critical Band

Noise and vibration signals are usually analyzed in frequency domain. According to different analyzed objects, the frequency analysis is usually divided into one-third octave analysis and narrow band analysis. However, in psychoacoustics analysis, according to the characteristics of human hearing, the frequency domain is rearranged and is divided into 24 special frequency bands (1–24), each representing a critical band. Within one critical band, the auditory perception of the human ear is assumed to be the same. The frequency unit is Hz, while the critical band unit is Bark which comes from the name Von Barkhausen, a famous acoustics scientist.

Lower than 500 Hz, there are 5 Bark bands, and the bandwidth of each critical band is 100 Hz. Above 500 Hz, there are 19 Bark bands and the bandwidth of each critical band is $0.2f_c$. f_c is the center frequency of a bandwidth. Table 7.1 gives the frequency range of the 24 Bark critical bands.

The critical bandwidth is a filter of the psychological sound, which can effectively represent the characteristics of the human brain. Its range covers all the hearing frequency. The psychoacoustic objective evaluation indexes, such as loudness, sharpness, fluctuation, and roughness, are calculated in the critical band domain.

7.2.4 Loudness

Loudness is a psychoacoustic index that quantifies the perceived intensity of a sound. The psychological researches show that the human auditory system has the following characteristics: its sensitivity to the sound frequency is nonlinear, its masking effect

Table 7.1 Bark critical band.

Bark band #	Center frequency	Bandwidth	Start frequency	Stop frequency
1	50	100	0	100
2	150	100	100	200
3	250	100	200	300
4	350	100	300	400
5	450	110	395	505
6	570	120	510	630
7	700	140	630	770
8	840	150	765	915
9	1000	160	920	1080
10	1170	190	1075	1265
11	1370	210	1265	1475
12	1600	240	1480	1720
13	1850	280	1710	1990
14	2150	320	1990	2310
15	2500	380	2310	2690
16	2900	450	2675	3125
17	3400	550	3125	3675
18	4000	700	3650	4350
19	4800	900	4350	5250
20	5800	1100	5250	6350
21	7000	1300	6350	7650
22	8500	1800	7600	9400
23	10500	2500	9250	11750
24	13500	3500	11750	15250

on different frequency sound is different, and it has a temporary masking effect. Traditionally, the sound characteristics are evaluated by the magnitude and energy of the sound, such as sound pressure level and sound power level, but these parameters cannot really reflect the human auditory perception. Therefore, the loudness, an index with psychological features, can more effectively reflect people's subjective auditory perception, and it is a linear measure of the magnitude of auditory sensation.

The loudness level is defined as the SPL of a pure tone at 1000 Hz, and its unit is phon. The loudness level is closely related with frequency and sound pressure level, as shown in Figure 7.5. Each curve represents the same loudness level at different frequencies. The figure shows that the auditory perception at middle frequencies is higher than that at low and high frequencies, and the most sensitive frequency is at 4000 Hz.

The loudness is an auditory sensation index of sound magnitude. Table 7.2 lists the relationship between loudness and sound pressure level. When the SPL changes 3 dB, people

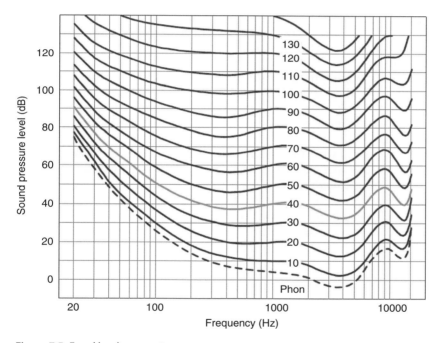

Figure 7.5 Equal loudness contours.

Table 7.2 Relationship between loudness and sound pressure level.

Change in sound pressure level (dB)	Change in perceived loudness
3	Just perceptible
5	Noticeable difference
10	Twice (or 1/2) as loud
15	Large change
20	Four time (or 1/4) as loud

just have the perceptible change of the loudness; when the SPL changes 5 dB, people feel loudness changes significantly; when the SPL increases 10 dB, the loudness is doubled.

The loudness is proportional to the hearing sensation. The loudness unit is sone. The loudness of 1 sone refers to a 1000 Hz pure tone with a SPL of 40 dB (or loudness level of 40 phons). The relationship between the loudness (N) and loudness level (L_N) can be expressed as

$$N = 2^{\frac{L_N - 40}{10}} \tag{7.1}$$

Figure 7.6 shows the relationship between the "sone" and the "decibel". A loudness of 2 sones is perceived as double loud of 1 sone. When the loudness increases from 1 sone to 2 sones, the SPL increases 10 dB at 1000 Hz, but only 5 dB at 50 Hz. From the plot, it is clearly to see that sound pressure levels increase more at high frequency than at low frequency when the loudness increases the same amount, which indicates that the loudness is more sensitive for low-frequency sound, i.e. the loudness can more effectively represent low-frequency characteristics. When the loudness level increases 10 phons, the loudness is doubled.

There are several methods to calculate the loudness, such as Stevens method (ISO532A), Zwicker method (ISO532B), etc. Among these methods, the Zwicker method is the most widely used. Zwicker was a German acoustic scientist who pioneered electric-acoustics and achieved outstanding contribution on loudness research.

For a steady-state sound, Zwicker considered the influence of three factors: nonlinear frequency sensitivity; masking effect at critical band; and sound field (free field and diffuse field) effects. Regarding loudness, namely for the sound with the same sound pressure level, it is perceived louder in the diffuse field than in the free field. For an unsteady-state sound, in addition to above three factors, the characteristic of the loudness varying with time, i.e. temporal effect of the sound, should be considered.

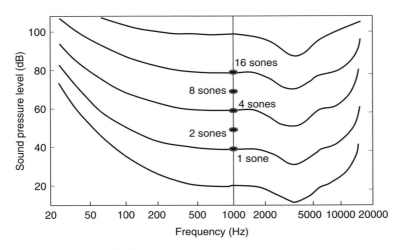

Figure 7.6 Relationship between the loudness and the sound pressure level.

The loudness changes with the critical band, which can be plotted as a graph. Specific loudness refers to the loudness graph varying with frequency. The loudness is a summation of the specific loudness at each critical band, and can be expressed as follows:

$$N = \int_{0}^{24\,Bark} n'(z)\,dz \tag{7.2}$$

where N is the overall loudness; $n'(z)$ is the specific loudness; z is the critical band.

Figure 7.7 shows SPL (a) and corresponding specific loudness curve (b). Because the sound masking effect is considered, the curves of the loudness and the SPL look different.

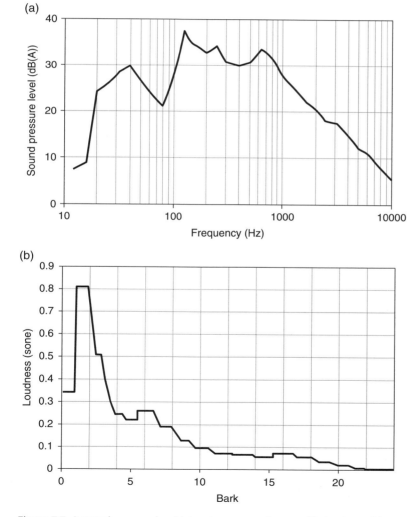

Figure 7.7 A sound pressure level (a) and corresponding specific loudness (b).

The loudness is much higher at low frequencies than at medium and high frequencies, which indicates that more effect at the low frequency is considered.

7.2.5 Sharpness

The sharpness is a psychoacoustic index to measure the high frequency components of a given sound. The sharpness can be interpreted as the ratio of the high-frequency components to the overall sound level. It can also be understood as the "center of gravity" of a spectrum on frequency scale. The higher the c.g., the sharper the sound, and vice versa. The sharpness can be expressed as

$$SHARP = \frac{\int_0^{24\,Bark} n'(z)z\,dz}{N} \tag{7.3}$$

The sharpness unit is acum. 1 acum is defined as a narrow band (less than 150 Hz bandwidth) noise where the center frequency is 1000 Hz and SPL is 60 dB. Figure 7.8 shows two sounds with the same sound pressure level, 44.3 dB(A). However, the sharpness corresponding to the two sounds are 0.275 acum (Figure 7.8a) and 4.4 acum (Figure 7.8b), respectively. In the spectral distributions of the sound pressure levels, Figure 7.8b has more high-frequency components than Figure 7.8a, so the sharpness for Figure 7.8b is higher than Figure 7.8a.

The sharpness expressed in Eq. (7.3), defined as the "center of gravity," is not weighted sharpness. Because the sharpness is also an important index of psychoacoustics, the subjective sensation has an impact on it. Some scholars modified the sharpness and obtained some weighted sharpness calculation methods, such as Aures method, von Bismark method, and the Zwicker method. Among them, the Aures sharpness and von Bismark sharpness methods are most widely used.

The Aures sharpness can be expressed as follows:

$$S_{Aures} = K_1 \frac{\int_0^{24\,Bark} n'(z)g_1(z)\,dz}{\log\left(\frac{N}{20}+1\right)} \tag{7.4}$$

where K_1 is the modified coefficient and $g_1(z)$ is the weighted coefficient, which is expressed as

$$g(z) = e^{0.171z} \tag{7.5}$$

The von Bismark sharpness can be expressed as follows:

$$S_{Bismark} = K_2 \frac{\int_0^{24\,Bark} n'(z)zg_2(z)\,dz}{N} \tag{7.6}$$

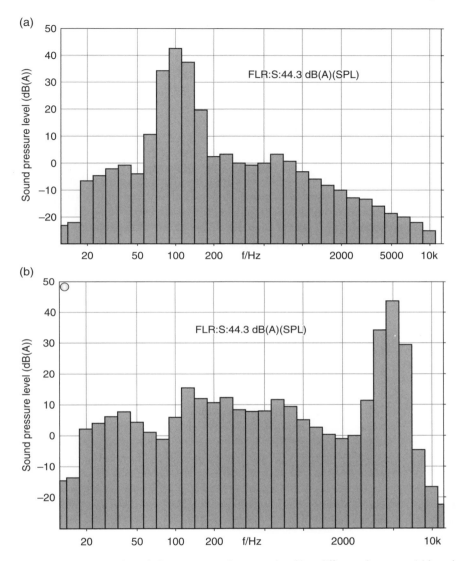

Figure 7.8 Two sounds with the same sound pressure level but different sharpness: (a) low sharpness; (b) high sharpness.

where K_2 is the modified coefficient and $g_2(z)$ is the weighted coefficient, which is expressed as

$$g_2(z) = \begin{cases} 1, & z < 14 \\ 1 + \dfrac{3}{1000}(z - 14), & z \geq 14 \end{cases} \tag{7.7}$$

The sharpness values calculated by different methods are different. The values calculated by the Aures method are relatively large, so the sharpness difference between the various noises is large as well. This method is beneficial for noise source identification, but is not consistent with the subjective sensations. The values calculated by von Bismark method have smaller differences for different noise, but they are close to the subjective perceptions. At present, the sharpness calculation is not standardized, so these methods are applied according to different cases.

7.2.6 Modulation, Fluctuation, and Roughness

Modulation, fluctuation, and roughness are the same concept, but in different descriptions. Modulation is the superposition effect of two waves with different frequencies and/or different amplitudes. Because the different modulation frequencies provide different subjective sensations, the modulation is divided into fluctuation and roughness. When the modulation frequency is low, people perceive the superimposed sound or vibration as similar to riding on an ocean wave, so this kind of modulation is called *fluctuation*. When the modulation frequency is high, people perceive that the sound or vibration is very rough, like riding on a rough road, so this kind of modulation is called *roughness*.

7.2.6.1 Modulation
When two pure tones with different frequencies and/or magnitudes are superimposed, a modulation is generated. Assume the two pure tone signals are

$$x_1(t) = A_1 \cos \omega_1 t \tag{7.8a}$$

$$x_2(t) = A_2 \cos \omega_2 t \tag{7.8b}$$

The superimposed signal can be expressed as

$$x(t) = A_1 \cos \omega_1 t + A_2 \cos \omega_2 t = A \cos(\omega t + \theta) \tag{7.9}$$

where A and ω are the amplitude and angular velocity of the superimposed signal, respectively, expressed as

$$A = \sqrt{A_1^2 + 2 A_1 A_2 \cos(\omega_2 - \omega_1)t + A_2^2} \tag{7.10a}$$

$$\omega = \frac{\omega_1 + \omega_2}{2} \tag{7.10b}$$

In order to illustrate the superposition, an example is provided. Assume $A_1 = 10$, $f_1 = 20Hz$, $A_2 = 15$, $f_2 = 23Hz$. Figure 7.9 shows the time-domain and frequency-domain plots of the two pure tone signals.

According to Eq. (7.9), the superimposed signal can be plotted, as shown in Figure 7.10.

After the two signals are superimposed, the superimposed signal with the new amplitude and frequency fluctuates. This phenomenon is called *modulation*. The maximum and minimum amplitudes of the modulated signal are $A_{\max} = A_1 + A_2$, $A_{\min} = |A_1 - A_2|$,

(a)

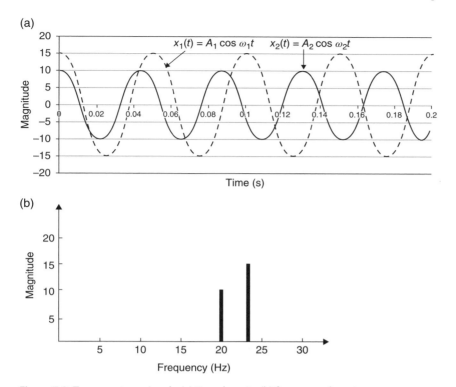

Figure 7.9 Two pure tone signals: (a) time domain; (b) frequency domain.

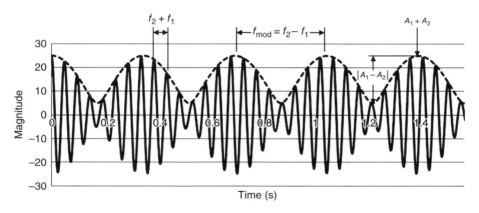

Figure 7.10 Superimposed signal of two signals with different frequencies and magnitudes.

respectively. The modulated signal fluctuates between A_{max} and A_{min}. The difference between the maximum value and the minimum value of the modulation is called as the modulation depth, represented by D:

$$D = A_{max} - A_{min} \tag{7.11}$$

The modulated signal periodically moves from the maximum to the minimum, then from the minimum to the maximum. The periodic change can be represented by modulation frequency f_{mod}, expressed as

$$f_{mod} = f_2 - f_1 \tag{7.12}$$

Let's consider two special cases. The first one is that the two signals have the same frequency, but different magnitudes, i.e. $f_1 = f_2$, $A_1 \neq A_2$. The superimposed signal is plotted in Figure 7.11. Due to the same frequency, it is still a pure tone signal and its amplitude is the superposition of two signal amplitudes. For this case, there is no modulation phenomenon.

The second case is that the two signals have the same magnitude, but different frequencies. For the above example, the superimposed curve is plotted, as shown in Figure 7.12. Due to the same amplitude, the minimum amplitude of the modulation is zero; that is, $A_{min} = |A_1 - A_2| = 0$, and the maximum amplitude is twice that of a single signal.

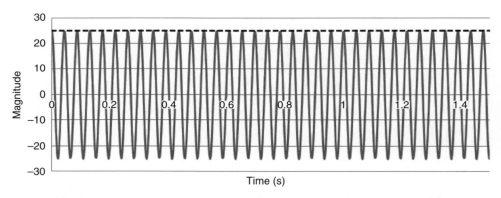

Figure 7.11 Superimposition of two signals with the same frequency but different magnitudes.

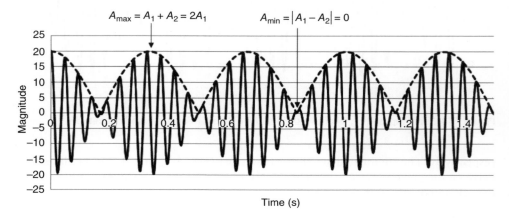

Figure 7.12 Superimposition of two signals with the same magnitude but different frequencies.

According to the modulation frequency, the modulation is divided into fluctuation and roughness. When the modulation frequency is less than 20 Hz, the superimposed sound or vibration performs fluctuation, and the corresponding modulation is fluctuation. When the modulation frequency is 20–300 Hz, the superimposed sound or vibration performs rough, the corresponding modulation is roughness. When the modulation frequency is higher than 300 Hz, people can distinguish the two frequency sounds and regard them as pure tone signals.

7.2.6.2 Fluctuation

The fluctuation describes the hearing or vibration sensation on variations of a sound or vibration with a modulation frequency between 0.5 and 20 Hz. For example, two pitch-forks with different frequencies are put together. Two pure tones are generated when they strike one another. When the frequencies are close, the people hear a mixed sound of the two pure sounds, which periodically changes from strong to faint and then from faint to strong, and sounds like a wave moving ups and downs.

As another example, a vehicle driving on an uneven pavement is used to illustrate the vibration fluctuation. The vehicle moves vertically at a frequency. Assume the road surface profile is an undulating sine wave, which has a certain frequency. The vertical movement the occupants perceive is the superimposition of the vehicle movement and the road surface "movement," so a modulation effect is formed by the two movements with different frequencies. Because the frequencies of the undulating road surface and the vehicle vertical movement are low, and their frequency difference is relatively small, the occupants traveling on this road feel like riding on wave ripples. They are moved up and down, so they really perceive the "fluctuation."

The fluctuation level can be expressed by fluctuation strength. The unit of fluctuation strength is vacil. 1 vacil is defined as a tone with sound pressure 60 dB at 1000 Hz, which is modulated by 4 Hz modulation frequency and 100% modulation depth. The fluctuation strength can be calculated as

$$F = 0.008 \frac{\int_0^{24} \Delta L \, dz}{f_{mod}/4 + 4/f_{mod}} \tag{7.13}$$

where f_{mod} is the modulation frequency and ΔL is the temporal masking depth, which can be obtained by calculating the signal response spectrum. The masking depth represents the human sensation change on the sound amplitude. The masking depth and modulation depth are two different concepts, and the masking depth can be obtained from the temporary masking effect.

The factors influencing the fluctuation strength include: fluctuation frequency, SPL, and fluctuation depth. When the modulation frequency is 4 Hz, the fluctuation strength has the largest value. When the modulation frequency is lower than 4 Hz, the fluctuation strength increases with the frequency increase. When the modulation frequency is higher than 4 Hz, the fluctuation strength decreases with the frequency increase. Figure 7.13 shows the fluctuation strength varying with frequency. The more strength the fluctuation, the more discomfort the people experience.

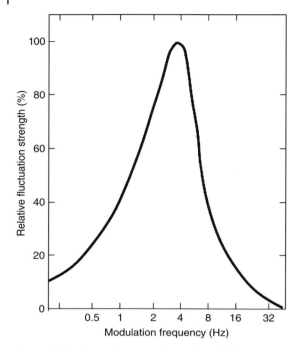

Figure 7.13 Fluctuation strength varying with frequency.

7.2.6.3 Roughness

When the modulation frequency is higher than 20 Hz, people cannot distinguish the change of intensity of a sound or vibration; instead, they perceive a "rough" sound or vibration. For example, a vehicle passing a speed-bump zone consisting of many strips is used to illustrate the vibration roughness. The surface profile "frequency" of the speed bumps is much higher than that of the vehicle vertical movement. Due to a big difference of the two frequencies, the "modulation frequency" is high. Driving on this road, people do not perceive the up and down movement, but they experience coarseness of the road.

Roughness is the sensation of rapidly modulated sound or vibration. The roughness frequency range is from 20 to 300 Hz, and its unit is asper. A pure tone at 1000 Hz and with 60 dB SPL is modulated by 70 Hz modulation frequency and 100% modulation depth. The roughness of the modulated tone is defined as 1 asper. Roughness can be calculated as

$$R = 0.3 \int_{0}^{24\,Bark} f_{mod} \Delta L dz \tag{7.14}$$

From Eq. (7.14), the factors affecting the roughness include the modulation frequency and the masking depth. The roughness has a maximum value at 70 Hz. The greater the roughness, the more uncomfortable the people perceive the sound or vibration.

The roughness is not limited to the pure tone modulation. Modulation for broadband sound or narrowband sound can also cause roughness.

7.2.7 Tonality

Tonality is used to describe a sound prominence among a noise. A pure tone or a sound within a bandwidth is much higher than those within the adjacent bandwidths. It is defined as tonality, and its unit is tu. A pure tone at 1000 Hz and with SPL of 60 dB is defined as 1 tu.

Figure 7.14 shows that within a Bark band, the sound power spectral density at a frequency is much larger than those of the other frequencies, which indicates that a prominence peak exists. If the power spectral density of the prominence peak (W_t) is 6 dB higher than those (W_n) excluding the prominence peak within the band, the prominence peak represents a tonality.

If the power spectral density of a Bark bandwidth is 7 dB higher than those of the adjacent bandwidths, the sound within the bandwidth can be clearly heard, so the sound within the bandwidth is called sound prominence. Figure 7.15 shows that the power spectral density of a bandwidth (W_e) is 7 dB higher than those (W_l and W_u) of bands on either side, so the sound is prominent.

7.2.8 Articulation Index

An articulation index (AI) is an index to describe the amount of sound that is audible in a noisy environment. AI is a measurement of the intelligibility of speech between

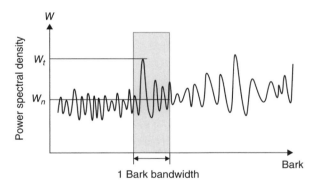

Figure 7.14 A prominence tonality within a Bark band.

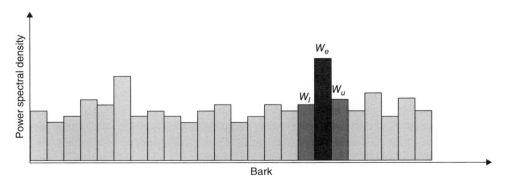

Figure 7.15 Prominent sound within a Bark bandwidth compared with adjacent bandwidths.

people. AI is expressed as a percentage. 100% AI means the sound is fully audible, and 0% AI indicates the sound is completely inaudible. When the background noise is louder than the speech voice, the voice cannot be heard clearly. When the noise is 12 dB louder than the voice, the voice is completely inaudible, which provides an upper limit of the noise. When the noise is 30 dB lower than the upper limit, the voice is fully audible, which provides a bottom limit of the noise.

The upper limit, $UL(f)$, and the bottom limit, $LL(f)$, of the noise can be expressed as follows:

$$UL(f) = H(f) + 12 \tag{7.15}$$

$$LL(f) = UL(f) - 30 \tag{7.16}$$

where $H(f)$ is the speech signal.

The most sensitive frequency range for the human hearing is 200–6300 Hz, and the sensitivity varies with the frequency, so a weighting coefficient, $W(f)$, is introduced, as shown in Figure 7.16. The weighting coefficient and the noise signal are included simultaneously to calculate the articulation index, expressed as

$$AI = \sum W(f)D(f)/30 \tag{7.17}$$

where $D(f)$ is determined by the noise signal, $N(f)$, and the upper and the bottom limits:

$$D(f) = \begin{cases} 0, & N(f) > UL(f) \\ UL(f) - N(f), & LL(f) < N(f) < UL(f) \\ 30, & N(f) < LL(f) \end{cases} \tag{7.18}$$

AI is a very useful index to identify the noise between 200 and 6300 Hz. For example, the transmission rattle is usually within the frequency range. The SPL of a rattle noise is usually much lower than the overall SPL, so the rattle problems cannot be identified by the overall sound pressure level. Instead, the rattle noise can be clearly recognized by the articulation index signals.

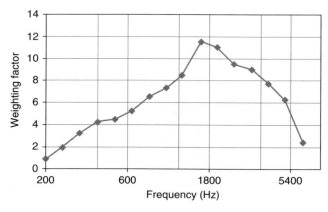

Figure 7.16 AI weighting coefficient.

7.2.9 Sound Masking

When two sounds A and B are present simultaneously, if sound B is masked by sound A, only sound A can be heard. This phenomenon is called the *sound masking effect*. Masking is the listening capability reduction to hear one sound due to the presence of another sound. For example, on a quiet street, we can clearly hear neighbors talking outside, but if suddenly, there is an extremely loud sound coming from a nearby horn, we cannot hear the voices anymore, because the voices are masked by the horn sound.

Assume sound A is 50 dB at 1200 Hz and sound B is 70 dB at 800 Hz. If only sound A exists, we can hear it clearly. But when sound B appears, we cannot hear the sound A because sound A is masked by sound B. The sound masking other sounds is called *masking sound*, while the sound being masked is called *masked sound*. If sound A is tuned up to 60 dB from the original 50 dB, which can be just heard, the increase amount is 10 dB. The increased decibel for a sound that can be just perceived is the amount of masking. In this example, the amount of masking of sound B to sound A is 10 dB. For sound B, the SPL of sound A that is just heard is called the critical point of sound B at 1200 Hz. For sound B, the critical points at different frequency sound are different. The critical points are connected to form a critical line for sound B at 800 Hz. The critical line is also called the masked threshold.

The masking effect depends on frequency components and temporary relations of two sounds, so it is divided into frequency masking and temporary masking. The frequency masking is further divided into broadband masking (or white noise masking), narrowband masking, and tone masking. Figure 7.17 shows a white noise masked threshold line. For each SPL, the masking effects are different for different frequencies, and the amount of masking increases with the frequency increases.

Figure 7.18 is the masked threshold for 60 dB sound pressure levels at each narrow-band frequency. The masking effects for each narrowband on other frequencies are different. The farther away from the center frequency of the sound, the greater the masked amount.

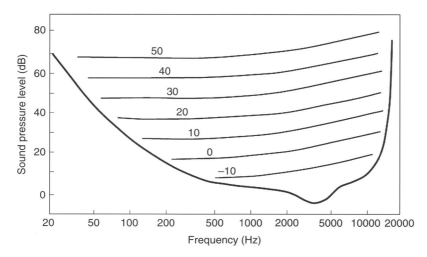

Figure 7.17 A white noise masked threshold.

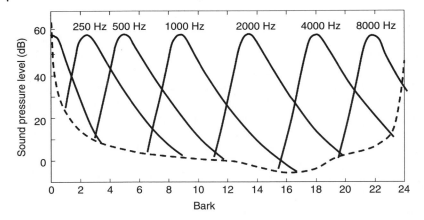

Figure 7.18 Masked threshold for 60 dB sound pressure levels at each narrowband frequency.

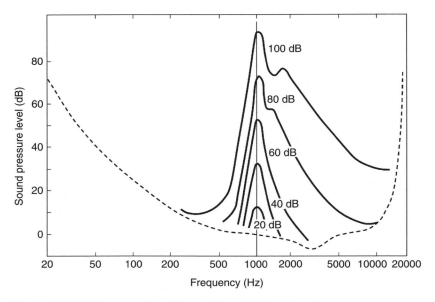

Figure 7.19 Masking curves for different SPLs at 1000 Hz center frequency.

For narrowband sounds with the same center frequency, the masking curves are different for different SPL sounds. Figure 7.19 shows masking curves for different SPLs at 1000 Hz center frequency.

From these curves, it is clearly to see that the masking effect not only relates to frequency but also depends on the sound pressure level. The masking phenomenon for several sounds existing simultaneously is called simultaneous masking or frequency masking. There is another masking phenomenon; that is, several sounds are not present simultaneously, and instead, one sound follows another sound in a short period of time, which introduces the temporary masking. The phenomenon where one sound is masked by later generated sound is called premasking. The phenomenon where one sound is masked at the beginning of its generation is called postmasking.

The masking effects widely exist for automobile noise. For example, when a vehicle runs at $130\,km\,h^{-1}$ speed, the engine noise could be masked by the wind noise. Another example is idle noise. In a quiet environment, the $10\,dB$ oil pump whine cannot be masked by $40\,dB$ idle noise. The idle noise SPL is dominated by low and middle frequency components, but the oil pump noise frequency is several thousand hertz. The masking amount of the low-frequency sound on the high-frequency sound is over $30\,dB$, so the oil pump whine is still audible.

7.3   Evaluation Indexes of Automotive Sound Quality

Automotive sound provides a lot of information to the customers, such as luxury, comfort, powerfulness, cheapness, etc. This section will briefly introduce the classification and evaluation indexes of the automotive sound quality.

7.3.1   Classification of Automotive Sound Quality

Every system and component in an automobile could induce NVH problems, so sound quality is related to these systems and components. Usually, the automotive sound quality is divided into three categories: power train sound quality, electrical sound quality, and body closure sound quality.

7.3.1.1   Power train Sound Quality

Power train sound quality involves engine, transmission, driveshaft system, intake and exhaust systems, and power plant mounting system. The power train sound quality is determined by its orders and frequency components. The sound dominated by the firing order and corresponding harmonic orders sounds is perceived as comfortable and soft, while the sound uniformly contributed by both the firing orders and half orders is perceived as sporty and powerful. Eliminating resonance can improve the sound quality and comfort, while appropriately increasing resonance can enhance the sensation of powerful or sporty sound.

The orders and frequency components of the engine noise, which are determined by the combustion characteristics, combustion pressure fluctuation, manifold design, etc., are cores of the power train sound quality.

Intake system and exhaust system are the key systems to tune the power train sound quality, which can be achieved by the systems' design and the manifold design. First, acoustic tuning elements are applied in the intake and exhaust systems design to eliminate unwanted noise and enhance the desired frequency sound. Second, the exhaust manifold and intake manifold are specially designed to achieve order components of the orifices' noise. Combination of the desired frequencies and orders of the orifice sounds provide different sound sensations, such as sporty sound, powerful sound, luxury sound, comfortable sound, etc.

Power plant mounting system also influences the power train sound quality. The engine vibration is transmitted to the vehicle body through the mountings, forming the structural-borne noise, which is the major source of interior low frequency noise.

The low-frequency components of the interior noise can be tuned by the mounting design. In addition, by modifying the mounting bracket design, the unwanted resonances can be eliminated in order to improve the comfortable sound quality. Sometimes, some brackets are specially designed to enhance resonances in order to achieve sporty or powerful sound quality.

Transmission and driveshaft system impact the sound quality as well, but they do not have functions to tune the sound quality. Take the transmission as an example; meshing gears generate whine and unmeshed gears induce rattle, which deteriorate the sound quality. In order to eliminate the whine and rattle and to improve the sound quality, the transmission must be modified.

7.3.1.2 Electrical Sound Quality

There are many electrical devices in an automobile, such as generator, fuel pump, and motors (wiper motor, heater motor, etc.). There is a tendency to use more electrical devices in today's vehicles. The electrical devices combine with other components to form functional systems, such as the cooling system, HVAC system, etc. The running electrical devices significantly influence the automotive sound quality. In addition, the standstill electrical components, such as turn light indicator, horn, and seat belt reminder, also impact the sound quality.

The electrical sound quality can be divided into three categories: sound quality generated by rotary machines, sound quality induced by motor electromagnetic noise, and sound quality of the standstill sound generators.

The rotary machine noise is generated by friction between its blades and the air, and its internal friction, such as friction between bearings and their supporting. The noise includes not only fundamental frequency responses of the blades and the bearings but also their harmonic frequency responses. The harmonic frequency responses concentrated in the high-frequency range coincide with the most sensitive frequency range of the hearing. Even though the sound pressure levels at middle- and high-frequency ranges are low, people still perceive them as annoying, resulting in sound quality problems. For example, in a quiet evening at an underground parking garage, a noise peak at 2000 Hz induced by a fuel pump is only 10 dB(A), but it is clearly audible and could make drivers or passengers uncomfortable.

Similar to the rotary machine noise, the electromagnetic noise also includes fundamental and harmonic responses, and the harmonic responses are concentrated at high-frequency areas, which are also annoying.

The standstill sound generators usually emit a single frequency tone or a sound combining just several frequency components. Today, the customers are more and more sensitive to sounds. Sometimes, according to the heard sounds, they can not only perceive the sound quality but also judge whether the vehicles are luxury ones or cheap ones.

7.3.1.3 Body Closing Sound Quality

Body closing sound quality mainly refers to the sound quality connected to opening and closing the body closure parts, including doors, hood, trunk lid, side windows, sunroof, and fuel tank cap. Opening and closing these parts also induce sound quality problems.

Among the body closing sound quality, opening and closing door sound quality is the most important, so the remainder of this chapter will focus on this sound quality analysis.

7.3.2 Indexes Used to Describe Automotive Sound Quality

As already discussed, automotive sound quality description includes subjective description and objective descriptions. The subjective description refers to the sound quality evaluation by language description and scoring. The language description is to use special words or terminologies to describe the customer perception of sound quality. For example, the words or terminologies to describe power train sound quality include sporty, powerful, exciting, pleasant, responsive, effortless, smooth, reliable, and so on. The scoring is to evaluate sound quality by scoring it. The subjective description and scoring can be plotted in a spider diagram, where it is quick and convenient to see the sound quality features and quantized scorings. Figure 7.20 shows a spider diagram of a power train sound quality, including the subjective description and scoring.

The objective description measures and analyzes the sound quality parameters, including automobile sound characteristic indexes, such as engine order, and psychoacoustic indexes. The automobile sound characteristic indexes include:

- Components of engine firing orders, half orders, and other integer orders
- Characteristics of the orders varying with the engine speed
- Order components of rotary machines
- Linearity of overall sound level varying with engine speed
- High-frequency components of sound at high engine speed
- Resonant frequencies and magnitudes

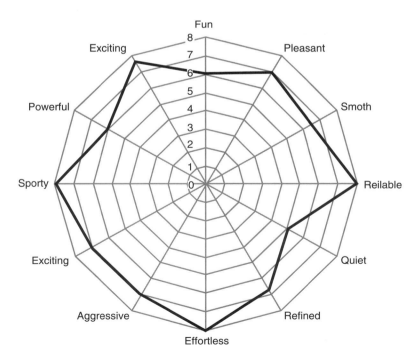

Figure 7.20 A spider diagram of a power train sound quality including subjective description and scoring.

The psychoacoustic indexes include:

- Loudness
- Sharpness
- Fluctuation and roughness
- Tonality
- Masking

The indexes to describe the overall sound quality and system sound qualities are different. It is still a hot and attractive topic for NVH engineers and scientists to explore effective methods to obtain the consistency between the subjective evaluation and the objective testing.

7.3.3   Indexes Used to Describe System Sound Quality

According to the classification of the automotive sound quality, the evaluation can be divided into three categories: power train sound quality evaluation, electrical sound quality evaluation, and door closing sound quality evaluation. The automobile sound characteristic indexes and psychoacoustics indexes are combined to obtain the evaluation indexes for these systems:

1) *Evaluation of power train sound quality*. Objective evaluation indexes of power train sound quality include:

- Combination of order components
- Order components varying with engine speed
- Linearity of overall sound level varying with engine speed
- High-frequency components of sound at high engine speed
- Fluctuation and roughness
- Articulation index

2) *Evaluation of electrical sound quality*. Objective evaluation indexes of electrical sound quality includes:

- Order components of rotary machines
- Loudness
- Sharpness
- Fluctuation and roughness
- Tonality
- Articulation index

3) *Evaluation of door closing sound quality*. Objective evaluation indexes of door closing sound quality include:

- Loudness
- Sharpness
- Ring-down

7.4   Evaluation of Door Closing Sound Quality

7.4.1   Importance of Door Closing Sound Quality

Some systems and components installed on the body can be opened and closed, such as door, engine hood, trunk lid, glove box, and sunroof. Opening and closing the systems or components can be operated either manually, such as a door, or electrically, such as sunroof.

When the systems or components are opened and closed, sound is generated, which leads to the problems of the body closing sound quality. The body closing sound quality includes: opening and closing door sound quality, opening and closing trunk lid sound quality, opening and closing hood sound quality, opening and closing glove box sound quality, and opening and closing sunroof sound quality. Among these systems or components, the door is the most frequently used system, so this chapter only describes the door sound quality. Moreover, the door closing sound quality is more prominent than door opening sound quality, so this chapter only introduces the sound quality of the closing door.

The first contact between the customers and automobiles is sight-based. A vehicle's profile, styling, curve, beauty, painting, etc. will provide the first impression. Beyond that, the first physical contact between the customer and the automobile is usually to open the door. Many people are used to opening the door first, and then closing it, listening to the sound generated by the closing and opening actions. At the moment when they hear the door closing sound, they could have their first instinctive impression on the automobile quality.

The solid, one-impact, and pure sound of closing door provides the customers good impressions on a vehicle, and could stimulate their interests to further know more about the vehicle. On the contrary, multi-impact, sharp, fragmented, metal percussion sound will make them uncomfortable, even left a bad memory. Some picky customers will speculate that the vehicle has design or manufacture problems, even reliable and safety problems.

In summary, the customers can "read" reliability, safety, comfort, luxury, quality, and other information from the door closing sound. Thus, door closing sound quality is very important for creating an impression of automobile quality, and this has become an important branch of NVH.

7.4.2   Subjective Evaluation of Door Closing Sound Quality

7.4.2.1   Subjective Evaluation Methods

Subjective evaluation relates to how customers or a specific group of people express their feelings on vehicle sounds, such as like or dislike, and then provide scores on the sounds. Subjective evaluation can be processed by driving vehicles or by listening to recorded sounds. The jury who participates in the subjective evaluation are usually the following groups:

- *Customers*. A group of customers are organized to evaluate the developing vehicles and benchmarking competitive vehicles; or a group of specific customers, such as young customers, are organized to subjectively evaluate some specific model vehicles. Through these reviews, the automobile companies got the feedback on their products, and then developed the vehicles to satisfy the customers.

• *Product development engineers or professional evaluators.* In each phase of the development processing, they can evaluate the sound quality of the vehicles under development in order continuously to find problems. They identify the sound quality problems in detail, and their evaluation results can be used to guide the design.

• *Company senior management team.* At key milestone phases of a vehicle development, the company senior management team will drive the prototype vehicles and provide their opinions, which will directly influence the direction of the vehicle development, and the sound quality improvement as well.

The door closing sound quality evaluation can be divided into on-site evaluation and sound quality room evaluation. In on-site evaluation, people close and open the door in person. First, they stand outside of the evaluated vehicle, then close and open the door. Second, they sit inside the vehicle and close the door and push to open it. From the sounds of the closing and opening, they give their comments, such as "The sound is solid, one-time impact, low level, comfort," "The sound is scattered, chirp, multi-impact, metal-like echo," and so on. Usually they score the opening and closing sounds and provide their overall impression.

A sound quality room is a special semi-anechoic chamber, as shown in Figure 7.21. Hi-fi sound processing and playback systems are installed in the room. NVH engineers put the on-site recordings into the playback systems. A group of evaluators listen to the recording sound by a public speaker, or each listen to the sound using headphones. The evaluation and scoring are similar to those of in-site evaluation.

Figure 7.21 A sound quality room.

7.4.2.2 Subjective Scoring System

In addition to the impression and description on the evaluated sounds, the evaluators should score the sound quality. Currently, two scoring systems are used in the automobile industry: a 7-point system and a 10-point system. The 10-point system is much more widely used than the 7-point, so the 10-point system is adopted in this book.

The 10-point system means that the human subjective sensation is divided into 10 levels, scoring 1, 2, …, 10, as shown in Table 7.3. Score 1 represents that the sound quality is terrible and completely unacceptable. Score 10 indicates that the sound quality is very good and impeccable, and the sound brings great satisfaction to the customers. The other eight levels are scored between 1 and 10. The higher the score, the more satisfied the evaluators. Usually, scores 1–4 show the sound quality is bad and unacceptable by the vast majority of customers. Scores 5–6 are in a transitional stage, which means that the sounds are acceptable by some customers, and unacceptable by others. The scores above 7 represent that the sound quality is good and accepted by the majority of customers.

For different levels of vehicles, the customer satisfaction on scoring is different. For example, for economy cars, the customers are very satisfied with sound quality with scores of 7 points. However, luxurious car owners are not satisfied with 7 points; instead, what they desire is over 8 points.

7.4.2.3 Description of Subjective Evaluation

Each person has his or her own sensation on the door closing sound, and each person's scoring could be different from others. In order to reach scoring consistency, the evaluators must be trained to fully understand the scoring system. On the other hand, they should describe their comments or impressions on the door closing sound. The scoring from a group of people could have bias, but usually their subjective descriptions are consistent because they should have some common identity on the sound quality.

Nice sound of closing door is low level, one-impact, solid, and clean sound. Bad sound is loud, multi-impact, sharp, fragmented, and metal percussion sound. In addition, the bad sound will linger for a while.

7.4.3 Objective Evaluation of Door Closing Sound Quality

7.4.3.1 Testing Methods

In order to objectively evaluate the door closing sound quality for different vehicles, an objective and repeatable test method must be established that includes two issues. The first one is to objectively test the sound quality, and second one is to find appropriate indexes to evaluate the sound quality.

Table 7.3 Ten-point subjective evaluation system.

Rating scales	1	2	3	4	5	6	7	8	9	10
	Unacceptable				Borderline acceptable			Acceptable		
Conditions noted by	All customers	Average customers			Critical customers			Trained observers		Not perceptible

Figure 7.22 An artificial head used to measure door closing sound.

The objective evaluation should be processed in low background noise environments (such as anechoic chamber, open and quiet place). The signals are tested by the high-precision acoustic artificial head, and then the tested data are analyzed to give the corresponding objective indexes that are used to evaluate the sound quality.

The customers standing outside or sitting inside the vehicles open and close the doors. The door closing sound quality for the both cases should be measured. The interior sound testing is relatively simple because the acoustic artificial heads can be placed on the occupants' seats. However, the outside sound testing is a little more complicated because the locations of the acoustic heads should be defined.

The locations for different customers standing outside a vehicle are different. In order to make sure the tested data are reproducibility and comparability, the locations of the acoustic heads must be determined. The artificial head is placed outside, as shown in Figure 7.22, which is parallel to the door, and the top of the head is 1.7 m from the ground. The handle center of the tested door is used as a reference point, and the horizontal distance between the artificial head nose and the reference point is 0.3 m.

Size, weight, sealing effect, and closing force for different doors are different, so a consistent closing door action must be determined. Usually, the minimum speed to close the doors is regarded as a control standard. The reason to use the minimum speed as the standard is that the customers hope the doors will be completely closed when they slightly push them. A photoelectric sensor is placed at 0.1 m from the door's edge and is used to measure the closing door speed.

7.4.3.2 Objective Evaluation Indexes

Closing door phenomenon is complex, and many indexes can be used to identify the closing sound. Usually four indexes – loudness, sharpness, ring-down and time-frequency spectrum – are used to evaluate the closing door sound quality.

Loudness is an indicator of sound intensity or sound energy, where the sound frequency distribution and the sound masking effect on the human ears are included simultaneously. The greatest contribution of loudness comes from low-frequency components.

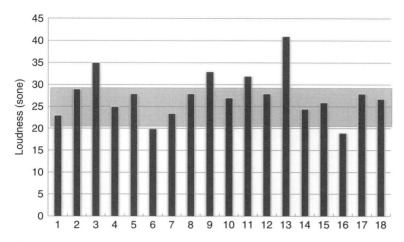

Figure 7.23 Loudness of closing doors of a series of vehicles.

Loudness not only represents the magnitude of closing door sound but also reflects solid sensation of the sound. Figure 7.23 lists loudness of closing doors of a series of vehicles. In general, the lower the loudness, the better the closing door sound. The lower loudness indicates that the closing door sounds light and handy. The loudness between 20 and 25 sones sounds very comfortable. The loudness between 25 to 28 sones sounds comfortable. The loudness between 28 and 30 sones is acceptable, but the loudness larger than 30 sones will make the people uncomfortable.

Sharpness is an indicator representing the high-frequency components in a sound. The higher sharpness represents more high-frequency components in the sound of the closing door, and corresponding sounds sound like scream, chirp, metal percussion, etc. The sharpness reflects the crisp perception of closing door action. The low sharpness sounds are clear and sweet. Figure 7.24 shows sharpness value of closing doors of a group of vehicles. Generally, the sharpness should be between 2 and 2.5 acums. The sharpness higher than 2.5 acums sounds uncomfortable.

Ring-down refers to the lingering sound of a sound generated by collision of two objects. There are many examples of the ring-down phenomena in our daily life; for example, a bell generates loud sound after impacted by a hammer. After the impact, the bell sound will last for a period of time, and then disappear slowly. Similar to impacting the bell, after a door is closed, the sound generated by collision between the door and body and by impact of the internal structure doesn't stop immediately; instead, the sound will last for a while. The "lingering" sound includes multiple impact sounds, chirp, and so on. The high-frequency ring-down sounds like a bell "lingering sound," while low frequency ring-down sounds like "buzzing sound."

The ring-down effect of closing the doors can be represented by curves in time domain, as shown in Figure 7.25. The highest peak in the figure stands for the sound produced by the first collision of the closing action, and after that, the curve continues to oscillate and decay. The magnitudes of the second and later peaks represent the sound attenuation level. The decaying tendency of the oscillated curve and the time the sound disappearance symbolize the ring-down attenuation speed.

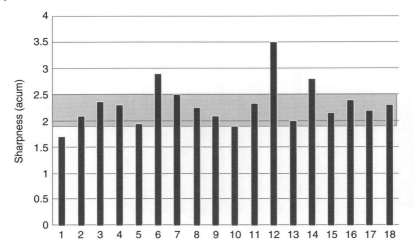

Figure 7.24 Sharpness values of closing doors of a group of vehicles.

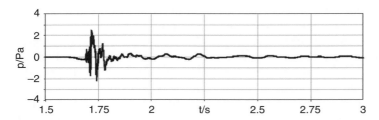

Figure 7.25 A time-domain curve of closing door action.

Time–frequency spectrum refers to a color map representing the varying of a signal with time and frequency. Time–frequency spectrum is obtained by processing a sound signal by wavelet analysis, as shown in Figure 7.26. The horizontal axis represents time and the ordinate is frequency, and the color indicates the sound intensity. From the time axis, the sound attenuation characteristics can be identified, such as number of collisions, collision strength, ring-down, and so on. For example, in Figure 7.26, it can be seen that there are three impacts, impact sound is mainly contributed by the high frequency components, and the maximum sound intensity is between 2000 and 5000 Hz. The sound between 50 and 200 Hz lasts longer.

In the frequency axis, the sound intensity at each frequency and their relation with the structure modal characteristics can be recognized. In general, the low frequency sound around 50 Hz is mainly contributed by the door structure because the door modal frequencies are usually around 50 Hz. The sound between 50 and 200 Hz is mainly contributed by the body panel radiated sound. The middle frequency sound mainly comes from the impacts between the doors and B/C pillars. The high-frequency sound mainly comes from the impact between the latch and the striker, and between the latch and the pawl.

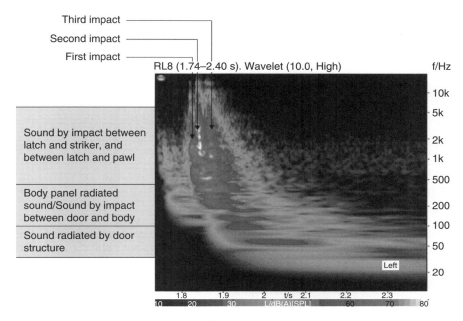

Third impact
Second impact
First impact

RL8 (1.74–2.40 s). Wavelet (10.0, High)   f/Hz

Sound by impact between latch and striker, and between latch and pawl

Body panel radiated sound/Sound by impact between door and body

Sound radiated by door structure

10k
5k
2k
1k
500
200
100
50
20

Left

1.8   1.9   2   t/s   2.1   2.2   2.3
10   20   30   L/dB(A)[SPL]   60   70   80

Figure 7.26 Time–frequency spectrum of a closing door sound.

7.4.4 Relation between Subjective Evaluation and Objective Evaluation

Subjective evaluation is to describe subjective sensations of closing door actions and provide scores. Objective evaluation is to judge the door closing sound quality by analyzing the tested data. The subjective evaluation can be easily and quickly processed and the phenomenon can be directly identified by the evaluators. The objective evaluation can quantitatively provide sound quality characteristics in time domain and frequency domain where the engineers can identify the relation between the sounds and the body structures, which is beneficial to guiding the body design. If the relationship between subjective evaluation and objective testing is established, the door closing sound quality can be comprehensively evaluated, which is helpful for more precisely describing and improving the sound quality.

Through a large number of synchronized sound quality subjective evaluation and objective tests, the two methods can be intrinsically linked. The subjective evaluation and objective analysis can be synchronized in laboratories or in sound quality rooms.

In laboratory testing, after the tested vehicle is inspected, the engineers place the artificial heads and sensors, and then record the data. After processing the tested signals, the loudness and sharpness can be analyzed. Simultaneously, the evaluators participating in the subjective evaluation score the door closing actions and provide corresponding description.

In a sound quality room, the speakers or headphones are used to playback the recorded sounds. Since the sounds are recorded by the artificial acoustic heads, the signals are very fidelity. For nonprofessional participators, they score the overall closing door sound quality from 1–10. For professional evaluators, in addition to provide the overall scorings, they also need to score for loudness, sharpness, and ring-down

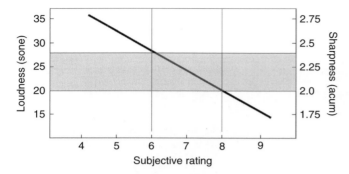

Figure 7.27 Objective tested loudness and sharpness values and subjective scorings.

of each door closing action. After the scores from all evaluators are collected, the statistical results will be obtained by statistical analysis of these data. At the same time, the engineers process the tested data and obtain corresponding loudness, sharpness, ring-down, and color map. After interpretation of these data and analyzing the reasons generating the sounds, the relationship between the sound and the door structure will be identified.

After the subjective evaluation results and objective data are compared and statistically analyzed, they are plotted on a graph where their relation is clearly presented. Figure 7.27 shows that the objective tested loudness and sharpness values and subjective scorings are plotted on a graph. From the figure, it is clearly to see that the loudness values between 20 and 28 sones correspond the score between 8 and 6, and the sharpness values between 2.0 acums to 2.35 acums correspond to scores between 8 and 6.

Similarly, the relationship between the ring-down, multiple collision time and subjective scoring can be plotted on a graph. From the figure, it is also clearly to see the relationship between subjective and objective data.

7.5 Structure and Noise Source of Door Closing System

7.5.1 Structure of Door Closing System

In order to design a door with good closing sound quality, it is necessary to understand the system structure influencing on the closure sound quality. From the perspective of sound quality, the door closing system is divided into three structures: body and door structure, door locker structure, and sealing structure.

The door closing is the impact processing for the door to impact the body frame, so the stiffness and modes of the body frame influence the closing force, impact time, impact sound and lingering sound, and so on.

A door structure includes doorframe, outer panel, internal panel, glass, and accessories (such as a speaker, glass regulator, etc.), as shown in Figure 7.28. Lock, door check and side mirror are installed on the frame. Side-impact beam, reinforcement adhesive, damping layer (mastic sheet) are placed on inner surface of the outer panel. Speaker and glass regulator are mounted on the internal panel. Sound-absorptive material is placed between the outer and inner panels as well.

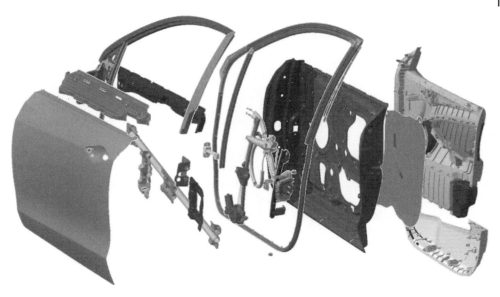

Figure 7.28 A door structure.

(a)   (b)   (c)

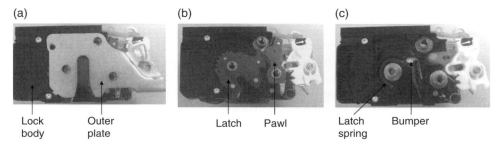

Lock   Outer   Latch   Pawl   Latch   Bumper
body   plate   spring

Figure 7.29 A door lock structure: (a) lock outside structure, (b) lock main structure, (c) lock internal structure.

Figure 7.29a shows a door lock structure, which consists of a lock body and the outer plate. After the outer plate is removed, the lock main structure can be seen clearly, which consists of latch and pawl, as shown in Figure 7.29b. Furthermore, the latch and the pawl are moved away, it is clearly to see the lock internal structure including latch spring, bumper and other parts, as shown in Figure 7.29c. The lock body is mounted on the doorframe, and the striker is installed on the body.

Sealing includes main sealing and secondary sealing between door and body, and belt seal and cutline seal between the doorframe and glass. The detailed sealing materials are introduced in Chapter 6. The sealing influences not only the wind noise but also the closing force and closure sound quality.

All three structures affect the door closing sound quality. Understanding the interaction between these structures and modal parameters, we can deeply analyze the relationship between the sound quality and the structural parameters, and provide a guidance to change the structure design and enhance the sound quality.

7.5.2   Noise Sources of Door Closing

There are four noise sources for the door closing: impact sound between the door and the body, radiated sound by the internal and outer door panels, sound generated by impact inside a lock, and sound induced by collision between glass and door.

7.5.2.1   Impact Sound between Door and Body

The impact between the door and the body includes the impact between the seals, and between the door and an over-slam bumper.

The impact sound is generated at the moment of closing the door. The sudden pressure pulsation is generated at the moment of closing door action, which pushes the air to move through the apertures, inducing the flow noise. When the seals are pressed, an internal cavity is formed and energy is stored. As long as the door is opened, the energy is released and an undesirable "pop-off" sound is generated, which is the major noise source for opening door action.

The purpose for using the over-slam bumper is to evenly distribute the closing door force and to avoid the impact between the body metal and door metal. Of course, noise will be generated by the impact between the bumper and the door or body.

The impact sound of the door and the body is dominated by the low- and middle-frequency components, so it is the major contributor of loudness of the door closing sound.

7.5.2.2   Radiated Sound by Internal and Outer Door Panels

When the door collides with the body, the impact force is transmitted to the internal and outer panels, causing them to vibrate. Since the internal and outer panels are thin-walled plates, they radiate noise when subjected to the excitation.

The mechanism of radiation noise of the thin plates is introduced in Chapter 3. The duration of noise radiation by the panels is relatively longer, which is one of the reasons to generate the ring-down. The radiation noise is dominated by middle-frequency components.

7.5.2.3   Sound Inside Door Lock by Impact

When the door is closed, we usually only hear a one-time impact sound. In fact, three impacts happen between the latch and the striker inside the lock. The first impact happens between the latch and the striker. The pawl pushes the latch to rotate to a semilocked position, then the latch impacts with the pawl. The third impact happens between the latch and pawl again when the latch is rotated to the fully locked position. The whole impact processing between the latch and the striker can be tested, as shown in Figure 7.30. The door lock is installed in a specially designed lock impact device, and an artificial acoustic head is used to record the impact sound.

Figure 7.31 shows a tested sound in time domain of the impact between the latch and the striker. From the curve, it is clear to see the three impact peaks and corresponding attenuations. Due to the very short time interval between the impacts, usually what people hear is a one-time impact sound. However, if the internal lock structure is poorly designed or manufactured, the interval between impacts becomes longer, and several impact sounds will be heard.

The impact between the latch and the striker could happen between the metals or between hard rubbers. The impact sound includes both low-frequency and high-frequency components. Compared with the lower frequency sound generated between

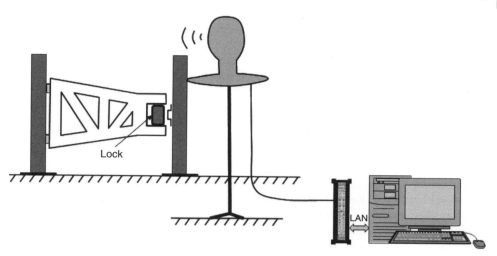

Figure 7.30 Test setup of lock impact.

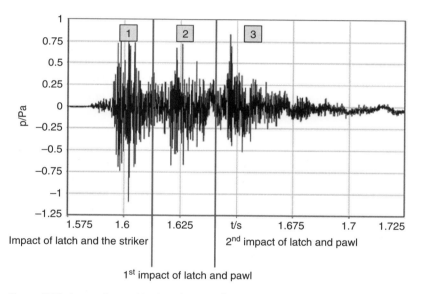

Figure 7.31 A tested sound in time domain of impact between latch and striker.

the door and the body impact, the sound level induced by the impact between the latch and the striker is much lower, so its contribution to the loudness is too low to negligible. But, its major contribution is the high-frequency sound. In addition, this impact will bring the ring-down sound, which will last a long time.

7.5.2.4 Sound Induced by Impact between Glass and Door

Window glass running inside the cutline seals and glass run seals could shake and generate sound. The sound pressure is low but its frequency is high, and it makes a fragmented sound, which mainly affects the ring-down and the sharpness.

7.6 Control of Door Closing Sound Quality

From the viewpoint of the structural design, only good component design and system design can guarantee good sound quality. Control of the structures, including body and door, locks, and seals will be described as follows.

7.6.1 Control of Door Panel Structure

7.6.1.1 Overall Stiffness of the Body and the Door

Body stiffness influences not only the overall vehicle NVH performance but also the door closing sound quality, and squeak and rattle. Closing door action is a process of the door impacting the body, so the body must have sufficient stiffness, especially the stiffness of the pillars and frames. Please refer to Chapter 2 and Chapter 8 regarding body frame stiffness.

The door overall stiffness is one of the most important parameters influencing the door closing sound quality. If the door stiffness is insufficient, the impact between the door and the body cannot achieve one-time solid impact sound; instead, the sound is multi-impact and sloppy.

The door overall stiffness is determined by doorframe, panels, and accessories. The frame is the foundation. Insufficient frame stiffness could lead to the door distortion at moment of closing action, and even influence the air leakage. The upper frame, which is supported only by the glass, is the weakest part on the door. Insufficient upper frame stiffness could induce obvious deformation of the door and generate fragmented and chirp sound.

7.6.1.2 Control of Outer Panel

The door outer panel is a thin metal plate, which is similar to a piece of paper or a drumhead. Sounds are easily generated by gently tapping the drumhead, or slightly waving the paper. Similarly, the outer panel is excited and radiates sound when it is subjected to the door closing force. The vibration characteristics of the outer panel can be tested by a laser measurement system or calculated using the finite element method. After the vibration response of the outer panel is obtained, the sound radiation can be predicted by Green's function.

Suppressing the panel vibration can reduce its sound radiation. The methods to control its vibration include increasing its stiffness and damping and adding additional mass on it. The outer panel shapes are determined by the design styling, so increasing its stiffness cannot be achieved by punching ribs on the surfaces; instead, the stiffness is increased by adding side-impact beams or reinforcement adhesives on the inside surface of the outer panel.

There are two ways to increase the stiffness of the outer panels: supporting beam and reinforcement adhesive. The first method is to glue one or several beams on the inner surface of the outer panels. These beams have two functions; one is to protect collision and the other one is to increase the panel stiffness. The beam is generally called a side-impact beam, as shown in Figure 7.32a. The second method is to place the reinforcement adhesive on the inner side of the panel. The reinforcement adhesive is a flexible sheet. After being baked at a high temperature, the sheet becomes very hard, like a steel plate, as shown in Figure 7.32a.

(a)

(b)

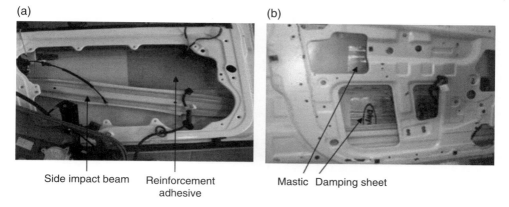

Side impact beam   Reinforcement   Mastic Damping sheet
                       adhesive

Figure 7.32 (a) Side-impact beam, reinforcement adhesive, and (b) damping sheet used on inner side of outer door panel.

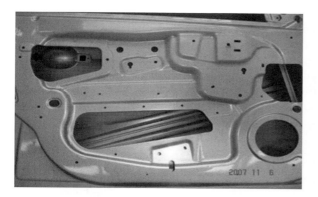

Figure 7.33 Complicated reinforced structure of an inner panel.

The increasing damping method is to place viscoelastic damping material on the inside surface of the outer panel in order to reduce the vibration response. Now, the widely used damping sheet is the constrained mastic damping material, as shown in Figure 7.32b.

7.6.1.3 Control of Internal Panel

The internal panel is a thin-walled metal plate, and lock, glass regulator, and speaker are installed on it. These concentrated masses reduce the modal frequency of the inner panel, so the panel must be reinforced for it to have enough high modal frequencies. The local structures should be stiff enough, especially the locations to install door locks, glass regulator, and speaker; otherwise, these locations could have big low-frequency vibrations, inducing booming noise.

Design of the inner panel is not limited by body design styling, so it can be designed as a complicated reinforced structure to improve its stiffness, as shown in Figure 7.33. Its stiffness can also be improved by using supporting beams.

Figure 7.34 Over-slam bumper on the door.

Over-slam bumper

7.6.1.4 Design of Over-Slam Bumper

The over-slam bumper is mounted on the door, as shown in Figure 7.34. When a door is slightly closed, generally, the door is not in contact with the bumper. Only when the movement of the door exceeds the normal seal compression do the door and the over-slam bumper affect each other. The bumper should have high loss factor and low stiffness in order to achieve the optimal touching between the door and the body and reduce the door's vibration.

7.6.2 Control of Door Lock

7.6.2.1 Control of Door Lock Body

During the closing door action, multi-impacts happen inside the lock, and corresponding forces are generated, including: impact force between the lock sleeve and the door, impact force between latch and striker, impact force between latch and pawl, impact force between latch and bumper, impact force between pawl and bumper. Control of these impact forces is the key to reducing the impact sound and enhance the sound quality.

The locker sleeve must be securely fastened with the door. The sleeve is recommended to be wrapped up by soft materials (such as rubber, foam etc.) in order to reduce the impact force. The soft materials also have function of noise insulation. The locker sleeve should be avoided to open apertures as little as possible in order to prevent the internal noise from transmitting to the outside.

The latch is a rotational part. The factors affecting its impact forces include: the moment of inertia, angular acceleration, and material. The latch impact force can be decreased by reducing its moment of inertia and angular acceleration, and by using damping material. The factors influencing the latch's moment of inertia are its mass, material density, and distance between its mass center and the rotation center. The latch's acceleration depends on the spring stiffness. In order to reduce the latch impact force, non-metal damping material can be used to make the latch, or the metal latch can be wrapped up by the damping materials. If the above-mentioned factors are controlled,

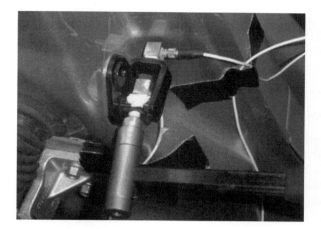

Figure 7.35 IPI testing on a striker.

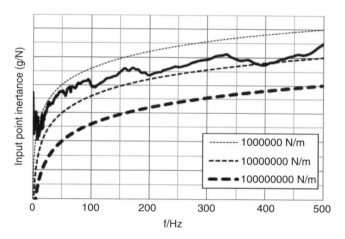

Figure 7.36 A tested IPI curve of a striker.

the impacts between the latch and the striker, between the latch and the pawl, and between the latch and the bumper can be controlled, so the corresponding generating sound is reduced.

Similar to the latch, the pawl is also a rotational part. The factors affecting its impact forces also are the moment of inertia, angular acceleration, and material. The control methods are similar to those for the latch control.

The striker is mounted on the door, and it is impacted by the latch, so the striker influences the impact sound. If a damping material is wrapped up on the striker, the impact force will be significantly reduced.

7.6.2.2 IPI of the Striker

Since the striker mounted on the body is subjected to the force by the latch, so the input point inertance (IPI) of the impacted point on the striker must be high enough. The importance of the IPI is the same as a vehicle body IPI.

An accelerometer is placed on the striker, and then a hammer strikes the same point, as shown in Figure 7.35, so the IPI will be obtained by processing the tested acceleration and force. Figure 7.36 shows a tested IPI curve of a striker, which characteristics are similar to those of body IPI.

7.6.3 Control of Sealing System

Seals are very important for both static sealing and dynamic sealing, and they are also very important for the door closing sound quality. From the perspective of door closing sound quality, the seals have following functions. The first one is to avoid direct metal impact between the door and the body, reducing the noise when the door is slammed. The second one is to make the closing force distributed as evenly as possible throughout the entire sealing contact surface. The third one is to attenuate the high-frequency noise generated by impact between the latch and the striker and internal collisions.

Some of the energy generated when a door is slammed is wasted and some energy is stored. The wasted energy is converted to heat energy or sound energy. The slamming door energy is stored when the seals are compressed, which is similar to a spring. The compressed over-slam bumper also stores the energy. When the door is opened, these "stored" energies are released, generating the "pop-off" sound.

The most important parameters for the seal design are compression load deformation (CLD), deformation uniformity on the entire compressed seal, and stored energy. If the CLD value remains constant, that is, the closing action is not affected by the compression deflection, the good sensation of closing door action and sound quality can be achieved. Uniformly distributed forces will generate the same sound at each part of the seal, so the superimposed sound sounds well. Small apertures on the seal help to reduce the stored energy, which is beneficial for reducing the "pop" sound when the door is opened.

7.7 Design Procedure and Example Analysis for Door Closing Sound Quality

7.7.1 Design Procedure for Door Closing Sound Quality

The design of door closing sound quality is based on customer demands and brand positioning. Depending on market positioning of the product and brand strategy, positioning of the closing door sound quality should be determined. According to market positioning, the market demands are converted to engineering targets in engineering development departments, and the targets are cascaded to every system which will is used to guide designs. When the design is completed, the prototype vehicles are produced, and subjective evaluation and objective tests can be processed. If the design goals are met, the development processing of the door closing sound quality completes, otherwise, the design must be modified and the prototypes should be re-evaluated, until the targets are reached. Figure 7.37 shows this procedure.

The detailed works for each phase of the procedure are introduced as follows.

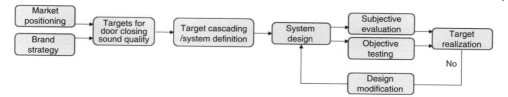

Figure 7.37 Design procedure of door closing sound quality.

7.7.1.1 Market Positioning and Brand Strategy

As already noted, the first physical contact most customers have with a vehicle is to open and close the doors. The sounds of slamming doors usually leaves a first impression, and customers judge the vehicle as good or bad according to these sounds. The vigorous, strong, and short time impact sound is good, while sharp, fragmented, and multi-impact sound is bad.

The closing door sound quality of different levels of vehicles varies a lot. The doors of luxury cars make a solid, clear, and one-time-impact sound, which delivers the customers a sensation of comfort and luxury. However, for many economy cars, in addition to the first loud impact sound of slamming the door, there could exist multiple low-level-impact sounds, even ring-down, connected to the sensation associated with the cheap and affordable vehicles. The sound quality is related to the designs of the body, the doors and the seals, and cost. Thus, market positioning is the basis to analyze and design the door closing system and to set up the sound quality targets. Similarly, brand positioning also affects sound quality development.

7.7.1.2 Target of Door Closing Sound Quality

The benchmarking vehicles can be chosen after the market positioning of the developed vehicle is determined. After studying the door closing features and sound quality of the benchmarking vehicles, and combined with the brand positioning, the targets of the door closing sound quality can be determined, including subjective scoring, loudness, sharpness, ring-down decay time, the time interval between multiple collisions, and so on.

7.7.1.3 Target Cascading and System Definition

After the targets of door closing sound quality are determined, they can be cascaded to each subsystem, forming the subsystem targets that include:

- *The door targets.* Torsional stiffness and bending stiffness of the doorframe, the outer panel stiffness, the inner panel stiffness, accessory support (such as speaker stand) stiffness
- *The body targets.* IPI of the connecting points between the body and the door, the deformation of the body closure opening, IPI of the striker.
- *The lock targets.* Vibration isolation level of the lock sleeve wrapper, the impact interval time between latch and pawl and between latch and striker, and corresponding sound pressure levels.
- *The seal targets.* Seal CLD value, opening area.

7.7.1.4   System Design

Subsystem targets are the indicators to guide the system design. Only when the subsystem targets are achieved, can the overall targets of the door closing sound quality be reached. For example, if the stiffness of a door outer panel does not meet its target, the stiffness can be increased by adding side-impact beam and/or reinforcement adhesive.

7.7.1.5   Subjective Evaluation/Objective Test/Achievement of Targets

The subjective evaluation and objective testing are used to check whether the design goal of door closing sound quality is achieved. If the goal is not achieved, the reasons must be analyzed and the design must be modified. For example, in the color map of Figure 7.26, the reason to cause the long ring-down time is the low stiffness of the outer door panel. After a side-impact beam is added, the panel stiffness increases, and the ring-down time is shortened, so the goal is achieved.

7.7.2   Analysis of Factors Influencing on Loudness, Sharpness, and Ring-Down

Loudness, sharpness, and ring-down are three indexes to objectively evaluate door closing sound quality. Among the three indicators, the sharpness is the most important, because it makes the customers the most uncomfortable. The ring-down also gives the customers a perception of fragmented door structure. The least important one is the loudness. Compared with the sharpness and ring-down, the loudness on people's sensation has much less impact. If the sharpness and ring-down are well controlled, and even the loudness is high, the customers usually do not complain the noise.

The design of the door structure is influenced by the three indexes. The relationship between the indexes and the door structure will be briefly described in the this section.

7.7.2.1   Sharpness and Door Structure

The first task to control the door closing sound quality is to reduce the sharpness. The moving components influencing the sharpness include:

- Impact components inside the lock
- Door seal and body seal
- The body and over-slam bumper
- The accessories attached on the door, such as the glass regulator, speaker, inner panel
- The door outer panel

Among the structures, the lock is the largest contributor to the sharpness. The multi-impacts among the latch, striker, and pawl generate board band noise, especially, the impact between the latch and striker induces very high sharpness.

In order to reduce the sharpness, it is extremely important to reduce the magnitude and rising slope of the impulse force of the slamming door. The impulse force is determined by the moment of inertia of the latch and the spring force, so it is important to control the latch's moment of inertia, angular acceleration, and spring stiffness. The impulse force can also be reduced by using the damped latch, and/or wrapping the striker with shock-absorbing material.

7.7.2.2 Ring-Down and Door Structure

The ring-down is induced by impact between structures, lingering constantly and then decaying gradually. The factors influencing the ring-down include: the stiffness of the outer panel, damping processing on the door structure, stiffness of inner panels and trimmed parts.

The modes of metal panels significantly influence the ring-down. When the thin outer panel is excited, the radiated sound will last a long time. By adding an anti-impact beam or reinforced beam on the outer panel, the panel stiffness will be increased and its ring-down time will be reduced as well. The ring-down time can also be shortened by pasting damping material on the outer panel.

In summary, the ring-down can be reduced by adding reinforced structures and/or pasting damping material on the panels, increasing the stiffness at the attached points, properly arranging the trimmed parts, and reducing the slamming force caused by excessive sealing compression.

7.7.2.3 Loudness and Door Structure

The loudness magnitude is determined by magnitude of slamming force and the door structure stiffness. In general, the higher the slamming force, the louder the loudness. Factors affecting the slamming force include the sealing system and the body ventilation system. The seals contact each other at the moment of slamming door, so the force is determined by the seal compression deflection and ventilated amounts. Smooth air ventilation will aid in achieving lower loudness.

The factors related to the door structure stiffness influencing the closing door loudness include door geometry, stiffness of the door inner, and outer panels, IPI at the striker, stiffness of the over-slam bumper, and so on.

7.7.3 Example Analysis of Door Closing Sound Quality

An example is used to illustrate the characteristics of the door closing sound quality and improvement methods. Figure 7.38a and Figure 7.39a show a color map and time-domain curve of a door closing sound quality. In the color map, the sound can be divided into three sections in the frequency range. The first section is the range of 30–50 Hz. After the first impact happens, intensity of the lingering sound is still high, and the door's modes are usually within the frequency range, so the ring-down is induced by the door's vibration.

The second section is the range of 100–200 Hz. After the first impact, the strong intensity sound lasts a short period of time. The frequency of radiation sound of the door panels is usually within the range, so it can be inferred that the sound is generated by the door panels.

The third section is in the range of 400–2000 Hz and the SPL is very high. Observed in the time domain, there exist three impact peaks. The sound intensities for the three impacts are moderate strength, high strength, and moderate strength, respectively. The time intervals for the impacts are very short, so it can be inferred that the impacts happen inside the lock.

From the time-domain curve in Figure 7.39a, the three peaks indicate three impacts. In addition to the peaks, the fluctuated curve represents low-frequency sound wave.

(a)   (b)

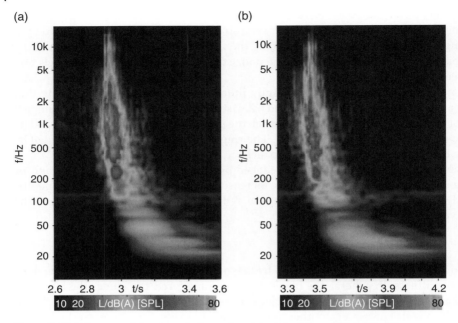

Figure 7.38 Color map of the door closing action: (a) original door; (b) modified door.

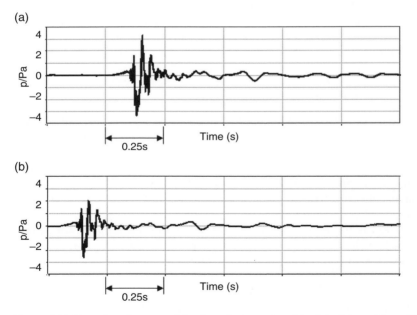

Figure 7.39 Time domain curve of the door closing action; (a) original door; (b) modified door.

Based on these characteristics of the door closing sound quality, the door system is modified in order to improve the sound quality. First, the lock is replaced. The latch is wrapped with damping material in order to reduce the impact sound. The lock body is wrapped with sound insulation material to further reduce the sound transmission from the inside to outside. Second, the reinforced adhesive is pasted on the inner side of the outer panel in order to increase the panel stiffness and reduce its sound radiation.

Figure 7.38b and Figure 7.39b are the color map and the time domain curve of the door closing sound quality of the modified door, respectively. The sound qualities of the original door and the modified door are compared from the three frequency sections. In the range of 30–50 Hz, there is almost no change. The doorframe is not modified, so its modes remain the same and the corresponding sound quality does not change. In the range of 100–200 Hz, the sound intensity is reduced, which results from the reinforced door panel. In the frequency range of 400–2000 Hz, sound intensity is greatly reduced due to the replaced lock.

7.8 Sound Quality for Other Body Components

In addition to the door closing sound quality, there are many other sound quality issues related the body:

- Door open sound quality
- Truck lid open and closure sound quality
- Locking and unlocking sound quality
- Sunroof open and closure sound quality
- Mirror rotation sound quality
- Adjustable seat moving sound quality
- Wiper moving sound quality
- Horn sound quality
- Turn signal sound quality
- Safety belt warning sound quality

The listed sound quality can be divided into three categories: mechanical impact sound quality, electrical motor sound quality, and sound quality for warning sound generator.

Similar to the door closing sound quality, mechanical impact sound quality refers to the sound generated by impact of two mechanical structures, such as opening and closing the trunk lid.

The electrical motor sound quality refers to the sound generated by a motor to move components such as the sunroof, mirror, and seat. For example, sound will be generated when a mirror is driven by a motor to rotate. This will create either a good or bad perception of the sound quality.

The sound quality of warning sounds refers to the warning or reminding sound generated by a buzzer or a loudspeaker, such as turning signal sound, safety belt warning sound, etc. The sound is usually a tone or combination of several tones. The tones with different magnitudes and frequencies will bring the customers different sensations, such as annoying or sweet sound.

The mechanism of the mechanical impact sound quality is similar to that of the door closing sound quality, so the book will not describe these issues again. In addition, this book does not cover the electrical motor sound quality and the sound quality for warning sound generators, so if readers are interested in these topics, please refer to related references.

Bibliography

AVL. (2010). Sound Quality Training. November.

Beniwal, R. and Wu, S.F. (2007). System Level Noise Source Identification and Diagnostics on a Vehicle Door Module. SAE Paper; 2007-01-2280.

Beranek, L.L. (1996). *Acoustics*. Acoustical Society of America.

Bhangale, R. and Mansinh, K.S. (2011). Investing Factors Affecting Door Slam Noise of SUV and Improved Performance by DFSS Approach. SAE Paper; 2011-01-1595.

B&K. (2008). Psychoacoustics –a qualitative description.

Blommer, M., Yang, B., and Vandenbrink, K. (2005). Detecting and Classifying Secondary Impacts in Door Closing Sound. SAE Paper; 2005-01-2471.

Blommer, M., Amman, S., Gu, P. et al. (2003). Sound Quality Aspects of Impact Harshness for Light Trucks and SUVs. SAE Paper; 2003-01-1501.

Brandl, F., Biermayer, W., and Thomann, S. (2001). Efficient Passenger Car Sound Engineering Using New Development Tools. Styrian Noise, Vibration & Harshness Congress; 22.-23.10.2001, Graz, Austria.

Brandl, F.K. and Biermayer, W. (1999). A New Tool for the Onboard Objective Assessment of Vehicle Interior Noise Quality. SAE Noise and Vibration Conference; May, Traverse City, Michigan.

Bray, W.R. (2009). Bi Measurements in an Information Technology Acoustic Program. Inter-Noise 2009; Ottawa, Canada, August 23–26.

Bray, W.R., Blommer, M., and Lake, S. (2003). Sound Quality Workshop. SAE Noise and Vibration Conference and Exhibition; Traverse City, Michigan.

Cerrato, G. (2009). Automotive Sound Quality – Powertrain, Road and Wind Noise. Sound and Vibration April: 16–24.

Cerrato-Jay, C. (2007). Sound/Vibration Quality Engineering, Part 1 – Introduction and the SVQ Engineering Process. Sound and Vibration; April, pp. 2–9.

Clapper, M. and Blommer, M. (2003). Masking Perception Analysis Software (MPAS) for Tonal Level Setting in Powertrain NVH. SAE Paper; 2003-01-1500.

Crocker, M.J. (2007). *Handbook of Noise and Vibration Control*. Wiley.

Everest, A.F. (2001). *The Master Handbook of Acoustics*. McGraw-Hill.

Fastl, H. and Zwicker, E. (2007). *Psychoacoustics: Facts and Models*. Berlin, Heidelberg: Springer-Verlag.

Gallo, M., Anthonis, J., van der Auweraer, H., et al. (2010). Evaluation of an Active Sound Quality Control System in a Virtual Car Driving Simulator. Inter-Noise 2010; Lisbon, Portugal, June 13–16.

Genuit, K. (2008). Product Sound Quality of Vehicle Noise – A Permanent Challenge for NVH Measurement Technologies. SAE Paper; 2008-36-0517.

Head Acoustics. (2009). Objective Classification of Subjective Perceived Sound Quality.

Howard, D.M. and Angus, J.A.S. (2009). *Acoustics and Psychoacoustics*. Focal Press.

Hu, X.L. (2009). Sound Quality – Some basic but important concepts and practical notes. International Conference on Automotive NVH Control Technology; 10–12 November, Chongqing, China.

Kang, K.T. and Byun, U.S. (2010). Sound quality development for passenger vehicle. Inter-Noise 2010; Lisbon, Portugal, June 13–16.

Kumbhar, M., Chavan, A., Nair, S. et al. (2008). Investigations of Factors Affecting Door Slam Noise Quality in SUV. Inter-Noise 2008; Shanghai, China, October 26–29.

Kumbhar, M., Edathute, S., and Chavan, A. (2003). Investigation of Factors Influencing Vehicle Audio Speaker Locations for Better Sound Quality and Spread. SAE Paper; 2007-01-2318.

Lee, H., Kwon, O.J., and Lee, J. (2007). Modeling of Door Slam Noise Index by using Sound Quality Metric. SAE Paper; 2007-01-2394.

Lee, S.-K., Chae, H.C., and Park, D.C. et al. (2003). Booming Index Development for Sound Quality Evaluation of a Passenger Car. SAE Paper; 2003-01-1497.

Lee, S.-K., Kim, B.S., Chae, H.C. et al. (2005). Sound Quality Analysis of a Passenger Car Based on Rumbling Index. SAE Paper; 2005-01-2481.

Musser, C.T. and Young, S. (2005). Application of Transient SEA for Vehicle Door closing sound quality. SAE Paper; 2005-01-2433.

Narayana, N. (2011). A Finite Element Method for Effective Reduction of Speaker-Borne Squeak and Rattle Noise in Automotive Doors. SAE Paper; 2011-01-1583.

Noumura, K. and Yoshida, J. (2003). Perception Modeling and Quantification of Sound Quality in Cabin. SAE Paper; 2003-01-1514.

Pang, J., Sheng, G., and He, H. (2006). *Automotive Noise and Vibration- Principles and Application*. Beijing Institute of Technology Press.

Petniunas, A., Otto, N.C., and Amman, S. et al. (1999). Door System Design for Improved Closure Sound Quality. SAE Paper; 1999-01-1681.

Pohlmann, K.C. (2013). *Principles of Digital Audio*. McGraw Hill.

Saha, P. and Roman, P.J. (2003). Polce III C.T. A Design Study to Determine the Impact of Various Parameters on Door Acoustics. SAE Paper; 2003-01-1430.

Sarrazin, M. and Janssens, K. (2012). Auwereaer H.V.D. Virtual Car Sound Synthesis Technique for Brand Sound Design of Hybrid and Electric Vehicles. SAE Paper; 2012-36-0614.

Scholl, D. and Yang, B. (2003). Wavelet-Based Visualization, Separation, and Synthesis Tools for Sound Quality of Impulsive Noises. SAE Paper; 2003-01-1527.

Schulte-Fortkamp, B. and Genuit, K. (2005). Exploration of Associated Imaginations on Sound Perception A Subject-centered Method for Benchmarking of Vehicle. SAE Paper; 2005-01-2263.

Sontacchi, A., Holdrich, R. and Girstmair, J. et al. (2012). Predicted Roughness Perception for Simulated Vehicle Interior Noise. SAE Paper; 2012-01-1561.

Terazawa, N. and Wakita, T. (2004). A New Method of Engine Sound Design for Car Interior Noise Using a Psychoacoustic Index. SAE Paper; 2004-01-0406.

Tousignant, T., Govindswamy, K., and Leibling, C. (2011). Evaluation of Source and Path Contributions to Sound Quality Using Vehicle Interior Noise Simulation. SAE Paper; 2011-01-1685.

TSTech. (2010). Vehicle Sound Quality.

Vergara, E.F. (2004). Sound Quality Tools in the Design Process of Electro-Hydraulic Steering System. SAE Paper; 2004-01-3275.

Ver, I.L. and Beranek, L.L. (2006). *Noise and Vibration Control Engineering: Principles and Applications*. Wiley.

Zeitler, A. (2012). Psychoacoustic Requirements for Warning Sounds of Quiet. SAE Paper; 2012-01-1522.

Zhang, L. and Champagne, A. (2003). Toward an Objective Understanding of Perceived Glovebox Closure Sound Quality. SAE Paper; 2003-01-1499.

8

Squeak and Rattle Control in Vehicle Body

8.1 Introduction

8.1.1 What Is Squeak and Rattle?

From perspective of noise existing period, the automotive noise can be divided into two categories: continuous noise and transient noise. The continuous noise refers to the noise lasting relative long time or existing all the time, such as engine noise, wind noise. The transient noise refers to the noise coming and then disappearing in a relative short period. The transient noise is an abnormal sound, which should be avoided during vehicle operation.

This chapter focuses on the transient noise of squeak and rattle (S&R). Squeak is generated by the friction between two objects. Rattle is induced by the impact between two objects. In some references, buzz, another transient noise, is included in the scope of squeak and rattle, and they are called BSR.

Buzz is the sound generated by the structural resonance, which is similar to the radiation sound induced by a panel structural resonance. The S&R refers to the sound generated by action between two adjacent parts, while buzz refers to the sound generated from a single part excited by external excitation. Therefore, for most cases, buzz can be separated from the S&R.

Both the S&R and continuous noise are unexpected sounds, but they have differences. In general, the continuous noise is the regular, relatively long duration sound. For example, vehicle occupants can hear regular engine noise during the whole driving period. The engine noise has its own characteristics such as orders, speed, and resonance, which can be analyzed by mature theories and methods.

However, the S&R does not have specialized regularities. For example, some S&R lasts for a very short duration, such as only 10 ms; some S&R happens suddenly, only once or several times; some S&R intermittently appears and then disappears, and after a while might appear again. The S&R occurs randomly, and its repeatability is uncertainty. It is extremely hard to find good methods to process the transient random signals.

Rattle is induced by the impact of two objects and corresponding frequencies are relatively low, while the squeak is generated by the friction between two objects and corresponding frequencies depend on their materials, surface pressure, and other factors.

Noise and Vibration Control in Automotive Bodies, First Edition. Jian Pang.
© 2019 China Machine Press. All rights reserved. Published 2019 by John Wiley & Sons Ltd.

8.1.2 Components Generating Squeak and Rattle

Most S&R problems are induced by the road excitation. The squeak and rattle could be generated in many components of a vehicle, especially driving on a concrete road, rough road, brick road, etc. Today, the automobiles become more and more complex, and more and more accessories are installed, which increases the possibility to generate the squeak and rattle. The components that could induce the squeak and rattle include instrument panel (IP), seats, glove box, electrical modules, doors, body, seat belt retractor, airbags, HVAC, seals, and so on. According to the statistical data, the major components where the squeak and rattle occur are the following areas:

1) Instrument panel
2) Steering wheel/column/supporting system
3) Seats
4) Body closure parts
5) Underbody

8.1.3 Importance of Squeak and Rattle

Customers do not want to hear squeak and rattle. They dislike and complain about the noise; sometimes they even doubt the quality of a vehicle due to the existence of S&R.

Traditionally, the engine noise, road noise, and wind noise are relatively high, which mask some low level squeak and rattle. With development of technology and control methods of automotive noise and vibration, however, the traditional sources and transmission paths of noise and vibration have been well suppressed, which makes low-level squeaks and rattles become much more prominent. In addition, automotive lightweight tendency could also induce the squeak and rattle. Today, the squeak and rattle control is a big challenge for automotive noise and vibration control.

The squeak and rattle for high mileage vehicles have great influence on their brands. For some new vehicles, there are no squeaks and rattles, but after running tens of thousands of kilometers, the squeak and rattle can be perceived. Thus, the customers could complain or even doubt the vehicles' quality and reliability. If this phenomenon becomes well known, the sales of that model could be seriously influenced. For a good-quality vehicle, even after it runs for a 100,000 kilometers, the possibility for occurring squeaks and rattles should be no greater than from a new vehicle.

S&R is one of the most important factors in creating a negative impression of product quality. In the J.D. Power survey, squeak and rattle noises were commonly noted, making S&R an important factor affecting the J.D. Power rankings of vehicle brands. Vehicles with lots of S&R problems are gradually losing their market competitiveness. In addition, customers send their vehicles having the S&R problems to be repaired in 4S (sales, spare parts, service, and survey, i.e. feedback) shops, which greatly increases the quality warranty costs after market.

8.1.4 Mechanism of Squeak and Rattle

Relative motion does not necessarily generate the squeak and rattle, but the squeak and rattle are absolutely induced by the relative motion. The dynamic characteristics and acoustic mechanism of the squeak and rattle are very complex, because it is a highly nonlinear physical phenomenon. Factors influencing the squeak and rattle include material and structural features. The material features include its friction, impact, temperature, and humidity, etc. The structural features include the structural dynamic characteristics, dynamic stiffness of the contact point, the characteristics of the impact force or frictional force, etc. Therefore, S&R is a very complex nonlinear dynamic problem. It is very difficult to establish mathematical models for the squeak and rattle, and it is almost impossible to get an analytical solution. The actual state of the squeak and rattle is much more complex than the mathematical models, and many factors are involved; therefore, even if the models can be solved, the results are far away from the actual situation.

The reasons to induce the squeak and rattle can be summarized as: insufficient structural stiffness, incompatibility of the material friction pair, and inappropriateness of structural design or manufacture. Insufficient stiffness of the structures makes components easily deformed during the vehicle running, which causes the impact of the adjacent components, so the rattle happens. The friction pair is incompatible for two contacted surfaces, and their frictional coefficients change with environmental temperature and humidity, generating the squeak. Improper structural design, such as a gap between the two parts, improper manufacturing error, etc., will also generate the squeak and rattle.

8.1.5   Identification and Control of Squeak and Rattle

Historically, identification and control of the squeak and rattle experience three stages. In the first stage, the S&R problems are identified after vehicle sales, and then solved according to the customers' requirement. In the second stage, targets for the S&R are set up during the vehicle development process, and the problems are identified and solved by testing a large number of prototype vehicles. In the third stage, the S&R control is moved to the early stage of vehicle development and controlled by CAE and other analytical methods.

The first stage, the identification and control of the S&R, is a "discover problem – solve the problem" process. After the vehicle sales, the problems are found by the market surveys or the customers' feedback. Then, the engineers analyze the problems, identify the problems' exact locations, and provide the solutions. The engineers drive the vehicle on different roads, such as rough road, cobblestone road, brick road, etc. or put it on a four-poster simulator. They subjectively evaluate the problems or test the data, and then provide the solutions after finding the reasons.

The second stage, the S&R identification and control are processed, on the prototype vehicles. The S&R targets are set up during the vehicle development process. Based on the benchmarking results of competitive vehicles in the market, the targets are set up. Once the prototype vehicles are ready, they are put on the four-poster simulator or driven on the road to be tested. After the S&R problems are identified and solved, the results are fed back to the design team. The design engineers modify the design based on the feedback and then provide a new design. After the new prototype vehicles are assembled, they are tested again. After several rounds of testing and design, it is possible to effectively control the squeak and rattle in the design stage; however, the disadvantage is that the testing costs are very high.

In the third stage, squeak and rattle control is moved up to early design stage. In recent years, with development of computer-aided engineering (CAE) technology, the control in the early design stage is possible by CAE analysis and other analytical methods, such as using finite element method to analyze the modal shape, deformation, and sensitivity. In addition, the factors affecting the squeak and rattle, such as materials, friction coefficient of the friction pair, and gap between adjacent parts can be analyzed in the early stage, which provides the data for squeak and rattle control. If the analysis in the early stage is well done, the possibility of the squeak and rattle occurring for the prototype vehicles will be significantly reduced compared with the traditional methods. The early analysis brings two benefits. One is that the control is moved to the early stage of the vehicle development process and the other one is that the cost is greatly reduced.

8.2 Mechanism and Influence Factors of Squeak

The squeak is generated by the friction between two objects, and its mechanism is very complex. For two objects in contact with each other and subjected to external forces, the friction force is induced on the contact surfaces and relative motion between them could happen. If the friction coefficients of the two surfaces are not compatible, a squeak will be generated.

8.2.1 Mechanism of Squeak

Figure 8.1a shows that object A is placed on the top of object B, and object A is subjected to a normal force (F_N) and a horizontal force (F). The friction force exists on the contacted surfaces of the objects, as shown in Figure 8.1b.

The friction force F_f is be expressed as,

$$F_f = \mu F_N \tag{8.1}$$

Where, μ is friction coefficient. When object A and B are relatively standstill, μ is static friction coefficient, and noted as μ_s; when they have relative motion, μ is dynamic friction coefficient, and noted as μ_d. The friction force can be expressed as

$$\begin{aligned} F_d &= \mu_d F_N, \quad \dot{x} > 0 \\ F_s &= \mu_s F_N, \quad \dot{x} = 0 \end{aligned} \tag{8.2}$$

where, x and $\dot{x}$ are the displacement and velocity of object A, respectively.

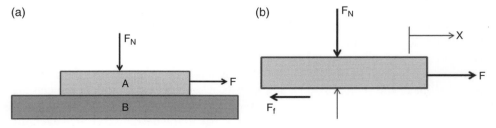

Figure 8.1 (a) Two contacted objects subjected to a normal force and horizontal force; (b) friction force on the contacted surfaces.

Generally, the static friction coefficient is larger than dynamic friction coefficient, so the static friction force is larger than dynamic friction force.

Two contacted objects subjected to external forces moving from standstill condition will experience three processes. The first process is that at the standstill condition, the stick–slip friction effect on the contacted surfaces results in the squeak. In the second process, the relative motion induces the self-excited vibration that generates the squeak. In the third process, when the natural frequencies of two or more moving objects are close, the coupled vibration is induced by the friction force, which causes the squeak. The squeak mechanisms generated by the three vibrations will be described in the following material.

8.2.1.1 Squeak Induced by Friction Stick–Slip Effect

Not all surface frictions will generate noise. Only the instable stick–slip movement occurs on the friction surfaces can the squeak be generated. When the static friction force is greater than the dynamic friction force, there is no relative movement between the objects, but there exists the stick–slip friction effect on the contacted surfaces. From the microscopic viewpoint, the two surfaces are uneven. Driven by the external forces, although the subjects do not overcome the static friction, the two surfaces are alternatively pulled and pushed, forming an alternatively pressed and released movement. The stick–slip effect induces elastic deformation on the surfaces, and the energy is stored. When the movement is released, the deformation disappears and the stored energy is released. The pressed and released movement induces the objects to vibrate and generate sound. The stick–slip motion cycle frequency is low, but the induced vibration could make the contacted surfaces to generate 200–10 000 Hz squeal.

8.2.1.2 Squeak Caused by Self-Excited Vibration

Typically, the friction coefficient is not a constant, as shown in Figure 8.2; instead, it changes with time. The friction force is not constant as well; instead, it varies with the speed of movement.

A single degree of freedom (DOF) system is used as an example to illustrate the friction movement, as shown in Figure 8.3. The system dynamic equation can be written as follows:

$$m\ddot{x} + c\dot{x} + kx = \mu(u - \dot{x})F_N \tag{8.3}$$

where, u is the belt speed, which is a constant speed; $(u - \dot{x})$ is the slip movement speed.

The friction force changes with the movement speed of the object, as shown in Figure 8.4. The friction force decreases with the speed increase.

Figure 8.2 Friction coefficient varying with time.

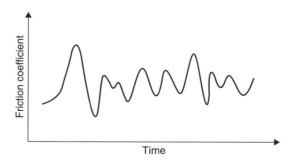

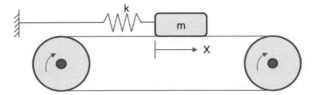

Figure 8.3 Single DOF friction system.

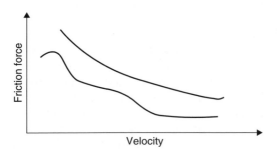

Figure 8.4 Friction force decreases with object movement speed increase.

The friction force is expanded by the Fourier series as follows:

$$\mu(u-\dot{x})F_N = \mu(u)F_N - \frac{\partial\mu(u-\dot{x})}{\partial u}\dot{x}F_N + \frac{\partial\mu(u-\dot{x})}{\partial u^2}\dot{x}^2 F_N - \cdots \tag{8.4}$$

Substituting Eq. (8.4) into Eq. (8.3), we obtain

$$m\ddot{x} + F_N\left(\frac{c}{F_N} + \frac{\partial\mu(u-\dot{x})}{\partial u}\right)\dot{x} + kx = \mu(u)F_N + \frac{\partial\mu(u-\dot{x})}{\partial u^2}\dot{x}^2 F_N - \cdots \tag{8.5}$$

In Eq. (8.5), if the damping force is negative, i.e.

$$\frac{c}{F_N} + \frac{\partial\mu(u-\dot{x})}{\partial u} < 0 \tag{8.6}$$

the system is instable. The case only happens when the negative slope of the friction force is very large. The friction force with large negative slope will cause self-excited vibration of the system, which induces the squeak. Especially when the self-excited vibration frequency is the same as or close to the natural frequency of the moving object, the object vibration will reach its maximum value, and corresponding, it generates loud noise.

8.2.1.3 Squeak Induced by Coupled Vibration Excited by Friction Force
The friction force between two moving objects also induces a coupling movement. Especially, when the natural frequencies of the two objects are the same or close, the coupling movement is intense. Even resonance could happen, which results in the squeak.

8.2.2 Factors Influencing Squeak

Two contacted surfaces form a friction pair. The compatibility of their friction coefficients is the main factor to affect the squeak. Material compatibility and the squeak characteristics of the friction pair can be obtained by testing. Assume that one of two contacted objects is fixed and the other one moves. A normal pressure is applied on the fixed object to force it to press the moving object, and simultaneously, a reciprocating load is applied on the moving object, as shown in Figure 8.5. The two objects are placed in a special environment cabin, where the squeak induced by the friction in different temperatures and humidity conditions can be measured.

Factors affecting the squeak include: temperature, humidity, excitation frequency, surface normal pressure and load time-history, sliding speed, contact area and contact situation, material characteristics of two contacted parts, and so on. The main factors affecting the squeak are described as follows.

8.2.2.1 Influence of Temperature

Squeak is generated by the relative motion between a friction pair varies with the temperature change. Figure 8.6 shows that the squeak sound pressure levels (SPL) vary with the temperature for two sets of friction pairs. In some temperature ranges, SPL increases with the temperature increase, while in other temperature ranges, they decrease with the temperature increase, which indicates that there is no regularity between the squeak sound level and the temperature. The relationship between the squeak sound level and the temperature is nonlinear; however, each friction pair has its own temperature range for the loudest sound level. Understanding the regularity of squeak sound level varying with temperature for a friction pair is helpful to guide the choice of the friction materials in order to avoid the temperature range where the loudest sound will be generated.

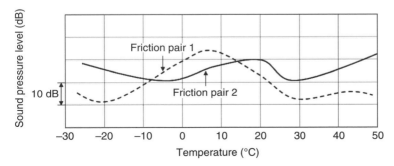

Figure 8.5 Squeak testing of a friction pair.

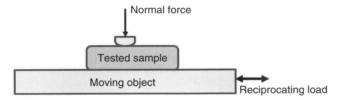

Figure 8.6 Squeak sound pressure level varying with temperature.

8.2.2.2 Influence of Humidity

The squeak generated by the relative motion of the friction pair changes with the humidity. Figure 8.7 shows the change of the squeak sound pressure levels with the humidity for two sets of friction pairs. As the humidity increases, the squeak SPL for one friction pair increases, but the other pair decreases. Thus, the humidity influence on different friction materials is different. Understanding the influence is helpful for choosing materials of a friction pair.

8.2.2.3 Influence of Normal Pressure

Figure 8.8 shows the change of the squeak sound pressure levels with the normal pressure for two sets of friction pairs. The larger the normal pressure, the deeper the contacted depth of the two objects, the stronger the tearing force on the uneven surfaces, and thus, the louder the squeak. The squeak induced by the friction can be attenuated by reducing the normal pressure.

8.2.2.4 Influence of Excitation Frequency

The excitations applying on the friction pairs include power train excitation, road excitation, etc., and the excitations have wide frequency range. Figure 8.9 shows the change of the squeak sound pressure levels with the excitation frequency for two sets of friction pairs. Generally, the squeak level increases with the excitation frequency.

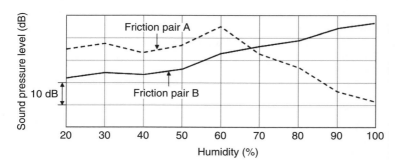

Figure 8.7 Change of squeak sound pressure levels with humidity.

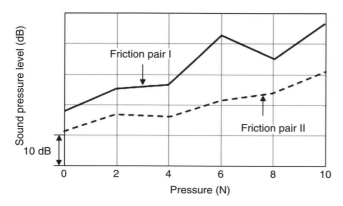

Figure 8.8 Change of squeak sound pressure levels with normal pressure.

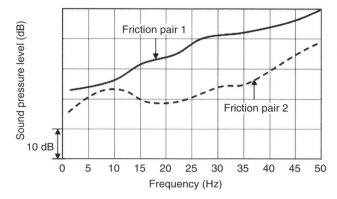

Figure 8.9 Change of squeak sound pressure levels with excitation frequency.

(a)   (b)

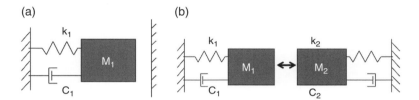

Figure 8.10 (a) A moving object hits a standstill surface; (b) two objects impact each other.

8.3 Mechanism and Influence Factors of Rattle

8.3.1 Mechanism of Rattle

Two adjacent objects subjected to external forces impact each other, and then separate. During the repeatedly impact and separation, rattle sound is generated. For a vehicle body, road excitation is main excitation source to induce the rattle. The reasons generating the rattle include small gap between the adjacent components, insufficient stiffness of the components, and structure loosening. Figure 8.10a shows that a moving object hits a standstill surface, and the object is impacted. Figure 8.10b shows two objects collide with each other, and both are impacted.

The impact time is very short. The most extreme impact occurs at one moment, which can be expressed by δ function, as follows:

$$\begin{cases} \delta(t - t_0) = 0, t \neq t_0 \\ \int_{-\infty}^{\infty} \delta(t - t_0) dt = 1 \end{cases} \tag{8.7}$$

This function is called a $\delta(t - t_0)$ function. Figure 8.11a shows its graph in time domain, which is a vertical line. After the δ function is transferred by the Fourier transform into the frequency domain, the function is expressed as follows:

$$F(\omega) = \int \delta(t - t_0) e^{-j\omega t} dt = 1 \tag{8.8}$$

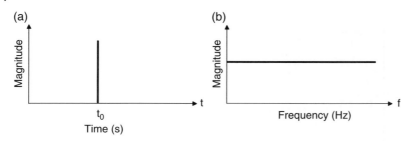

Figure 8.11 δ function: (a) graph in time domain; (b) graph in frequency domain.

In the frequency domain, it is a horizontal line, as shown in Figure 8.11b. The amplitude is the same for all frequencies, so it is a white noise function. If an excitation force was a δ function, the responses for all frequencies of the component would be excited.

Assume an excitation is a half-sine function, as shown in Figure 8.12a, and its expression is

$$y = \begin{cases} \sin\dfrac{\pi t}{t_0}, & 0 < t < t_0 \\ 0, & t > t_0 \end{cases} \tag{8.9}$$

After transferred by the Fourier transform, the expression in the frequency domain is

$$F(\omega) = \int_0^{t_0} \sin\frac{\pi t}{t_0} e^{-j\omega t} dt = \frac{e^{-j\omega t_0}\pi\cos(\pi t_0) + je^{-j\omega t_0}\omega\sin(\pi t_0) - \pi}{(\omega^2 - \pi^2)t_0} \tag{8.10}$$

Figure 8.12b shows the frequency response curves for several half-sine functions. It is clear to see that the shorter the half-sine time, the wider the corresponding spectrum.

The δ impulse excitation and the half-sine excitation are two special functions, which can be written by mathematical expressions. Actually, such special excitations do not exist for the body impact; however, most excitations for body impact can be assumed between the two special excitations, and corresponding spectrum is also in a wide frequency range.

The body is subjected to wideband frequency excitations, so the rattle noise is also in wide frequency range. The noise frequency responses depend on the excitation, structural vibration characteristics, and sound radiation efficiency. Most of the body rattle noise is in the frequency range of 200–2000 Hz.

8.3.2 Factors Influencing Rattle

Factors influencing the rattle include: space design, structure design, material pairing, manufacture, and assembly.

8.3.2.1 Space Design

Insufficient space or gap between two parts is a common cause of rattle noise. In a standstill condition, a gap exists between the two parts. But when the relative displacement of the parts' movement exceeds the gap, impact between the parts is not

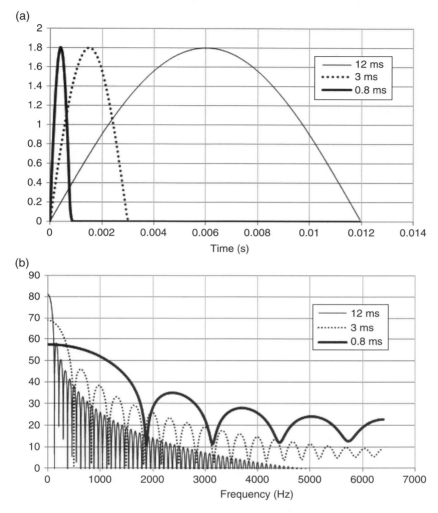

Figure 8.12 Graphs of a half-sine function: (a) in time-domain; (b) in frequency domain.

avoided, so rattle happens. For example, two tubes are very close, but there is a gap between them. When their movement displacement is beyond the static gap, they impact each other.

Even for two contacted parts at standstill condition, they could impact each other when their movement displacements are inconsistent. For example, two tubes contact each other at standstill condition, but when the vehicle moves, their movements are different, so they impact each other and generate rattle noise.

In the early phase of body design and system space layouts, the movement gaps for the adjacent systems and components must be clearly determined. The gaps can be determined by analysis of the systems' extreme movement envelope. For example, by analyzing extreme movement tracks of a power train, the distance between the power train and the body can be determined.

8.3.2.2 Structure Design

Inappropriate structure design also generates the rattle noise. Structure design includes the body overall stiffness and modes, local structural stiffness and modes, components' deformations, etc. For example, when a body is subjected to external excitations, the insufficient overall stiffness and local stiffness could induce the whole-body vibration and rattle noise.

Another example is a CD player mounted on an IP. The insufficient IP bending stiffness and/or the insufficient stiffness of the CD bracket will make these components have large movement displacements, which easily cause impact between the CD box and surrounding components. If the stiffness is increased by adding additional brackets on the IP and/or the CD box, the box movement displacement will be reduced and the impact possibility will be reduced as well or even eliminated.

8.3.2.3 Material Pairing

Not all impacts generate rattle noise. The impact between metal parts absolutely generates sound, but the impact between a metal part and a soft rubber may not generate sound, so the material pairing is very important. For example, latch, pawl, and striker inside a lock impact one another. If these parts are made of metals, the impact sound is not only louder but also includes more high-frequency components, resulting in poor sound quality. If some parts are wrapped by insulation materials, the impact sound level and the high-frequency components will be significantly reduced, and corresponding sound quality will be improved.

For the parts that have potential to be impacted, a good method to reduce the rattle noise is to carefully choose the appropriate material pairing.

8.3.2.4 Manufacture and Assembly

From the design viewpoint, the rattle noise can be controlled by appropriate space layout, structure design, and material pairing. However, even if the design is reasonable, but the poor quality control during manufacture and assembly processing will cause the vehicle body to generate rattle noise. For example, poor quality components, material deformation, and big dimensional tolerance will result in rattle noise.

Many components of the body are assembled using fasteners or bolts. Improper installation, missing installation of the fasteners, inadequate bolting force, parts loosening, etc. could induce the rattle noise.

8.4 CAE Analysis of Squeak and Rattle

CAE methods can be used to analyze and predict the squeak and rattle problems in vehicle early development stage. Today, many companies have begun to use CAE tools to realize the early control of the body squeak and rattle in order to save development time and cost. Due to complex and highly nonlinearity of the squeak and rattle problems, it is extremely difficult to find the precise calculation methods and the evaluation criteria. Some companies had developed a number of in-house software to analyze the squeak and rattle, but there is no mature commercial software in the market. However, based on the features of the squeak and rattle, and characteristics of structural vibration, the engineers have been trying the traditional methods to analyze and control the squeak and rattle, and have achieved good results. The finite element method (FEM) is the most widely used method.

Most squeak and rattle happen in the areas of the closure parts and instrument panel. The squeak and rattle in these areas are related to the body stiffness, deformation, and sensitivity of the response to the external excitation. Therefore, according to the relationship between the structural features of these components and their relative motion, conventional finite element analysis can be used to process following four analyses:

1) Analysis of stiffness, modes and deformation of body and door
2) Modal analysis of subsystems
3) Sensitivity analysis of body squeak and rattle
4) Response analysis of whole vehicle squeak and rattle

8.4.1 Analysis of Stiffness, Mode, and Deformation of Body and Door

Driven down the road, due to the road excitation, the vehicle body will produce bending and torsional deformations. The insufficient body stiffness will result in large deformation, which not only induces noise, vibration, and harshness (NVH) problems but also generates the squeak and rattle. The body modes depend on its stiffness as well. The statistic data shows that the higher the body modal frequency, the lower the probability to generate the squeak and rattle, the higher the subjective scoring, as shown in Figure 8.13.

Body-in-white (BIW) stiffness is the foundation of the whole vehicle overall stiffness, so the stiffness is very important for the squeak and rattle control. BIW overall stiffness depends on its frame stiffness and the joints' stiffness. See Chapter 2 for more on the overall stiffness and modal analysis.

The deformation of a BIW doorframe is one of the main factors affecting the body squeak and rattle. Subjected to the road excitation, the body is deformed. The deformation amount depends on the stiffness of the body frame, i.e. the joint stiffness and the body frame stiffness. Figure 8.14 is a side view of a BIW. Four diagonal lines of the doorframe, L1, L2, L3, and L4, are plotted, respectively.

Similar to the measurement and calculation of the body torsional stiffness, after the boundary conditions and applied load are determined, the diagonal lengths of the body closure opening are measured, and then compared with the dimension at unload condition. The deformation amounts (δ_D) of the body closure opening can be defined and expressed as follows,

$$\delta_D = \frac{L_i - L}{L} \tag{8.11}$$

Figure 8.13 Relationship between body modal frequency and subjective scoring.

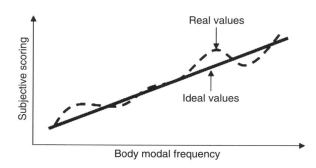

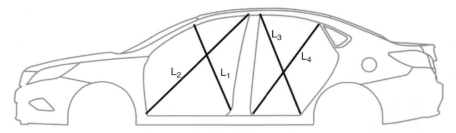

Figure 8.14 Body frame and four diagonal lines of body closure openings.

where, L is diagonal length of the body closure opening at unloaded condition; L_i ($i = 1$, 2, 3, 4) is the diagonal length of the body closure opening after applying load.

The body deformation can be reflected by the body closure opening deformations. Large deformation of the closure opening represents large deformation of the body, which easily leads to the interference and impact between the body and the door when the vehicle is moving; and vice versa. Figure 8.15 shows the deformation values of the body closure openings of five cars. In general, the deformation value should be less than 0.03. The body deformation can be controlled by limiting the closure opening diagonal deformation.

The door stiffness and modes are also very important. High stiffness and high modal frequencies of the doorframes are beneficial to reducing the vibration during slamming the door, the diagonal deformation of the doorframe, sound radiation of the door panels, and the squeak and rattle generated during the door closing. In order to measure the door modes, the accelerometers are placed on the doorframe. Figure 8.16 shows a tested first bending mode and a first torsional mode of a door. Figure 8.17

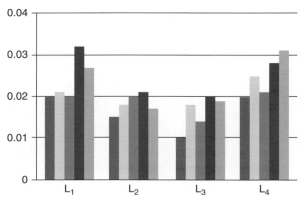

Figure 8.15 Deformation values of several body closure openings.

(a)   (b)

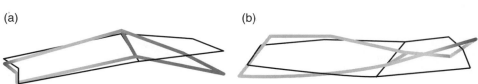

Figure 8.16 A door-in-white: (a) first bending mode; (b) first torsional mode.

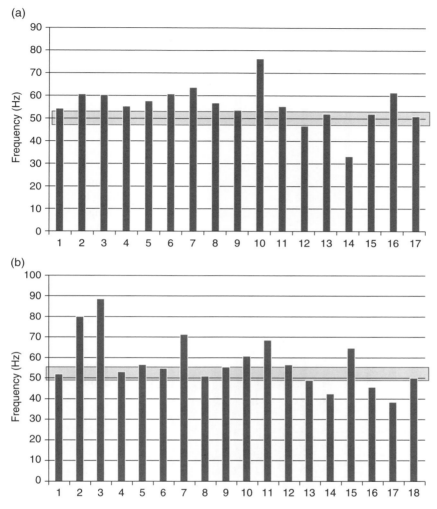

Figure 8.17 Bending modal frequencies of a group of door-in-white: (a) front doors; (b) back doors.

shows bending modal frequencies of a group of door-in-white, which are usually around 50 Hz. Figure 8.18 shows torsional modal frequencies of the doors-in-white, which are usually around 60 Hz.

8.4.2 Modal Analysis of Body Subsystems

A subsystem, a part of a body, is a collection of many components. The subsystem is connected with a body by welding or riveting or using brackets. The major subsystems generating the squeak and rattle include trim parts, instrument panel, steering supporting system, seats, etc.

There are three reasons for a subsystem to generate the squeak and rattle. The first reason is that the gap between the subsystem and the peripheral parts is too small.

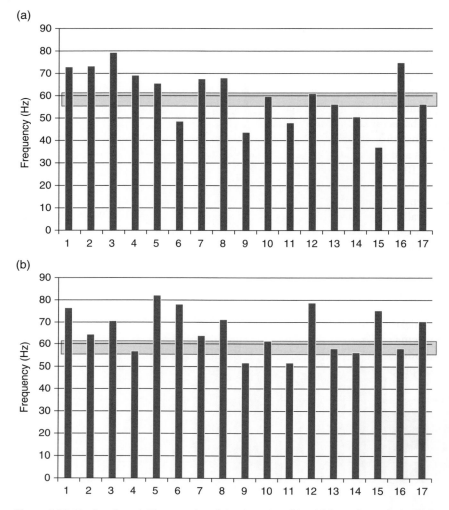

Figure 8.18 Torsional modal frequencies of the doors-in-white: (a) front doors; (b) back doors.

When the subsystem moving displacement exceeds the gap, impact is not avoided, so the S&R is generated. The second reason is that when the resonances between the subsystem and the body or between the subsystem and surrounding parts happen, the S&R will be induced. The third reason is that when the number of connecting points between the two parts are not enough or the connections are loose, the two parts will impact or slide, generating the squeak and rattle.

The above three reasons are associated with the subsystem structural modes. In the first reason, the large displacement is due to low modal frequencies of the subsystem, resulting in large displacement at the low frequencies. Taking a CD box placed on a bracket as an example to illustrate the problem, if the bracket stiffness is low and the corresponding modal frequencies are low as well, the bracket will have large displacement when subjected to external excitation. In the second reason, the subsystem modes are directly coupled with the body modes or modes of the surrounding parts. In the third reason, the number of connecting points and the connection tightness between the two parts directly affect the subsystem boundary conditions, therefore, directly affect its modes.

The purpose for the subsystem modal analysis is to understand the modal frequencies of each subsystem, the modal frequency separation between the subsystems and the body, and to find the corresponding control methods. Therefore, the subsystem modal analysis is the basis for the local squeak and rattle control.

Taking an instrument panel subsystem as an example to illustrate the subsystem modal analysis and the squeak and rattle analysis. The instrument panel include: cross car beam (CCB) or called IP reinforcement tube, ventilation ducts, defroster nozzle, glove box, CD box, passenger airbag assembly, finish panels, ashtrays, and so on.

The first step is to analyze its modal frequencies and shapes. The analysis goal is to obtain its overall modes and modes of its internal parts. Figure 8.19 shows an IP overall modal shape. Figure 8.20 shows the modal shapes of several internal parts. Figure 8.20a and b are modes of the glove box and the ventilation pipe, respectively. Comparing these modal frequencies with the external excitation frequencies and other modes of the adjacent parts, the resonated parts will be found. For example, the first modal frequency of the glove box is 25 Hz and the engine idle speed is 750 rpm and corresponding second-order frequency is 25 Hz. Thus, the glove box will be excited to resonate; therefore, the design of the glove box must be modified.

The subsystems can be controlled in the early stages of vehicle development by the subsystem modal analysis, without having to wait for testing the prototype vehicles. There are two purposes to do the subsystem modal analysis. One is to achieve the setup

Figure 8.19 An IP overall modal shape.

(a)   (b)

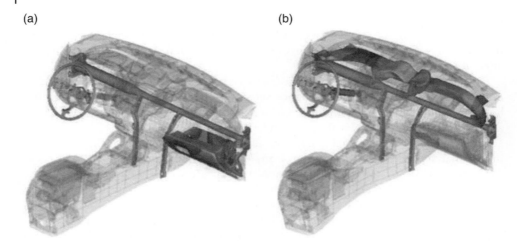

Figure 8.20 Local modes inside the IP: (a) a mode of the glove box; (b) a mode of the ventilation pipe.

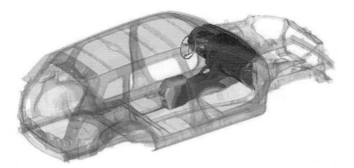

Figure 8.21 Modes of the instrument panel and the central console mounted on the body.

modal frequencies in order to realize the body modal separation. The other one is to control the modal number as fewer as possible within the analyzed frequency range (usually below 100 Hz).

The second step is to analyze the modal characteristics of the subsystem installed on the body. The purpose of this analysis is to identify the coupling status between the subsystem and the body, and to decouple the systems if they are coupled. Figure 8.21 displays the modes of the instrument panel and the central console mounted on the body, which are coupled with the body bending mode. There exist risks for resonance, so the design must be modified.

8.4.3 Sensitivity Analysis of Squeak and Rattle

The subsystems and the body are connected using fasteners, bolts, and welding glue. Figure 8.22 shows the two connection ways between the subsystems and the body. Figure 8.22a shows the instrument panel and the body are connected by fasteners, and Figure 8.22b shows the CCB and the A pillar are connected by bolts.

(a)

Connecting points

(b)

Connecting points

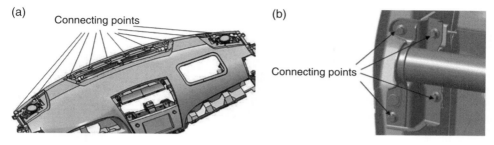

Figure 8.22 Connections between subsystems and the body: (a) IP and body; (b) cross car beam (CCB) and the A pillar.

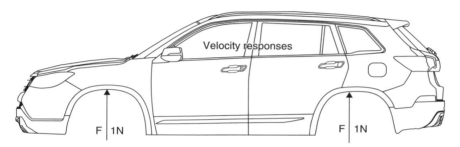

Velocity responses

F | 1N

F | 1N

Figure 8.23 Body S&R sensitivity: 1 N force is applied on four body points connected with suspension.

Subjected to the preloads and the external excitations, the fasteners or bolts could be loose, especially subjected to long duration. In general, the more intense the excitations, the easier loosening the connected parts. After the connected parts are loose, gaps or sliding between the originally fastened parts could happen, so the squeak and rattle will be generated.

The excitations applying on to the connecting points are mainly from the suspension system. It is very important to analyze the force transmission from the tire to the body points connected to the suspension, and the sensitivity between the responses at the fasteners or bolts and the excitations. Body S&R sensitivity refers to the ratio of the velocity responses at the fasteners/bolts to the external input force, and the unit is $mm\,s^{-1}\,N^{-1}$.

When analyzing the body S&R sensitivity, 1 N force is applied on the four body points connected with the suspension, respectively, as shown in Figure 8.23. The four excitation points are: left front suspension connected point, right front suspension connected point, left rear suspension connected point, and right rear suspension connected point.

Response points are: connected points of fasteners, bolts, welding points.

After calculating the velocities of connected points, the sensitivities between the velocities to the force are obtained, as shown in Figure 8.24. This figure shows the velocity sensitivities of an IP fastener in three directions to left rear suspension excitation. At 40 Hz, the resonances exist in all three directions, where the sensitivities have peaks and its magnitude at Z direction reaches to $0.41\,mm\,s^{-1}\,N^{-1}$, and the sensitivity magnitude at X direction at 62 Hz reaches $0.31\,mm\,s^{-1}\,N^{-1}$.

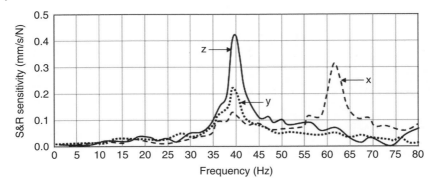

Figure 8.24 Body sensitivities at an IP fastener in three directions to left rear suspension excitation.

The same connected point has different velocity responses to different excitation points. For example, the response at an IP fastener to the left front suspension excitation is $0.13\,\mathrm{mm\,s^{-1}\,N^{-1}}$, but $0.05\,\mathrm{mm\,s^{-1}\,N^{-1}}$ to the right rear suspension excitation, which indicates that the connected point is more sensitive to the front suspension excitation than the rear suspension excitation.

The S&R sensitivity analysis includes two issues. One is to check whether a subsystem has resonances and the other one is to calculate its peaks that could induce the fasteners to loosen and the S&R. The subsystem modal analysis and resonant analysis can be analyzed simultaneously. The sensitivity peaks are closely related to the squeak and rattle. Under normal circumstances, the sensitivity at the connected point between a body trim part and a sheet metal should be less than $0.25\,\mathrm{mm\,s^{-1}\,N^{-1}}$. The sensitivity at a fastener inside the IP (such as the glove box, storage box, CD player, airbag, etc.) should be less than $0.4\,\mathrm{mm\,s^{-1}\,N^{-1}}$. If the sensitivities are lower than the mentioned values, the possibility to induce the squeak and rattle is relatively low. By the sensitivity analysis, its peak values, and corresponding frequencies can be used to judge whether the squeak and rattle could happen.

8.4.4 Dynamic Response Analysis of Squeak and Rattle

The vehicle response analysis refers to the analysis of responses at key points on a body installed on a whole vehicle subjected to the external forces. The input comes from the road surface, and the responses are at the connected points by fasteners.

The road excitation is transferred to the suspension system through the tires and the hubs, and then transferred to the body points connected with the suspension, and finally transferred to the connected points between the body and subsystems. The magnitude and frequency of the road excitation are closely related to the road surface profiles, so the profile displacement is chosen as the excitation input. The excitation of a rough road is larger than that of the smooth road, and the excitation of the cobblestone is even higher. The excitations with 1 mm displacement from 0 to 50 or 100 Hz are applied at the locations on the tires touching with the road, as shown in Figure 8.25.

The accelerations at the connected points are chosen as the responses, which are easily compared with the testing results. The magnitudes of the accelerations can be used to judge the occurring possibility of the squeak and rattle.

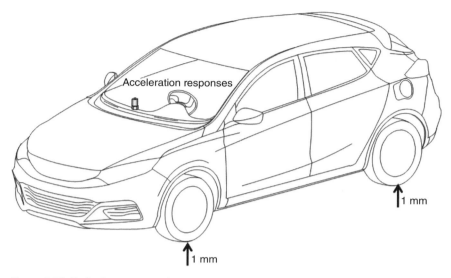

Acceleration responses

1 mm

1 mm

Figure 8.25 Excitation points and response points for whole vehicle response analysis.

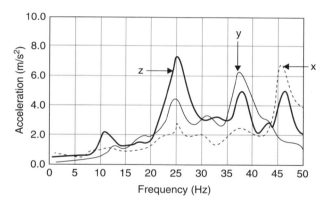

Figure 8.26 Acceleration in three directions at a fastener inside an IP.

Figure 8.26 shows the acceleration curves in three directions at a fastener inside the IP. It is clearly to see that there are three peaks at 25, 37, and 46 Hz, and the magnitude at 25 Hz in Z direction reaches $7.4\,\mathrm{m\,s^{-2}}$.

A lot of experiences show that if the accelerations at the connected points are smaller than $10\,\mathrm{m\,s^{-2}}$ when the tire is subjected to 1 mm excitation, the occurring possibility of the squeak and rattle is low, and vice versa.

8.5 Subjective Evaluation and Testing of Squeak and Rattle

In addition to using CAE methods to analyze and control the squeak and rattle in the early stage of product development, the subjective and objective methods are still widely used to identify the S&R problems. The methods are extensively applied to analyze the S&R problems in the mass produced vehicles, benchmarked vehicles and the developed prototypes.

8.5.1 Subjective Identification and Evaluation of Squeak and Rattle

Subjective evaluation refers that the S&R locations are identified by driving a vehicle on different roads and at different speeds, or by placing the vehicle on a four-poster simulator excited by various inputs.

8.5.1.1 Subjective Identification on Road

In a proving ground, there are many roads specially designed and constructed for NVH development, such as

- Asphalt road
- Smooth cement road
- Rough cement road
- Corrugated road
- Random shock road
- Cobblestone road
- Belgium road
- Brick road
- Bumper road
- Gravel road
- Dirt road

Figure 8.27 shows photographs of several roads. Some roads are used for riding comfort evaluation and/or for testing NVH performance, and some roads are specially designed for the S&R evaluation.

In addition to the proving grounds, street roads such as rough cement roads, deceleration zone, suburban bad roads, pavements of parking lots, etc., are usually used to evaluate the squeak and rattle as well.

(a)   (b)

(c)

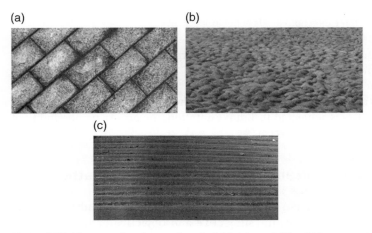

Figure 8.27 Photographs of several roads: (a) brick road; (b) cobblestone road; (c) bumper road.

Figure 8.28 Road in a special parking lot.

Some squeak and rattle can be heard on different roads, even on very smooth asphalt road. Some squeak and rattle can be perceived only on a special road. For example, abnormal sounds could not be heard on the impact road and the cobblestone road, but were identified on the Belgian brick road. Another example, a prototype vehicle is tested on different roads in a proving ground and on the street roads, and no S&R is identified, however, a rattle sound is perceived on a special road in a parking lot shown in Figure 8.28.

Driving speed also has an impact on the squeak and rattle. Some abnormal sound can be heard at all vehicle speeds, while others can be perceived at one speed or a speed range. At a special speed, the natural frequency of a subsystem is consistent with the road excitation frequency, so the subsystem is excited, inducing impact or sliding between the adjacent parts.

8.5.1.2   Subjective Scoring System

During the S&R evaluation in the proving ground or on street roads, in addition to finding the S&R occurring locations, the S&R severity should be scored. Two types of scoring systems are generally used: one is a "10-Points" system and the other one is a "Four-Points" system. The "10-Points" scoring system is the same as the one described in Chapter 7, which is used for sound quality evaluation. "1 point" represents very serious squeak and rattle, "10 points" stands for no squeak and rattle, and "6 points" indicate the abnormal noise can be accepted by most customers.

The S&R severity levels in the "4-Points" system are divided into: severe S&R, moderate S&R, slight S&R, and no S&R. The "severe" S&R indicates that the squeak and rattle are very bad, and the majority of customers can perceive it, which is completely unacceptable. The "moderate" S&R indicates the obvious presence of the squeak and rattle, which are unacceptable to many customers. The "slight" S&R indicates that the most customers cannot perceive the squeak and rattle, even if it is present. In the "4-Points" system, the scoring value is normalized and defined as S&R severity coefficient, that is, the scores for the "severe," "moderate," "slight" and "no" squeak and rattle are "1," "0.3," "0.1," "0," respectively. Table 8.1 lists the S&R severity coefficient, indicated by V_i.

Table 8.1 S&R severity coefficient (V).

S&R severity	S&R severity coefficient	Description
severe	1	The majority of customers can perceive
moderate	0.3	50% customers can perceive
slight	0.1	Most customers cannot perceive
no	0	No customers can perceive

Table 8.2 Weighting coefficient for different roads (R_i).

Road conditions	Weighting coefficients
smooth road	1
rough road	0.3
bad road	0.1

The S&R severity coefficient is different for the same vehicle driven on different roads. The coefficient is lower on the smooth road and higher on the rough road. Therefore, the road influence on the squeak and rattle must be determined. Usually, three typical roads are used to evaluate the S&R: smooth road, rough road, and bad road. The smooth road means a high-quality road, such as a good smooth asphalt road. The rough road means that there are obvious coarse particles on the road surface, such as coarse cement road. The bad road refers to various bad roads, such as pebble road, road with speed bumps, etc. If the subjective perceptions on the S&R for different vehicles are the same on the three different roads, the scorings for the vehicles will be definitely different. Taking two vehicles (A and B) as an example to illustrate the problem, the S&R severity for vehicle A driven on the smooth road is the same as that for vehicle B driven on the bad road, which indicate the squeak and rattle for vehicle A is much worse than vehicle B. Thus, it is necessary to provide a weighting coefficient, denoted as R_i, of the road influence on the squeak and rattle. The weighting coefficients for the smooth road, rough road, and bad road are "1," "0.3," and "0.1," respectively, as shown in Table 8.2.

From the above descriptions, in the squeak and rattle scoring, the vehicle S&R status and road conditions must be considered simultaneously. Thus, an index, called Squeak and Rattle Index (SRI), is introduced to represent both the vehicle status and road conditions, and is expressed as follows:

$$SRI = \sum_{n=1}^{N} SRI_n = \sum_{n=1}^{N} \sum_{i=1}^{3} R_i V \tag{8.12}$$

where N represents part number of a vehicle inducing squeak and rattle. SRI is the summation of the squeak and rattle contributed by all parts. The index obtaining from analysis or testing can be used to judge whether the S&R is acceptable.

Figure 8.29 A vehicle placed on a four-post simulator.

8.5.1.3 Evaluation by Vibration Simulator

In addition to identifying the squeak and rattle on the road, another way is to recognize the problem on a vibration simulator in a laboratory. Driving on the road, some locations inducing the S&R can be identified, but many times it is very hard to identify the exact locations. However, if the vehicle is tested on the simulator, the precise locations can be easily identified.

The vehicle is placed on a four-post simulator to be tested, as shown in Figure 8.29. The simulator consists of four individual exciters, and each exciter pushes one cylinder to move, where the four tires are placed on the corresponding cylinders. The vibration signals, including random signal, sine sweep, single-frequency input, and collected road spectrum, are input to each exciter. The four exciters can move in-phase or out-phase.

During the testing, the evaluators sit inside or stand outside of the vehicle, carefully listening to the squeak and rattle and to identify the locations. Compared with the road evaluation, the simulator evaluation is more precise to find the squeak and rattle locations. In addition, some professional tools or equipment are helpful for the evaluators to more accurately identify the squeak and rattle locations and corresponding causes. The professional tools include stethoscope, sound recording equipment and high-speed camera.

The stethoscope is basic equipment for doctors, which can also be used for preliminary diagnosis of NVH and S&R problems. The auscultation head of the stethoscope is removed, and a metal tube is connected, as shown in Figure 8.30, to form an NVH stethoscope. The end of the tube is equivalent to a probe, which can be moved around to approach the vehicle parts generating sound, so the near-field sound can be clearly heard. If the tube end touches the vehicle structure, the sound generated by the structural vibration can be perceived. By moving the probe of the NVH stethoscope to different locations, the precise locations generating the squeak and rattle can be identified. The advantage of this method is cheap and quick. Similar to an outstanding doctor, first, an excellent NVH engineer can use the stethoscopes and his or her rich experience to diagnose NVH and S&R problems, and then use the instruments to do measurements.

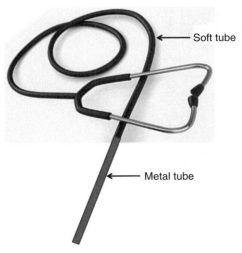

Figure 8.30 NVH stethoscope.

Soft tube

Metal tube

Figure 8.31 A door installed on a component S&R bench.

The squeak and rattle can be recorded by a sound recorder and can be repeatedly listened, which is helpful to diagnose the problems. The recording equipment can also be used to identify the squeak and rattle during road testing.

A high-speed camera can be used to record fast-moving objects or some motions that cannot be clearly seen with the naked eye. If the quick motions are recorded by the high-speed camera and then played back at normal speed, the motion can be clearly viewed and causes of squeaks and rattles more easily identified. For example, impact between an engine mount and its surrounding parts is often not clearly viewed, but if the motion is recorded by a high-speed camera and played back at normal speed, the impact locations, motion displacements, and root causes inducing the impact will be obviously identified.

Some systems can also be tested on a component S&R bench. Figure 8.31 displays a door installed on a component S&R bench. In this bench, the squeak and rattle inside the door system can be conveniently and clearly identified.

8.5.1.4 Examples

During driving a prototype (car M), an evaluator found four distinct S&R problems inside the instrument panel, including squeak between the IP and the A pillar, rattle inside the glove box, rattle inside the IP, and rattle from the airbag.

After the discovery of these S&R problems, the first work is to determine whether they are individual problems for this prototype or common problems for all the developed prototypes. If they are the common problems, the root causes in design and/or manufacture must be found out. If they are individual problems, the problems' sources must be analyzed, for example, to find out whether they are generated in manufacturing processing or in installation processing.

In order to determine whether the problems are common or individual problems, the evaluators needs to drive more prototypes. In this example, after five prototypes are driven on the roads, among the four problems, the first two are confirmed as common problems and the late two are the individual problems.

For the common problems, the design must be analyzed first. The evaluators also drive a competitive vehicle (car N) whose sales are outstanding in the submarket, and there are no S&R problems inside the IP and the surrounding parts. So, the competitive car is used as benchmark vehicle for the developed M car.

For the first problem, the connection ways between the IPs and A pillars, stiffness, and supporting of the steering systems for the two cars are compared, and significant differences between the two cars are found. The modal analysis shows that the modal frequency of M car's steering tube is low, which results in the large deformation of the tube and the IP. By strengthening the middle supporting of the steering tube, increasing the number of the connecting bolts between the tube and the A pillar, and increasing the bolts' preload, the modal frequencies of the tube are significantly increased and its deformation decreases, resulting in elimination of the squeak and rattle between the A pillars and the IP.

For the second problem, the root cause is that the glove box and the blower are interacting, and the glove box has significant vibration. The interference is avoided by shortening the length of the glove box, and the glove box vibration is reduced by adding an additional bracket at its end. After the designs are changed, the squeak and rattle disappear.

For the third problem, the long cable inside the IP and the loosened intermediate fasteners induce the cable to shake and affect the surrounding structures, so the rattle noise is generated. After the fasteners are tightened and the foam is wrapped on the cable surface, the rattle noise is eliminated.

For the fourth problem, the bolts used to install the safety airbag inside the steering column loosen, which generates the rattle noise. After the bolts are tightened, the rattle is eliminated.

8.5.2 Objective Testing and Analysis of Squeak and Rattle

The purpose of objective testing is to identify the S&R sources, characteristics and severity from the tested data. Because of the strong nonlinearity and irregularity of the S&R signals, it is very difficult to find suitable tested signals to evaluate and analyze the S&R characteristics. Nonetheless, similar to the CAE analysis, some approximation methods still can be used to evaluate the S&R.

The objective test can be implemented on road and/or on the four-poster simulator. In addition to a whole vehicle testing, the body subsystems can be tested as well.

8.5.2.1 Analysis of Acceleration Power Spectrum

The S&R is generated by the impact or friction between structural parts, which indicates that the S&R is related to the structural vibration, so the analysis of structural vibration characteristics is helpful for understanding the S&R characteristics.

The acceleration at the connected points between the components can be calculated by CAE analysis, but it is hard to obtain by the testing. Instead, the component surface acceleration is easily measured, and it can be used to estimate the vibration at the connected points. A lot of experiences prove that the higher the surface acceleration power spectrum, the larger the acceleration of the internal connected points, the higher possibility of the squeak and rattle occurring.

Assume a measured acceleration signal is $a(t)$, then the corresponding autocorrelation function is

$$R_{aa}(\tau) = \lim_{T \to \infty} \frac{1}{T} \int_0^T a(t) a(t+\tau) dt \tag{8.13}$$

The acceleration power spectral density is

$$S_{aa}(f) = \int_{-\infty}^{\infty} R_{aa}(\tau) e^{-j2\pi f\tau} d\tau \tag{8.14}$$

Accelerometers are placed on the body or subsystem surfaces, as shown in Figure 8.32. The vehicle is excited on a four-poster simulator or driven down the road, and then the acceleration signals on the body surface and subsystem surfaces are measured. The tested power spectral density can be used to deduce the S&R problems from two aspects. The first aspect is that the vibration of the internal connected points can be estimated by the surface power spectral density. Lots of tested data and experience show that the surface acceleration power spectrum is closely related to the connected points' accelerations, which determine the occurring potential of squeak and rattle. The second aspect is that an acceleration threshold of generating the S&R problems can be

Figure 8.32 Accelerometers placed on the body or subsystem surfaces.

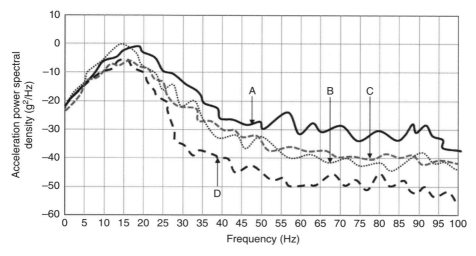

Figure 8.33 Acceleration power spectral density on the IP surfaces for the four vehicles.

obtained after comparing the acceleration spectra on the same locations for different vehicles where some have the S&R problems and some don't have.

Two examples are used to illustrate the relationship between the acceleration power spectra and the S&R. The first example is that three vehicles (A, B, and C) have the rattle problems inside the IPs and vehicle D doesn't have it. The tested curves of the acceleration power spectral density on the IP surfaces for the four vehicles are plotted, as shown in Figure 8.33. In the range of 25–100 Hz, the power spectral density for vehicle D is much smaller than other vehicles, while natural frequencies of the IPs and many internal components are within this frequency range. The surface power spectral densities of the four vehicles are well corresponded with the internal squeak and rattle, i.e. the higher the surface power spectral density, the louder the squeak and rattle, and vice versa.

The second example is that the rattle sound inside a vehicle glove box is eliminated by optimizing the structure. Figure 8.34 shows the tested curves of the glove surface acceleration power spectral densities for the original and optimized structures. In the frequency range of 25–50 Hz, the power spectral density for the optimized structure is significantly reduced compared with that of the original structure.

The above two examples prove that the surface acceleration power spectrum represents the S&R characteristics; therefore, the acceleration power spectral density can be used as an objective index to determine the S&R sensitivity.

8.5.2.2 Noise Analysis of Squeak and Rattle

When a vehicle is driven on road, it is difficult to distinguish the squeak and rattle from other noises by only using the SPL because they are mixed together. Sometimes, the S&R is not loud, but the human ear can tell its difference from other noises. For example, the "creak" sound induced by the seat leather friction is much lower than the engine noise, but the occupants can perceive it. Since frequency range of the S&R is wide, it is very difficult to use loudness, sharpness, and other indexes to quantitatively evaluate the squeak and rattle.

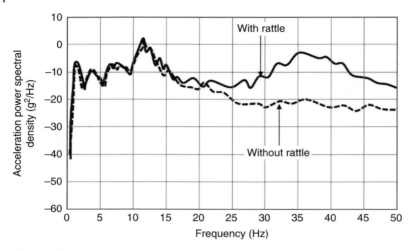

Figure 8.34 Comparison of glove surface acceleration power spectral densities with/without rattle.

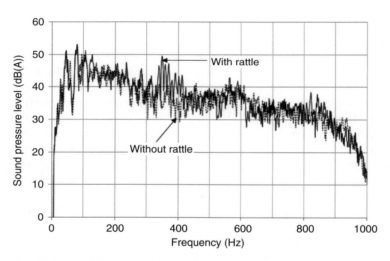

Figure 8.35 Near-field sounds of the mount with and without the rattle.

When a vehicle is excited on the four-poster simulator, there are no engine noise, road noise, and wind noise; instead, only the squeak and rattle exist. Thus, the physical indexes, such as loudness, can be used to evaluate the squeak and rattle. The near-field sound and interior sound are simultaneously tested, which can be used to analyze the S&R characteristics. For example, when a prototype vehicle is driven down the road, a rattle sound generated by a hydraulic engine mount is heard. The vehicle is placed on the four-poster simulator, and the near-field sound close to the mount and the interior sound are measured. The frequency range for the rattle mount is between 300 and 400 Hz. After disassembling the mount, the internal decoupler loosens, which is the root cause of the rattle. After the mount structure is modified, the rattle disappears. Figure 3.35 shows the near-field sounds of the mount with and without the rattle.

In addition to the laboratory testing of the squeak and rattle for the whole vehicle, the body subsystems and components can be tested on the component S&R bench to find their S&R problems. After the component squeak and rattle sound is recorded, the corresponding loudness, sound pressure level, and sharpness can be used to determine its locations and characteristics.

8.6 Control of Body Squeak and Rattle

8.6.1 Control Strategy during Vehicle Development

The three stages of S&R control have already been introduced in section 8.1. The first stage is a "discover problem – solve the problem" process. In the second stage, the S&R identification and control is moved forward to the prototype vehicles. In the third stage, the S&R control is further moved to the early stage of product development, and CAE and other analytical tools are used.

Today, the S&R control has been developed to the whole vehicle development process from the concept design to mass production. The S&R control strategy during the process will be introduced in the following materials.

8.6.1.1 Phase of Concept Design and Project Planning

The phase of concept design and project planning is a pre-product development stage. A vehicle project has not officially started, but staffs from the marketing department, R&D department, etc. work together to certify feasibility, market outlook, and other issues of the project. The task of the S&R engineers is to study the S&R cases for previous models, to collect and analyze the history data, to analyze the after-sales maintenance cost due to the S&R problems, and to analyze the benchmark vehicles. At the same time, they also need to predict some changes in the coming years, such as technology development, improvement of the competitive vehicles, and changes of market segments, customers' demands and regulations, and so on. Based on the above study and analysis, the engineers decide the S&R control strategy of the new vehicle development.

8.6.1.2 Early Design Phase

In the early phase of product development, the S&R control should be considered as a design target for each part and each system design, and be integrated with the whole vehicle design. Only in this way can the effective and low-cost S&R control be achieved.

The vehicle integrated design and CAE analysis play a central role in the early stage. The vehicle integrated design determines that the targets for every system and component are set up first, and then the optimal targets for the whole vehicle should be achieved after all systems and components are integrated. The S&R integrated design will be described in details in later materials of this section. At this stage, a large number of S&R CAE analysis, such as modal analysis of systems and components, body joint stiffness analysis, BIW door deformation analysis, BIP sensitivity analysis, whole vehicle dynamic analysis, and so on, must be implemented, which will predict the S&R locations and prevent the happening possibility of squeak and rattle. Meanwhile, robustness analysis can be used to forecast the potential squeak and rattle after high mileage.

8.6.1.3 Phase of Component and Prototype Verification

After the prototype components and systems are ready, they are tested on S&R simulators. For example, if an instrument panel was tested and several S&R locations were identified, the IP design should be modified.

After the prototype vehicles are ready, they are arranged to be tested on the four-poster simulators or on a road, so the locations generating the S&R can be identified. Simultaneously, the tested results can be used to verify the accuracy of CAE calculation.

8.6.1.4 Manufacture and Assembly Phase

The squeak and rattle of the prototype vehicles is well controlled by the conceptual design, CAE analysis, and prototype testing. However, in manufacture and assembly phase, the squeak and rattle could be found on some mass-produced vehicles. The S&R control during the manufacture and assembly processing involves three aspects: component quality check, quality control of the entire manufacture processing, and the vehicle performance check.

Component quality check is to determine whether the quality of all components meet the design requirements, especially for the key components and easily failure components. The key components, such as IP, seats, doors, etc. must be checked on the S&R benches.

Quality control of the entire manufacture process is to guarantee that the work at each station and each process meet the design requirements. The data show that many S&R problems are generated by poor quality control during assembly – for example, inappropriate installation (such as door misalignment) or missing installation of small parts (such as fasteners and bolts). These minor assembly errors will result in squeak and rattle problems, especially for the vehicles that have high mileage.

Vehicle performance check is to check whether the squeak and rattle could happen on the test track of the assembly plant. The test track includes some special roads, such as cobblestone roads, brick roads, impact roads, etc., where it is relatively easy to find squeak and rattle problems.

8.6.2 Body Structure-Integrated Design and S&R Control

Body S&R control should follow the principle of integration. The integration means that the S&R sources and transfer paths must be considered simultaneously as an integration that is input into the body design during the whole product development process. The sources causing the squeak and rattle come from the road and the engine, which are transferred to the body through various paths. The sources are transferred to the isolators or rubber pads, then to the connecting parts, such as brackets, and finally to the body. The body closure deformation and the system mode coupling affect the squeak and rattle. The above description can be plotted in Figure 8.36.

8.6.2.1 Isolation Efficiency and Isolator Stiffness

The connections between the body and other systems or components include rigid connection or flexible connection. The rigid connection refers that the systems and the body are directly connected by bolts, while the flexible connection means that the systems and the body are connected by isolators or isolation pads. For example, the power train is connected with the body by a flexible mounting system; the chassis is flexibly connected with the body by shock absorbers, bushings, etc.; the subframe is connected

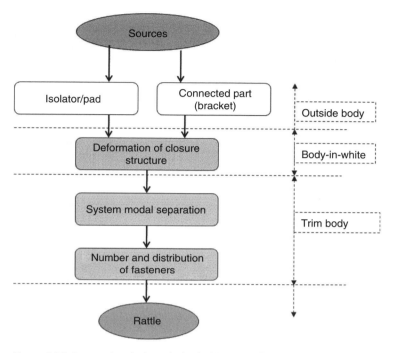

Figure 8.36 Integration design of a body S&R control.

with the body by either flexible bushings or rigidly bolts; the exhaust system is flexibly connected with the body by hanger isolators, and so on.

Vibration sources from the engine and the chassis are attenuated after passing through the isolators or bushings, and then transferred to the body. The isolation efficiency of the isolators and bushings determines the vibration attenuation, which affects not only the transmission of vibration but also the S&R control.

8.6.2.2 Design of Connected Part Stiffness

Many components are connected with the body by specifically designed connecting member, such as brackets. The power train vibration is transferred to the isolators, then to the bracket connecting with the body, as shown in Figure 3.88. The vibration from the exhaust system is transferred to the isolators, then to the hangers, and finally to the body. Battery, air cleaner box, fuse box, etc. are placed on the brackets connected with the body. There are many brackets in a vehicle body. Only their stiffness is high enough can the resonance between the brackets and the excitation sources be avoided, and the squeak and rattle could be avoided as well. For example, the bracket in Figure 3.88a is softer than that in Figure 3.88b, because there are more ribs on the bracket in Figure 3.88b and it is thicker, so it has higher modal frequencies.

8.6.2.3 Modal Separation

The principle of body mode separation has been introduced in the previous chapters. The S&R control should follow this principle as well. The body modal frequencies and the modal frequencies of the systems connected to the body must be separated in order to avoid resonance. In CAE analysis in the early development stage, modes and frequencies for each subsystem and each component must be analyzed in order to confirm that the modes of adjacent systems or components are separated.

8.6.2.4   Deformation of Closure Structure

The BIW will deform when subjected to external torques. If the deformation is large enough, the body closure opening deformation could be different from the door deformation, so, impact between the closure opening and the door is inevitable. Therefore, in the early design stage, it is necessary to analyze the closure structure deformation, and the deformation amount must be controlled in order to avoid the rattle noise.

8.6.2.5   Reasonable Gap Design between Parts

A gap exists between some adjacent components. When the displacement of their relative motion is larger than the gap, the impact is inevitable. Thus, gap control is very important. If a small gap is unavoidable, it must be isolated. For example, inside an engine compartment, there are many intersecting pipes and wires, as shown in Figure 8.37a. If the distance between the pipes and wires is too close, brackets or fasteners must be used to support them in order to avoid impact. In addition, sometimes, isolation materials, such as sponges, are used to isolate the pipes or the wires from the surrounding parts, as shown in Figure 8.37b.

Figure 8.38a shows disorganized arrangement of the wires that are not fixed with the body. When the vehicle moves, the wires will impact one another, and they impact with the surrounding objects as well, inducing the squeak and rattle. Figure 8.38b displays a set of bundled wires which are fastened with the body by fasteners, so the arrangement avoids the squeak and rattle.

8.6.2.6   Number and Distribution of Fasteners

There are many fasteners in a vehicle body. For example, the airbag is connected with the steering wheel by bolts, the trims and the door are connected by fasteners, the wires are connected with the body by fasteners, and the sound insulation mat is connected

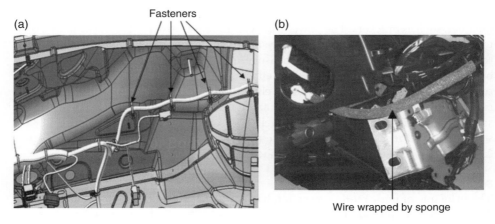

Figure 8.37 Isolation between adjacent components: (a) using fasteners; (b) using isolation material.

(a)   (b)

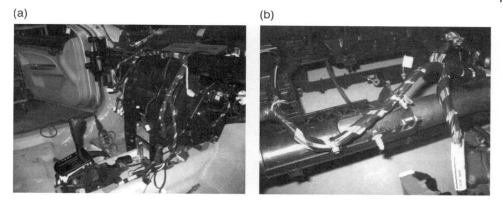

Figure 8.38 (a) Disorganized arrangement of the wires; (b) a set of bundled wires.

(a)   (b)

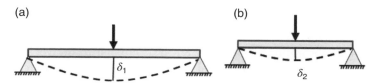

Figure 8.39 A section of a component connected by two bolts or fasteners can be regarded as a beam.

with the engine hook by fasteners. The squeak and rattle is influenced by the stiffness of the connected parts, installation location on the body, number and distribution of the bolts and fasteners, bolt torques, etc.

The stiffness of the connected parts and installation location on the body affect the body squeak and rattle. The fasteners connect the body and the connected part, so connected locations on both the body and the part should have sufficient stiffness. The insufficient stiffness will result in large deformation or large displacement motion, which has potential to induce them to impact or slide with the adjacent parts.

The number and distribution of the bolts or fasteners influence the body squeak and rattle. A section of a component connected by two bolts or fasteners can be regarded as a beam, as shown in Figure 8.39. If the "beam" span is too large, the vertical deformation at middle point will be large, as shown in Figure 8.39a. Excessive deformation will make it interfere or impact with the surrounding parts. Increasing the number of the fasteners will reduce the "beam" span, so the deformation at middle point will decrease, as shown in Figure 8.39b.

Figure 8.40 shows the connections between an IP upper portion and a body. Eight fasteners are used to connect the two objects shown in Figure 8.40a, and the S&R problem is found. After analyzing the problem, the large deformation of the IP between the fasteners is the root cause. After the fasteners are increased to 11 as shown in Figure 8.40b, the deformation reduces and the S&R disappears.

Therefore, the number and distance between the bolts or fasteners determine the deformations and accelerations of the connected parts, and further determine the possibility of the S&R occurring.

(a)

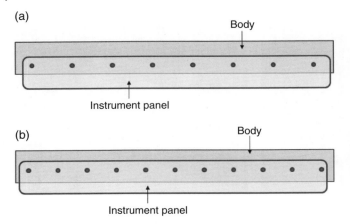

(b)

Figure 8.40 Connections between an IP upper portion and a body by different number of fasteners.

8.6.2.7   Sound Insulation and Sound Absorption Processes

Sound insulation and sound absorption reduce the S&R transmission to the vehicle interior. Some S&R happens in a specific location, and it is difficult to control its source, so the transfer path must be controlled. For example, a vehicle is running on a wet road and the water is splashed to the body, inducing the rattle noise. It is impossible to control the noise source, so sound absorption and sound insulation are placed in the wheelhouse area. A thick layer of damping material is splashed on the outer panels of the wheelhouse in order to reduce the water-splashing noise, and sound insulation can be improved as well.

8.6.3   DMU Checking for Body S&R Prevention

Digital mock up (DMU) checking refers to identification of NVH problems, such as structural stiffness, leaking, etc. in the designed digital vehicles and data. For example, the structural stiffness checking is to check the panel stiffness distribution and its influence on the local modes, damping distribution, etc. Checking for leaks is to discover the holes and apertures on the body and judge their influence on the leakage.

The S&R DMU checking is to determine whether gaps or connections between adjacent parts are reasonable; whether layout of wires and pipes brings interference problem; whether friction pair of nonmetallic parts is compatible; whether span between fasteners is acceptable; and whether resonance will happen for brackets and other structures.

The first task of the S&R DMU checking is to check whether a gap between adjacent components meets the design requirements. The gap must be larger than the displacement of their relative movement, otherwise, interference may occur.

The second task is to check connection status of wires and pipes with the body and interference between them. When they are not tightly connected with the body or there are no enough connection points, the squeak and rattle could happen. In addition, if the wires and pipes are interfered or too close one another, squeak, and rattle could happen as well.

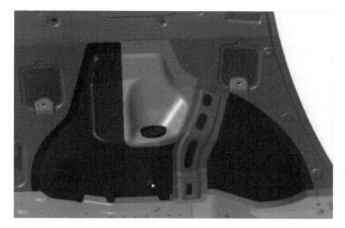

Figure 8.41 Wheelhouse consisting of several large flat plates.

The third task is to check the material compatibility characteristics of contacted parts. The nonmetallic parts on the body, such as seals, trimmed parts, etc., could be in contact with the other components. If the friction characteristics for the contacted parts are not compatible, a squeak could result.

The fourth task is to check the connection methods of the connected parts and the spans of the connected points. For example, the connection between the IP and the body using the fasteners in Figure 8.40, in order to reduce the displacement and to avoid the potential S&R problems, the spans between the fasteners must be less than a certain value.

The fifth task is to check whether the parts will cause resonance. For example, in Figure 8.41, the wheelhouse consisting of several large flat plates could be easily excited by road inputs, and then radiates noise. By the DMU inspection, this wheelhouse design is unreasonable, so, some ribs must be punched on the plates, similar to the structure shown in Figure 3.72.

8.6.4 Matching of Material Friction Pairs

Squeak is due to the relative friction movement on the material surfaces, but not all friction will generate the squeak. Squeak can be heard for a friction pair, but after a special material is coated on their surfaces, the squeak will be eliminated. Therefore, it is very important to understand the material characteristics of the contacted components and their friction characteristics.

It is unrealistic to use "no-squeak" materials or special coating for all components. Even though some parts are coated with a special material and the squeak is eliminated, but after the vehicle runs a period of time, the surface coating may be worn or reliability problems emerge, the squeak can be heard again.

The key to control the squeak is to study friction characteristics between the friction components and to find the friction pair that doesn't generate the squeak. For example, for materials A, B, and C, squeak is generated by friction between A and B, but there is no squeak for friction between A and C. A pair of materials where no squeak is generated when they slide against each other is called a compatible pair of friction materials.

So, the materials A and C are a compatible pair of friction materials. It is an effective method to control the friction noise by using the compatible pair materials.

In the selection of the matching pair materials, the factors such as temperature, humidity, external force, and its contacted angle should be considered, because the material friction characteristics are affected by these factors. For example, the temperature and humidity could cause the material to be deformed, especially for the elastic elements and flexible structures, which makes their friction characteristics different from those at room temperature. The materials meeting the requirements of the friction matching pair in various environments (such as different temperatures, humidity, etc.) are regarded as a good compatibility friction pair. Therefore, selecting materials having good compatible friction pair to make the contacted parts is a key to control the squeak.

8.6.5   Control of Manufacture Processes

Even if the S&R is well controlled in design phase, if there is no effective control during the manufacture and assembly, the squeak and rattle could be introduced for the mass-produced vehicles. The vehicle manufacture processing can be divided into three phases: parts entering to an assembly plant, process of manufacture and assembly, and vehicle inspection. All three phases should have strict S&R control measures in order to reduce the potential squeak and rattle for mass-produced vehicles.

The parts inspection is to check whether the components from the suppliers meet the design requirements. From perspective of S&R control, the inspection is to check the component quality, gaps between adjacent components, etc., such as the gap between an IP and a glove box and the connecting status by the fasteners, etc.

Assembly process control is to make sure that each process has right assembly operations in order to avoid the squeak and rattle due to the improper assembly. For example, if a door is improperly installed to a body, unacceptable impact and friction could happen for the door closing action, generating squeak and rattle. Therefore, it is very important to establish strict assembly processes and to train the workers in order to achieve good S&R control.

Vehicle inspection is to check whether the mass-produced vehicles have the S&R problems on the factory test track or on the four-poster simulator. Every produced vehicle must be driven on the track. Similar to the roads in a proving ground, the track includes a variety of roads. The experienced workers generally can identify the S&R problems after driving the vehicles on the track. Sometimes, for the special cases where it is hard to identify the S&R problems on the track, the vehicles can be placed on the simulator where the exact locations generating the squeak and rattle can be easily judged. In addition, using recorders to record the sound and playback in listening room, the squeak and rattle can be quantitatively determined.

The strict control of manufacture processes will significantly reduce the squeak and rattle, but the cost could be increased. In general, the luxury car makers have stricter control measures than the economy car makers.

8.6.6   Squeak and Rattle Issues for High Mileage Vehicle

In modern society, the automobile becomes a tool for daily life. The market competition becomes more and more fierce, and the customers have many choices. The customers are expecting the good performances not only for the new vehicles, but also for the vehicles running for several years, which induce a high mileage problem.

After a vehicle running for several years, its structural and material properties could change compared with a new vehicle. Due to the structure connection loosening, material aging, and fatigue, etc. the potential to generate the squeak and rattle for the used vehicle increases. For example, rubber component aging, vibration isolation effectiveness decreasing, loosening between the side window glass and seals, parts loosening inside the lock, etc. will increase the squeak and rattle. Another example is that missing of the fasteners and loosening of the bolts could induce the squeak and rattle as well. The third example is that the decreasing gap between the plates due to the body panels' deformations could induce the rattle noise.

8.6.7 Squeak and Rattle at High Mileage

High-mileage S&R evaluation can be implemented in different used miles, including subjective evaluation and objective testing. Figure 8.42 shows a subjective evaluation rating for three vehicles at new vehicle status and after using 20 000, 40 000, 60 000, 80 000, and 100 000 km, respectively. At 100 000 km, the rating for vehicle A reduces from 8.2 to 7.5 points, and the reduced score is 0.7; the rating for vehicle B reduces from 8.0 to 7.7 points, and the reduced score is 0.3; the rating for vehicle C reduces from 7.8 to 6.5 points, and the reduced score is 1.3. The vehicle C has the largest attenuation, then the vehicle A, and the vehicle B has the least attenuation, i.e. the vehicle B has the best high mileage performance.

Another index to evaluate high mileage S&R is the overall S&R indicator. Figure 8.43 shows the overall S&R indicators varying with the mileage (20 000, 40 000, 60 000, 80 000, and 100 000 km) for three vehicles. At 100 000 km, the indicator for vehicle A changes from 0.1 to 0.9, and increases 0.8; the indicator for vehicle B changes from 0.2 to 1.8, and increases 1.6; the indicator for vehicle C changes from 0.2 to 4.5, and increases 4.3.

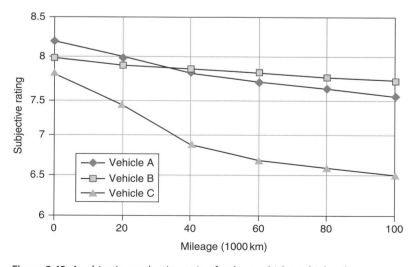

Figure 8.42 A subjective evaluation rating for three vehicles at high mileages.

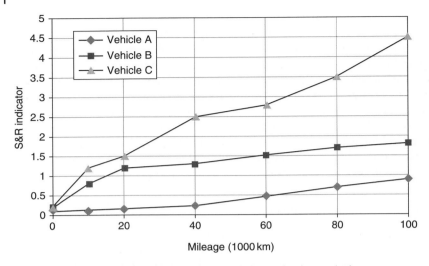

Figure 8.43 Overall S&R indicators varying with mileage for three vehicles.

Obviously, the S&R indicator increases the most for vehicle C, then the vehicle B, the least is vehicle A.

High-mileage S&R should be controlled in design, material selection and manufacture, etc. In the design, the major method to reduce the high mileage S&R is the robust design. Robustness refers to the ability of a system to maintain the designed performance under the change of the system internal conditions and external conditions. The high-mileage S&R robustness means that the S&R performance of a used vehicle is almost the same as the new one and still has a few squeak and rattle problems after running for several years and under the change of the body condition, such as structure loosening and deformation, rubber material aging, etc.

Many factors and their variations influence the S&R performance, such as, mounting rubber stiffness increasing or decreasing 30% from its designed value, the bolt preload reducing 50% to its original value. The performance targets of the robustness design include body modes, gaps between components, changing of the friction pair, etc. In the S&R robustness design, based on the influencing factors and their variable ranges, the S&R robustness performance attenuation is analyzed, and the probability of the high-mileage squeak and rattle problems can be predicted.

Bibliography

Bédouin, N., Hoguet T., and Pasquet, T. (2007). New Concept of Chassis Dyno Procedure for Squeal Noises Evaluation. SAE Paper; 2007-01-2265.

Boreanaz, G. Celiberti, L. and Falasca, V. (2007). In-Plant Fast Diagnostics of Vibration-Acoustic Quality of Cars. SAE Paper; 2007-01-2211.

Brines R.S., Weiss, L.G. and Peterson, E.L. (2001). The Application of Direct Body Excitation Toward Developing a Full Vehicle Objective Squeak and Rattle Metric. SAE Paper; 2001-01-1554.

Byrd, R. and Peterson, E.L. (1999). A Comparison of Different Squeak & Rattle Test Methods for Large Modules and Subsystems. SAE Paper; 1999-01-0693.

Cerrato-Jay, G. Gabiniewicz, J., Gatt, J. et al. (2001). Automatic Detection of Buzz, Squeak and Rattle Events. SAE Paper; 2001-01-1479.

Chen, F. and Trapp, M. (2012). *Automotive Buzz, Squeak and Rattle*. Elsevier.

Feng, J. and Hobelsberger, J. (1999). Detection and Scaling of Squeak & Rattle Sounds. SAE Paper; 1999-01-1722.

Grenier, G.C. (2003). The Rattle Trap. SAE Paper; 2003-01-1525.

Hunt K., and Rediers, B., and Brines, R. (2001). Towards a Standard for Material Friction Pair Testing to Reduce Automotive Squeaks. SAE Paper; 2001-01-1547.

Hyun, Y.W., Warden, G., Blenman, J. et al. (2012). A Displacement-Approach for Liftgate Chucking Investigation. SAE Paper; 2012-01-0217.

Ibrahim, R.A. (1994a). Friction-induced vibration, chatter, squeal, and chaos, part I: mechanics of contact and friction. *ASME Applied Mechanics Review* 47 (7): 209–226.

Ibrahim, R.A. (1994b). Friction-induced vibration, chatter, squeal, and chaos, part II: dynamics and modeling. *ASME Applied Mechanics Review* 47 (7): 227–254.

Jay, M., Gu, Y., and Liu, J. (2001). Excitation and Measurement of BSR in Vehicle Seats. SAE Paper; 2001-01-1552.

Juneja, V., Rediers, B., Kavarana F. et al. (1999). Squeak Studies on Material Pairs. SAE Paper; 1999-01-1727.

Kavarana F. and Rediers, B. (1999). Squeak and Rattle – State of the Art and Beyond. SAE Paper; 1999-01-1728.

Kim, U., Mongeau, L., and Krousgrill, C. (2007). Simulation of Friction-Induced Vibrations of Window Sealing Systems. SAE Paper; 2007-01-2268.

Kuo, E.Y. (2003). Up-Front Body Structural Designs for Squeak and Rattle Prevention. SAE Paper; 2003-01-1523.

Kuo, E.Y. and Mehta P.R. (2004). The Effect of Seal Stiffness on Door Chucking and Squeak and Rattle Performance. SAE Paper; 2004-01-1562.

Kuo, E.Y. and Mehta P.R. (2005). The Effects of Body Joint Designs on Liftgate Chucking Performance. SAE Paper; 2005-01-2541.

Kuo, E.Y., Mehta P.R., and Geck, P.E. (2002). High Mileage Squeak and Rattle Robustness Assessment for Super Duty Cab Weight Reduction Using High Strength Steel and Adhesive Bonding. SAE Paper; 2002-01-3064.

Kwon J. and Lee, H.S. (2005). A Study on the Evaluation Process of Rattle Noise Considering the Signal Characteristics in Frequency and Time Domain. SAE Paper; 2005-01-2543.

Lee, P., Rediers, B., Hunt K. et al. (2001). Squeak Studies on Material Pair Compatibility. SAE Paper; 2001-01-1546.

Maciel, E.L.F., Alves, L.H., and Rabi, E.R. (2002). Strategies and Proposals to Minimize Squeaks and Rattles – Strong Customers Enthusiasm Improvement Program. SAE Paper; 2002-01-3561.

Na, H.H., Park, H., Lee, H. et al. (2013). A Study on the Rattle Index from a Vehicle Door Trim under Audio System Inputs. SAE Paper; 2013-01-1914.

Neihsl, K.S. (2002). Performance of Thermoplastic Polyolefins in Automotive Roof-Pillar Covers Involved with Interior Head Impact and Roof-Rail or Side Air-Bag Deployment. SAE Paper; 2002-01-1040.

Padilha, P.E.F. and Nunes, A. (2002). A Brief Survey on Squeak & Rattle Evaluation Techniques at General Motors do Brasil. SAE Paper; 2002-01-3489.

Park, K.-H., Bae M.-S., Yoo, D.-H. et al. (2005). A Study on Buzz, Squeak and Rattle in a Cockpit Assembly. SAE Paper; 2005-01-2544.

Peterson, E.L. Sestina, M. (2007). Using Rumble Strips for Buzz, Squeak and Rattle (BSR) Evaluation of Subsystems or Components. SAE Paper; 2007-01-2267.

Peterson, C. Wieslander, C. and Eiss, N.S. (1999). Squeak and Rattle Properties of Polymeric Materials. SAE Paper; 1999-01-1860.

Shorter, P.J., Cotoni, V., and Chaigne, S. (2011). Predicting the Acoustics of Squeak and Rattle. SAE Paper; 2011-01-1585.

Sohmshetty R., Kappagantu, R., Naganarayana, B.P. et al. (2004). Automotive Body Structure Enhancement for Buzz, Squeak and Rattle. SAE Paper; 2004-01-0388.

Soine, D.E. Evensen, H.A., and VanKarsen, C.D. (1999). Threshold Level as an Index of Squeak and Rattle Performance. SAE Paper; 1999-01-1730

Trapp, M.A. and Peterson, E.L. (2007). A Systematic Approach to Preparing Drive Files for Squeak and Rattle Evaluations of Subsystems or Components. SAE Paper; 2007-01-2269.

Trapp, M.A. and Pierzecki, R. (2003). Squeak and Rattle Behavior of Elastomers and Plastics: Effect of Normal Load, Sliding Velocity, and Environment. SAE Paper; 2003-01-1521.

Trapp, M.A., and Pierzecki, R. (2005). Squeak and Rattle Behavior of Filled Thermoplastics: Effect of Filler Type and Content on Acoustic Behavior. SAE Paper; 2005-01-2542.

Trapp, M.A., McNulty, P., Chu, J. et al. (2001). Frictional and Acoustic Behavior of Automotive Interior Polymeric Material Pairs Under Environmental Conditions. SAE Paper; 2001-01-1550.

Weisch, G. and Stücklschwaiger, W. (1997). The Creation of a Car Interior Noise Quality Index for the Evaluation of Rattle Phenomena. SAE Paper; 97NV203.

Zhang, L., Sobek, G., Chen, L. et al. (2005). Improving the Reliability of Squeak & Rattle Test. SAE Paper; 2005-01-2539.

9

Targets for Body Noise and Vibration

9.1 Target System for Vehicle Noise and Vibration

9.1.1 Period for Vehicle Development and Targets

9.1.1.1 Product Development Period

The period to develop a brand new vehicle is usually three years. The "starting point" of the development period is the time when the company executives make a strategic decision to develop the new vehicle. The "ending point" is that beginning of the mass production in an assembly plant. Before the starting point, it takes about half a year for the market and research departments to do detailed analysis on the vehicle to be developed, including product positioning, market competition, technology support from internal company and outside sources, etc., which is called predevelopment work. Figure 9.1 shows the development period.

9.1.1.2 Target System and its Importance

Target is a desired result by a system, a person, or a unit. For automotive noise, vibration, and harshness (NVH), the target is to meet customer expectations and competitive demands of the market on noise and vibration. The customer language is abstract; for example, they wish a vehicle to be "quiet," "comfortable," "powerful at acceleration," and to have "less vibration." The market language is abstract as well; for example, a vehicle should have these characteristics: "competitive NVH performance," "best NVH performance among the segment-market," "NVH is an important part of the automotive DNA," "NVH is an attractive point for sales," etc.

The customers language and market input are abstract. After the marketing strategy is finalized, the requirements will be input to the development department. The abstract inputs from the market department will be studied by the development team and then translated into the engineering language. The engineering language is objective and measurable. For example, the inputs from the market department for a sedan to be developed are that the vehicle should be very quiet at idling for AC-On condition," and "its NVH performance is a leader among the segment-market." The requirements could be translated by the engineering team to be 37 dB (A) for idle noise. For another example, the input from the market department is that a vehicle to be developed should be the quietest one in its segment-market. The engineers translate the market language this way that they test three brand vehicles with the best NVH performances in the

Noise and Vibration Control in Automotive Bodies, First Edition. Jian Pang.
© 2019 China Machine Press. All rights reserved. Published 2019 by John Wiley & Sons Ltd.

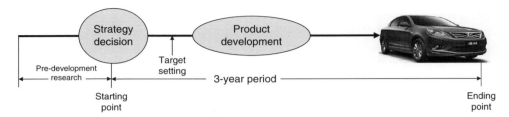

Figure 9.1 Development period of a brand-new vehicle.

segment-market and plot the tested curves together, and then set up its target that is lower than all the curves.

The set targets based on the inputs from the customers and the market are perceivable by the occupants, such as heard sound, perceived vibration by hand on steering wheel, by seat butt, and so on. The targets are called vehicle-level targets. A vehicle is a very complex structure, and almost all the systems and components involve NVH issues. During a vehicle development, each system and each component should have its own NVH targets. The system targets and component targets cannot be perceived by the occupants directly, but they affect the vehicle-level NVH performance. Thus, the vehicle NVH target system consists of three levels:

- First level: vehicle-level targets
- Second level: system-level targets
- Third level: component-level targets

The NVH target system will be introduced from the three level targets.

The target system is one of the most important contents in a vehicle development. The targets are set up in the early stage, and then the entire development process is based on the targets. The target setting is closely related to cost and technology; therefore, the target setting is not blindly to pursue the higher level. Once the targets are set up, the whole development process is deployed around the targets. Through computer-aided engineering (CAE) analysis and prototype testing, the set targets will ultimately be achieved. Setting a complete and reasonable target system is very important for vehicle development.

9.1.1.3 Execution of Target System

Execution of the target system is based on a development process. The process specifies the tasks, deliverables, and desired targets for each phase. The vehicle development process can be defined as a process to set targets, execute targets, and achieve the targets, as shown in Figure 9.2.

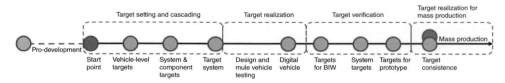

Figure 9.2 Processing of target setting, execution, and realization.

In addition to participating in "pre-development" work, an NVH team starts formal development work from the "starting point," and sets up the vehicle-level noise and vibration targets. The target setting is the most important work in the development process because all the later work is deployed around the "targets." The three-year development process can be divided into four phases: target setting and cascading, target realization, target validation, and target realization for mass production. From the NVH perspective, each phase is briefly described as follows:

1) *Target setting and cascading.* Based on market research and analysis of competitive vehicles, the vehicle-level targets are set, and then cascaded to each system and each component to form system targets and component targets, so a vehicle NVH target system is generated.
2) *Target realization.* By CAE analysis, mule-vehicle (or call hybrid-vehicle) testing, system and component design and analysis, and DMU inspection, the NVH targets for the developed digital prototype are achieved.
3) *Target validation.* After the set targets are achieved for the digital prototype, the targets for systems, body-in-white (BIW), trimmed body, and prototype should be verified to realize the prototype NVH targets.
4) *Target realization for mass production.* After the prototypes are tested and all the problems are solved, the set targets are finally realized; however, the NVH performance should be checked for the mass-production vehicles in order to guarantee consistent performance. The NVH issue feedbacks from the after-sale market must be followed and the countermeasures must be provided to solve the problems.

The four phases of the three-year development process are usually divided into more than a dozen milestones. Before each milestone, the target must be checked to make sure this milestone achieves the required NVH targets. If not achieved, risks for moving forward of the project should be assessed in order to clearly confirm the direction of the target execution.

9.1.2 Factors Influencing on Target Setting

The target setting is usually influenced by four factors: government regulations, customer demands, market competition, and company technology, shown in Figure 9.3:

1) *Government regulations.* Noise pollution is harmful to human health, especially hearing. Loud noise inside the vehicle makes conversation uncomfortable and is fatiguing for the occupants, which influences driving safety. Loud exterior noise impacts the surrounding residents as well, so, the World Health Organization

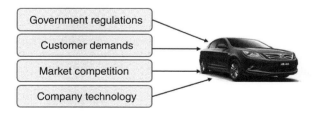

Figure 9.3 Four factors influencing NVH targets.

(WHO) and governments all over the world have enacted regulations to limit vehicle exterior noise. Under normal circumstances, there are two limits for the exterior noise: pass-by noise and standstill noise. The European Union has developed additional exterior noise standards.

2) *Customer demands.* The government regulations limit the exterior noise, but the customers concern more about the interior noise. Not only are customers concerned about sound and vibration levels but also they care about sound quality. The requirements on sound for different customers are not the same. Customers who drive economy vehicles expect that they will not be annoyed by the interior noise. Customers who own luxury vehicles desire not only low noise but also a comfortable sound. The customers who drive sports cars pursuit dynamic sound, even excitement and irritation.

3) *Market competition.* The vehicle market can be divided into many segment-markets, such as the SUV segment, the economy car segment, and so on. In a segment-market, there are many brand vehicles by different makers. Therefore, in order to develop a vehicle that can compete in the corresponding segment-market, benchmarking and target setting must be implemented. In addition, some institutions investigate the market and publish detailed reporters on vehicles performance, such as J.D. Power. The reports influence the customers decisions to purchase vehicles, company strategy to development vehicles, and target setting.

4) *Company technology.* Technology is a factor affecting the target setting. Even if the above three factors are analyzed and a best-in-class vehicle is developed, if the company technology cannot meet the desired requirements, the target setting and the vehicle development are just empty talk. The technology know-how is an important pillar of a company to support the vehicle development.

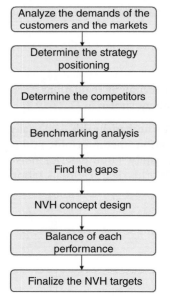

Figure 9.4 Principles and procedures of target setting.

9.1.3 Principles of Target Setting and Cascading

9.1.3.1 Principles of Target Setting

Figure 9.4 shows the principles and procedures of the target setting. In the premises of above four factors, following steps are followed to set the targets.

The first step is to analyze the demands of the customers and the markets. The market department specifies the requirements on the vehicles by the identified segment-market and target customers, and then lists the inputs.

The second step is to determine the strategy positioning on vehicle performances. The strategy positioning of performances refers to determining the competitiveness of the attribute performances of a vehicle to be developed in a specified segment-market. The performance competitiveness is divided into four categories as follows:

- *Leader.* The performance is best in a segment-market.
- *Leading group.* The performance is not the best, but among the best in a segment-market.

- *Competitor.* The performance is a little bit worse than the leading group but is competitive in a segment-market.
- *Follower.* The performance is not competitive, but acceptable by the customers in a segment-market.

An interior noise at wide-open throttle (WOT) is used to illustrate the relationship between the four roles. Figure 9.5 shows the interior noise curves at WOT for several best-sale vehicles in a segment-market and the set target lines. The leader's target line is lower than all the curves; the competitor's target line is 2–3 dB (A) higher than that of the leader; the target line for the leading group is between those of the leader and the competitor; the target line for the follower is about the same level as the loudest vehicle, and is about 5–7 dB higher than the leader's.

The third step is to determine the competitors. Once the customer demands and the strategy positioning of the performances are finalized, the competitors can be determined. Assume a vehicle to be developed is among the leading group, so the competitive vehicles can be picked up from the "leading group vehicles." For example, there are 20 vehicle models in the segment-market; after the "leader" is removed, the remaining three best-performance vehicles are chosen as the "competitors."

The fourth step is to do benchmarking analysis. The three chosen vehicles are used to be the benchmarking objectives. The NVH data of these vehicle are tested at various running conditions, such as idling, WOT, POT (partial-open throttle), etc. and then the NVH performance for the vehicle-level, system-level, and component-level are analyzed, so, their advantages and disadvantages will be fully understood.

The fifth step is to find the gaps. Through the benchmarking research, the NVH performances of the competitive vehicles are fully understood, and then the gaps between the competitors and the vehicle to be developed should be analyzed. Where are the gaps? Technology gaps? System gaps? Other gaps? How to bridge the gap and how to achieve the desired targets are important tasks for NVH development of the vehicle to be developed.

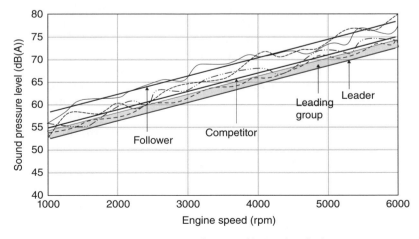

Figure 9.5 Interior noise curves at WOT for several best-sale vehicles in a segment-market and set target lines.

The sixth step is to do NVH concept design. In order to achieve the competition with competitors, NVH concept analysis and design must be implemented for each system. Then, the NVH target system is setup.

The seventh step is to analyze the balance of each performance. A vehicle includes many performances (attributes), such as NVH, collision safety, handling, fuel economy, and so on. For many cases, these performances are contradictory, for example, requirements on suspension bushings for handling and NVH are often opposing, higher bushing stiffness is benefit to the handling but worse for NVH. A good vehicle is the some that the balance is achieved among the performances. Of course, some vehicle could highlight some performances and scarify others.

The eighth step is to finalize the NVH targets. In considering the balance with other performances, the NVH target set in the sixth step is modified, and then the NVH targets are finalized.

9.1.3.2 Principles of Target Cascading

The vehicle-level NVH targets that can be directly perceived by the occupants are determined based on the customer and market demands. After the vehicle-level targets are set, they should be cascaded to each system and to each component. The "cascading" is not a one-way process from top to bottom, which means that the vehicle-level target is not simplify cascaded into the body target, engine target, etc. The target cascading should be two-way directions from the top to the bottom and from the bottom to the top, as shown in Figure 9.6.

In doing benchmark analysis, not only the characteristics of the vehicle-level NVH of the competitive vehicles, but also the features of each system and each component NVH, such as body modal frequency, intake orifice noise, muffler transmission loss, should be analyzed. The benchmarked results will be the references for the target setting of the vehicle to be developed. If the targets of each system and each component are determined, then, in turn, they can be "composed" into the vehicle-level targets. Of course, errors, such as phase errors, could exist in the "composed" process.

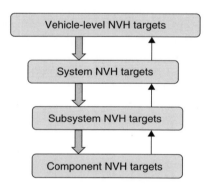

Figure 9.6 Two-way directions of target cascading.

The target setting of each system and each component is affected by vehicle design, technology, cost, and other factors. Thus, NVH targets must be adjusted to satisfy the constraints.

The target cascading is a "down" process from the vehicle level to system and component levels; simultaneously, it is also a "up" process from the system and component levels to the vehicle level. Only when these two processes blend well together can a perfect NVH target system be set up.

9.1.4 Principles of Modal Separation

When natural frequency of a system is consistent with an external excitation frequency, resonance happens. When natural frequencies of two neighboring systems are the same, if one system is disturbed, the other system will be affected. Therefore, for a system design, resonance must be avoided, and the principle of modal separation must be followed. The principle includes separation between a system natural frequency and an excitation frequency, and separation of natural frequencies of the adjacent systems.

For a vehicle to be developed, a modal table or chart must be established, as shown in Figure 1.40. The figure shows modal frequencies of all the systems and idling firing frequencies. It is obvious to see the modal frequencies of neighboring systems. If the modal separation of the adjacent systems does not meet the separation principle, the systems must be modified. Similarly, it is noticeable to see whether natural frequency of a system coincides with the idling firing frequency. If coincidence occurs, the system frequency or idling firing frequency must be modified. In the early stages of the vehicle development, the table together with the target system become the most important guidelines for the NVH development.

Similar to the vehicle modal table, and according to the modal separation principle, a detailed modal chart for each system can be established. For example, Table 2.2 lists mode description of all components on a body, including first and second body bending modes, first and second body torsional modes, panels (dash panel, roof, floor, trunk lid, hood, etc.) modes, modes of other systems connected with the body, excitation frequencies, etc. The table is a guideline for the body NVH development.

In addition to the vehicle modal table and the system modal table, an excitation source table or chart should be established, as shown in Figure 2.52. The main excitation sources, such as engine and rotating machineries, are plotted on this chart. It is clear to identify the excitation sources at each speed and every frequency on this chart.

9.1.5 Target System of Body NVH

The body NVH target can be divided into four levels:

- Vehicle-level body NVH target
- Trimmed body-level NVH target
- BIW-level NVH target
- Component-level NVH target

The vehicle-level body NVH target refers to the body NVH target when a body is installed in a full vehicle. For example, sound insulation is only determined by the body; so it is a vehicle-level body NVH target. For another example, the door closing sound quality is implemented and evaluated in a full vehicle, so it is a vehicle-level body target. The overall vehicle bending modes and torsional modes are substantially determined by the body structure, so the modes are closely related to the body targets.

The trimmed body NVH target refers to the one after the interiors, door, glass, windshields, seats, etc. are installed on a BIW. After components are installed, due to the change of the body stiffness and weight, its modal frequencies change. Because the trimmed components significantly determine the vehicle sound absorption and insulation, the targets of sound insulation, vibration transfer function, etc. can be set on the trimmed body.

The major NVH targets of a BIW include overall stiffness, overall modal frequencies, and local model frequencies.

The component target refers to the local mode target of each component such as mirror modal frequency, bracket modal frequencies, door modes, sound insulation coefficient, sound absorption coefficient, and so on.

This chapter will describe the body NVH target system in detail from above four aspects.

9.2 NVH Targets for Vehicle-Level Body

9.2.1 Vehicle-Level Body NVH Targets

The vehicle-level body NVH targets are closely related to the noise and vibration perceived by the occupants. The targets refer to those that can only be realized on the full vehicle but cannot be determined or conveniently determined on the trimmed body and BIW.

The overall vehicle modes and modal frequencies determined by the trimmed body are the basis of the vehicle NVH, whereas the trimmed body modes are determined by the BIW. So, from perspective of the vehicle body, the overall vehicle modes and modal frequencies belong to the body NVH targets.

The sound insulation and acoustic cavity modes can be controlled in the trimmed body, however, they are much more easily measured in the full vehicle, so they are classified to the vehicle-level body NVH targets.

Similarly, the target of the door closing sound quality can be determined by a trimmed body, but it is more conveniently measured and analyzed on the full vehicle, so it is attributed to vehicle-level body targets.

The local panel (such as dash panel) modal frequencies are set as targets in a BIW; however, their modes in the full vehicle could change. For example, some accessories (such as steering pump, etc.) and passing-throughs (such as air-conditioning pipe, shifter cable, steering column, etc.) are installed on the dash panel, which will force its modes to change. So, the local panel modes must be considered as one of the vehicle-level body NVH targets.

Some small accessories, such as mirrors, are installed in the full vehicle, so the vehicle-level body NVH targets should include them.

9.2.2 Vibration Targets for Vehicle-Level Body

9.2.2.1 Modal Frequency Targets for Vehicle-Level Body

The modal frequency targets for the vehicle-level body include:

- First bending modal frequency of the full vehicle
- First torsional modal frequency of the full vehicle

The low frequencies of the overall body modes are close to the range of engine idling frequency, easily resulting in the body resonance. The low frequencies also could induce the impact between components, generating squeak and rattle. Thus, the first bending and torsional modal frequencies should be included as the NVH targets, and they generally should be more than 30 Hz.

9.2.2.2 Modal Frequency Targets for Panels

The modal frequency of the dash panel is an NVH target; however, it is difficult to set a confirmed value. The principle is that the frequency is separated from the excitation frequency of the components connected with it – for example, an air conditioning pipe could transfer the vibration from the compressor to the dash panel, so the panel frequency should be separated from the compressor excitation frequency. In general, the higher the panel frequency, the better. The main method for increasing the panel frequency is to punch ribs and/or add reinforcement bars on it, and the ribs and bars should be connected to the surrounding components to form closed-loop structures. If a resonance cannot be avoided, damping materials or constrained layer damping structure can be used on the panel.

9.2.2.3 Vibration Targets for Accessories

The side mirrors are mounted on the door or on the body. Inadequate dynamic stiffness of the mounting point or low modal frequency of the connecting bracket could induce the mirrors to shake when a vehicle is driven down the road, which will make the driver or passengers see ambiguous objects on the mirrors. Therefore, the dynamic stiffness of the mounting points and the bracket modal frequencies must be among the NVH targets. Similarly, the dynamic stiffness of the mounting points and the bracket modal frequencies of the interior mirrors should be well controlled as well.

9.2.2.4 Vibration Targets for Instrument Panel

The instrument panel is mounted on the cross car beam (CCB). The CCB modal frequency not only determines the frequency of the steering wheel system but also affects the modal frequency of the instrument panel. The instrument panel with low modal frequency can be easily excited by external vibrations, for some severe cases, the naked eye can see the instrument panel trembling. The instrument panel has potential to impact with body structures, which will induce rattle and squeak problems. Therefore, the modal frequency of the instrument panel is set as a body NVH target.

9.2.3 Noise Targets for Vehicle-Level Body

The vehicle-level body noise targets include:

- Sealing
- Sound insulation
- Acoustic cavity modal frequency
- Door closing sound quality

9.2.3.1 Sealing

The body sealing is fundamental of sound package. It is extremely important to seal the openings and apertures on the body, such as the opening used for steering shaft and shifter cable on dash panel, in order to achieve good body sound insulation. Every opening and aperture must be sealed. If only partial openings are sealed, it is impossible to obtain an effective sound insulation. The body can be compared with a house. The house has eight windows, but if seven windows are closed and one is open, the outside sound will be transmitted through the open window into the house. In order to insulate the outside sound, all the eight windows must be closed. If the sealing is not good, no matter how good the sound insulation and sound absorption are used, the effective sound insulation cannot be achieved. Only when the sealing achieves its target can the sound insulation be effective.

The sealing can be quantitatively evaluated by leaking air per unit of time. The leaking amount is a target of the sealing.

9.2.3.2   Sound Insulation

Sound transmission loss is an index to evaluate the sound insulation performance of a vehicle body. The sound insulation can be measured in some special circumstances, such as reverberation room, as shown in Figure 5.58. The sound insulation can be expressed by the sound transmission loss or acoustic transfer function, and the targets are set based on the benchmarking results.

9.2.3.3   Acoustic Cavity Modal Frequency

The low frequency interior booming is usually induced by the coupling between an acoustic cavity mode and a body panel mode. The cavity modal frequency is determined by the vehicle internal dimensions, which is hard to be changed, but the mode and frequency should be monitored to realize the modal separation between the body structural mode and acoustic mode.

9.2.3.4   Door Closing Sound Quality

The targets for the door closing sound quality include:

- Loudness
- Sharpness
- Ring-down

The loudness and sharpness of door closing action of a series of vehicles are provided in Chapter 7, as shown in Figures 7.23 and 7.24. The loudness between 20 and 25 sones, between 25 and 30 sones, and over 30 sones, represents good, competitive, and bad door closing sound quality, respectively. The sharpness of door closing sound between 1.8 and 2.2 acum, between 2.2 and 2.5 acum, and over 2.5 acum is very good, competitive, and bad, respectively. The ring-down can be identified from the time domain and frequency domain of a sound signal. Good ring-down is that there is only one main impact peak and then the sound decays quickly.

9.3   NVH Targets for Trimmed Body

9.3.1   NVH Characteristics of Trimmed Body

The trimmed body is constructed based on a BIW. In addition to the body panels, it includes many sound-absorption and sound-insulation structures. When subjected to an external excitation, the trimmed body attenuates the noise and vibration transmission. When an excitation force is exerted on the body, the vibration transmission on the body is represented by the vibration–vibration transfer function, and transmission of the excitation vibration to the interior noise is represented by the sound-vibration transfer function. The transmission of the external noise excitation to the interior noise is represented by the sound-sound transfer function.

9.3.2 Vibration Targets of Trimmed Body

9.3.2.1 Modal Frequency Targets

The weight and stiffness of the trimmed body is different from those of a BIW, so their modal frequencies are different as well. In general, the influence of weight increase on bending mode frequency is more than the stiffness increase, so the bending modal frequency of the trimmed body is much lower than that of the BIW. However, for the torsional modes, the increased weight reduces the frequency, but the torsional stiffness is significantly increased due to the doors, windshields, etc. so the resulted modal frequency of the trimmed body will be close to the BIW. The modal frequency targets of the trimmed body include:

- First vertical bending stiffness
- First torsional stiffness
- First vertical bending modal frequency
- First torsional modal frequency

9.3.2.2 Vibration Transfer Function Target

The excitations from other systems connected with the body are transmitted to the body, and then to the steering wheel, seats, and floor. The vibrations are perceived by the hands, buttocks, and feet. Thus, the target to control the vibration transmitted to the occupants from the body point connected with the systems is the vibration–vibration transfer function that is expressed as VTF. The trimmed body VTF targets include:

- Vibration–vibration transfer function between the interior vibrations and the power plant mounting points
- Vibration–vibration transfer function between the interior vibrations and the exhaust hanger points
- Vibration–vibration transfer function between the interior vibrations and the shock absorber excitation points
- Vibration–vibration transfer function between the interior vibrations and the suspension spring excitation points
- Vibration–vibration transfer function between the interior vibrations and the suspension arm excitation points
- Vibration–vibration transfer function between the interior vibrations and the other system (air filter box, cooling module, etc.) excitation points

These targets are usually set based on benchmark testing and analysis.

9.3.3 Noise Targets for Trimmed Body

9.3.3.1 Sound–Sound Transfer Function

In addition to its basic functions, the trimmed body has functions of sound absorption and sound insulation. The trim parts together with the metal sheet form a barrier to insulate the outside sound. The index to evaluate a body sound insulation is the sound–sound transfer function that can be expressed by noise attenuation (NR) or acoustic transfer function (ATF). The noise targets of the trimmed body include:

- Sound–sound transfer function between the interior noise and engine radiation noise
- Sound–sound transfer function between the interior noise and exhaust orifice radiation noise
- Sound–sound transfer function between the interior noise and intake orifice radiation noise
- Sound–sound transfer function between the interior noise and tire radiation noise
- Sound–sound transfer function between the interior noise and other noise (wind noise, driveshaft noise, etc.)

The NR increases with frequency increases. The higher NR represents a better sound insulation, so the higher the NR, the better. However, ATF decreases with frequency increase. The lower ATF represents a better sound insulation, so the lower the ATF, the better.

The sound insulation target that closely relates to material cost, weight, etc. is set up by benchmark testing and analysis. Usually, the acoustic package design is based on the customer requirements and the market competition.

9.3.3.2 Sound–Vibration Transfer Function

The index to express the structural-borne noise transfer is the sound–vibration transfer function, also called the noise transfer function (NTF). The targets of the noise transfer functions include:

- Sound–vibration transfer function between the interior noise and the power plant mounting points
- Sound–vibration transfer function between the interior noise and the exhaust hanger points
- Sound–vibration transfer function between the interior noise and the shock absorber excitation points
- Sound–vibration transfer function between the interior noise and the suspension spring excitation points
- Sound–vibration transfer function between the interior noise and the suspension arm excitation points
- Sound–vibration transfer function between the interior noise and the other system (air filter box, cooling module, etc.) excitation points

Generally, the target is set to $55\,dB\,N^{-1}$.

9.4 NVH Targets for Body-in-White

9.4.1 NVH Characteristics of BIW

A BIW that is similar to a frame structure of a building consists of the frames and panels. The BIW is the foundation of a full vehicle, and many other systems are installed on it. The BIW NVH performance is critical to the full vehicle because it is subjected to all the excitation sources. Therefore, it is very important to set up the BIW NVH targets for vehicle development.

The frame stiffness and joint stiffness determines the BIW modal frequencies and S&R control. Low stiffness will induce the low body modal frequency and large deformation, resulting in the impacts among the adjacent components and generating squeak and rattle. The BIW modal frequencies determine the trimmed body and full vehicle modal frequencies.

The BIW includes many panels, such as dash panel, roof, floor, and so on. Due to their low modal frequencies, the panels are easily excited by external vibration sources and then radiate noise. In addition, some attached components and pass-throughs excite the panels as well. The panel frequencies could be coupled with the cavity mode frequencies. Therefore, it is very important to set targets to control the panels' vibration.

Stiffness of the connected points between the body and other excitation systems affects efficiency of vibration transmission inside the body. Therefore, the dynamic stiffness is set as a BIW-level NVH target.

9.4.2 Vibration Targets of BIW

According to the analysis in the previous sections, the BIW-level vibration target can be cascaded into three categories: overall modes, local modes, and driving point dynamic stiffness.

9.4.2.1 Target of Overall Modal Frequency
The targets of BIW overall modal frequency include:

- First vertical bending stiffness
- First torsional stiffness
- First vertical bending modal frequency
- First torsional modal frequency

There are two purposes to control first vertical bending mode frequency and first torsional mode frequency. First, the body modal frequencies must avoid the excitation frequencies of engine, road, etc. Second, if the targets are set, the squeak and rattle could be avoided.

9.4.2.2 Target of Local Modal Frequency
Chapter 3 lists many local modes, such as dash panel mode, roof mode, floor mode, hood mode, trunk lid mode, door panel mode, and so on, and also describes the noise and vibration problems generated by the local modes. Let us briefly review the problems. The panels are thin plates, so they are easily excited by the engine and road. Because of their relatively low frequencies, they could be possibly coupled with the cavity mode. The pass-throughs could excite the panels. Thus, it is very important to control the panel modal frequencies in order to reduce the body vibration and improve sound quality.

The targets of local modes include:

- Hood modal frequency
- Dash panel modal frequency

- Roof modal frequency
- Floor modal frequency
- Modal frequencies of other panels, such as side wall, trunk lid

The principle to set up the local modes' targets is to separate the frequencies of the local modes from the excitation frequencies and also from the frequencies of the connected components.

9.4.2.3 IPI Targets

Input point inertance (IPI) is a ratio of input force to acceleration response at the same point, and its unit is $N\,m^{-1}s^2$ or Ng^{-1}. The higher the IPI, the stronger the point.

For a vehicle body, low IPI represents that it is easily excited and then transmits structural-borne noise to the interior. Therefore, the IPI is an important BIW-level target, including:

- IPI (x-, y- and z-directions) at the connected points between body and power plant mounting
- IPI (x-, y- and z-directions) at the connected points between body and exhaust hangers
- IPI (x-, y- and z-directions) at the connected points between body and shock absorber
- IPI (x-, y- and z-directions) at the connected points between body and suspension spring

The IPI at a point has three-direction values. Traditionally, the stiffness at the loading direction should be higher than those in the other two directions. However, a lot of data indicate that the vibration magnitudes and loading direction are not necessarily related; that is, the vibration at a nonloading direction could be higher than that in the loading direction. Generally, $10^7 N\,m^{-1}$ is set as the target of driving point dynamic stiffness that is closely related to acceleration impedance, a reciprocal of the IPI.

9.4.3 Noise Target of BIW

In the absence of closure and trim parts, sound freely transmits through the BIW, so it is not necessary to set up the BIW noise target. However, in order to verify the leakage of the nonfunctional holes and manufacturing process holes of the BIW, the leakage is set as a BIW-level noise target.

In order to test the BIW leakage, a sturdy plastic sheet is used to cover the doors and windshields, and then the amount of leakage is measured and the leakage locations are identified. The timely discovery of the leakage on the BIW will provide a guide to control them as early as possible, which will lay down foundation for the full vehicle leakage control and also provide support for the body design and manufacture processing design.

9.5 NVH Targets for Body Components

In the previous introduction of the vehicle-level targets, trimmed body-level target, and BIW-level targets, some component targets are presented as well, such as side mirror, dash panel, and so on. This section will describe the targets of the components that have not yet been mentioned, including brackets, seats, hood and trunk lids, door, sound absorption material, and insulation material.

9.5.1   Component-Level Vibration Targets

Many accessories are mounted to the vehicle body by brackets. A bracket with low frequency can be easily excited by the attached component, which could result in interior resonance. Therefore, the bracket frequencies should be set as the control targets, such as

- Steering system bracket frequency
- Battery supporter frequency
- ECU supporter frequency
- Air cleaner box bracket frequency

Two principles should be followed to set up the bracket modal frequency targets. One is that the frequency avoids the frequency of external excitation, and the second one is that it avoids the modal frequencies of the adjacent components. For example, the airflow flowing through an air cleaner box will excite its supporter. If the airflow excitation frequency is consistent with the supporter frequency, the supporter will be resonated, transmitting the vibration to the body. When setting the target of the bracket modal frequency, the engine firing frequency and air cleaner box structure frequency should be avoided.

The seat targets include the frequencies of lateral and vertical modes; please refer to Chapter 3.

Modal frequencies of the hood and trunk lid are also included in the component-level NVH targets, and the purpose is to avoid resonance.

9.5.2   Component-Level Noise Target

Except for the body panels and the cavity barrier materials, the components to attenuate the body noise include sound insulation and sound absorption parts, and their basic characteristics are determined by coefficients of the sound absorption materials and sound insulation materials. Thus, the component-level noise targets include:

- Material sound absorption coefficient
- Material sound insulation coefficient
- Sound transmission loss of sound insulation structures

The material absorption coefficient can be measured in an impedance tube or in a reverberation chamber. The sound insulation can be measured in an impedance tube or in a combined suite of a reverberation chamber and anechoic chamber. The targets for the coefficients and sound transmission loss are setup based on benchmark testing and analysis.

9.5.3   Noise and Vibration Targets of Door

Door closing sound quality is the most important body sound quality. Chapter 7 introduces the factors influencing the sound quality and corresponding improved methods. This section briefly summarizes the NVH targets to control the sound quality of door closing action.

9.5.3.1 Modal Frequency Targets of Door Structure

The door structures influencing the door closing sound quality include the door overall structures of external and inner panels, thereby, the targets relating to the door structure are

- First bending modal frequency of overall door
- First torsional modal frequency of overall door
- External panel modal frequency
- Inner panel modal frequency

Chapter 8 lists the first bending and torsional modal frequencies for some doors, as shown in Figure 8.17 and Figure 8.18, respectively. Generally, the frequency targets for the bending and torsional modes for the white doors are 50 and 60 Hz, respectively.

9.5.3.2 IPI of Door Striker

The latch and striker instantly impact at the moment of door closing, so the striker IPI affects the door closing sound quality. The striker IPI should be the same as the one at other driving points on the body.

9.5.3.3 The Impact Targets for Latch-Striker-Pawl

There are multiple impacts among the latch-striker- pawl. If the time interval between two impacts is too long, several impact sounds will be heard, resulting in poor door closing sound quality. The targets of a locker include the time intervals, loudness and sharpness of the impacts induced by the latch-striker-pawl.

9.6 Execution and Realization of Body Targets

The first section introduces that the vehicle NVH development is a process of target-setting to target-execution. Only a reasonable target is setup, and then strictly executed at each phase of the development process can a satisfied vehicle be realized. This section will describe the execution process of the NVH target system from target setting and cascading to CAE analysis, DMU inspection, BIW control, trimmed body control, full vehicle body control, milestone check, etc.

9.6.1 Control at Phase of Target Setting and Cascading

The principles of target setting and cascading have been described in previous materials. After positioning of the segment-market and the NVH competitiveness of the vehicle to be developed are determined, three competitive vehicles are benchmarked and analyzed, and then the targets are set up based on the benchmarked results.

First, the vehicle-level noise and vibration are measured, including interior noise, perceived vibration on steering wheel, seat and floor, and noise and vibration data of each system, such as the tailpipe orifice noise, power train mounting vibration, etc.

Second, the full vehicle is dismantled to different systems, such as trim body, exhaust, power train, suspension, and so on. The NVH data are tested for these systems; for example, the first vertical bending mode and first torsional mode of the trim body are measured.

Third, each system will be further broken down into subsystems or components. For example, the trim body is broken down into BIW, doors, dash mat, etc. Then NVH data for each subsystem and each component are measured, such as first vertical bending mode and first torsional mode of the BIW, the sticker IPI, coefficient of sound absorptive material, and so on.

After the NVH targets for vehicle-level, system-level and component-level are obtained, according to the "top-bottom" and "bottom-top" two-way principle of the target setting and cascading, the set targets are modified based on the benchmarked results, so NVH targets are finalized.

After the NVH targets are determined, the achievable percentage of the targets at each development phase and DVP (Design and Verification Plan) must be established.

9.6.2 Target Checking at Milestones

The three-year development period is divided into about a dozen milestones. The purpose to set the milestones is to divide the development period into different phases, to set up targets for each phase and then to inspect whether the targets are realized.

The NVH milestone checking is to inspect the achievement of the NVH targets at a development phase. For a coming milestone, a vehicle NVH team "inputs" the finished work, the expert team leading by a company NVH chief engineer "checks" the input, and finally, they discuss the work and provide "output," as shown in Figure 9.7. The input includes the current status of NVH development, problems, achievable percentage of the NVH targets, etc. The checking is to inspect the NVH status and problems, to evaluate whether the requirements for the phase are achieved, and to determine whether the milestone is passed. If the existing problems do not influence the moving forward of the project, their risk and potential solutions must be proposed. The output is to provide a complete checking assessment and to assign the tasks from now to the next milestone.

The milestone checking is the most effective way to execute NVH target system. The checking breaks down the overall NVH target into each phase, so the targets at each

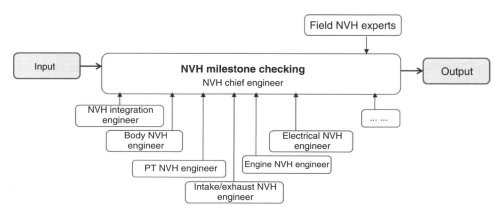

Figure 9.7 NVH milestone checking.

phase are more specific, clearer, and easier to achieve. If the targets are achieved at each phase, the ultimate targets will certainly be achieved.

9.6.3 CAE Analysis and DMU Checking

In addition to testing and analysis, CAE analysis and DMU checking are good means to control the NVH targets. They are closely used together with the testing and analysis to check the existing problems, to rapidly provide multi-designs and to verify the targets.

9.6.3.1 CAE Analysis

It is costly and time-consuming to do prototype testing. Sometimes, it is impossible to do the testing in early development phases. However, CAE is a good tool to do simulation because it is fast, cheap, and capable for multi-design analysis. In the early phases, CAE is used to analyze the strengths and weaknesses of the competitive vehicles, and to set up targets of the vehicle to be developed together with the benchmarking testing results. In the middle phases, CAE is used to analyze the existing problems, to do multi-design analysis, and to provide the optimal solutions. In the late phases, the prototype testing results can be used to verify the CAE analysis accuracy, which will help to improve CAE analysis capability.

For a BIW, the CAE modal analysis results usually are consistent with the testing results. However, for a trimmed body, after the doors, windshields, seats, etc. are installed on the BIW, the trimmed body become a nonlinear system and some parameters are hard to be simulated, so the CAE analysis has significant error with the testing results.

The acoustic calculation of a trimmed body generally uses the statistical energy method. Because of the difficulty of measuring the trimmed body parameters and the damping factors of structural parts, it is hard to control the calculation precision. The body acoustic calculation is not as mature as the body structural calculation. Nevertheless, for the trimmed body calculation, CAE analysis encounters great challenges, but the calculation is still beneficial for the multi-design in early phases and optimization in late phases.

9.6.3.2 DMU Checking

DMU represents a digital vehicle. Since the prototype body is available only at later phases, in the early phases, based on CAE analysis and body design, the digital prototype body will be generated. By checking this digital prototype, many NVH problems can be identified.

First, DMU checking looks for poorly designed structures. For example, the NVH engineers can identify whether the floor frames form a closed-loop structure, whether the local panels are too weak, whether the reinforced parts are reasonable, whether the damping materials are optimally arranged, and so on. By DMU checking, these problems can be found in the digital vehicle, and the corresponding structures can be modified before the prototype is manufactured.

Second, the goal is to discover opening and holes on the body. The checking for functional openings is to inspect whether they are reasonable or too big, whether vibration isolation and sound insulation are reasonably applied on the needed locations. For example, the openings on the dash panel for pass-through parts must be well sealed. The checking for manufacturing process openings is to inspect whether they are filled after the process is finished. The check for error-state openings is to inspect whether this kind of hole exists. For example, if the error-state holes are found, they must be filled.

Third, the sound package is checked. This evaluation includes locations, materials, and thickness of the used sound package, etc.

9.6.4   NVH Control for BIW

After a BIW is ready, its targets must be checked, including air leakage, overall modes, panel local modes, bracket modes, and IPI.

9.6.5   NVH Control for Trimmed Body and Full Vehicle

After a trimmed body or a full vehicle is ready, its targets must be checked, including air leakage, overall modes, local modes, sound transmission loss, vibration transfer function, and door closing sound quality.

Bibliography

Afaneh, A.H., Abdelhamid, M.K., and Qatu, M.S. (2007). Engineering Challenges with Vehicle Noise and Vibration in Product Development. SAE Paper; 2007-01-2434.

Williams, R., Henderson F., Allman-Ward, M. et al. (2005). Using an Interactive NVH Simulator for Target Setting and Concept Evaluation in a New Vehicle Programme. SAE Paper; 2005-01-2479.

Tousignant, T., Govindswamy, K., Tomazic, D. et al. (2013). NVH Target Cascading from Customer Interface to Vehicle Subsystems. SAE Paper; 2013-01-1980.

Pang, J., Sheng, G., and He, H. (2006). *Automotive Noise and Vibration – Principles and Application*. Beijing Institute of Technology Press.

Index

Noise and Vibration Control in Automotive Bodies, First Edition. Jian Pang.
© 2019 China Machine Press. All rights reserved. Published 2019 by John Wiley & Sons Ltd.